Environmental Effects on Volcanic Eruptions

From Deep Oceans to Deep Space

Environmental Effects on Volcanic Eruptions

From Deep Oceans to Deep Space

Edited by

James R. Zimbelman
Center for Earth and Planetary Studies
National Air and Space Museum
Smithsonian Institution
Washington, D.C.

and

Tracy K. P. Gregg
Department of Geology
State University of New York at Buffalo
Buffalo, New York

Kluwer Academic/Plenum Publishers
New York, Boston, Dordrecht, London, Moscow

Library of Congress Cataloging-in-Publication Data

Environmental effects on volcanic eruptions: from deep oceans to deep space/edited by
James R. Zimbelman and Tracy K. P. Gregg.
     p.   cm.
   Includes bibliographical references and index.
   ISBN 0-306-46233-8
   1. Planetary volcanism.   2. Volcanism.   3. Volcanism—Effect of environment on.   I. Zimbelman, James R.   II. Gregg, Tracy K. P.

QB603.V65 E58 2000
551′ .21′0999—dc21   99-056303

ISBN: 0-306-46233-8

©2000 Kluwer Academic / Plenum Publishers, New York
233 Spring Street, New York, New York 10013

10 9 8 7 6 5 4 3 2 1

http://www.wkap.nl/

A C.I.P. record for this book is available from the Library of Congress

All rights reserved

No part of this book may be reproduced, stored in a retrieval system, or transmitted in any form or by any means, electronic, mechanical, photocopying, microfilming, recording, or otherwise, without written permission from the Publisher.

Printed in the United States of America

To our fathers,

*Edwin Zimbelman (9/17/18–5/9/98)
and
Eugene D. Porter (5/21/44–).*

*Their encouragement to question and explore
will always permeate our lives.*

Contributors

Carlton C. Allen, Johnson Space Center, Mail Code C23, Lockheed Martin, 2400 NASA Road 1, Houston, TX 77058, USA
Jayne Aubele, Museum of Natural History and Science, 1801 Mountain Road NW, Albuquerque, NM 87104-1375, USA
Nathan T. Bridges, Jet Propulsion Lab., MS 230-235, 4800 Oak Grove Dr., Pasadena, CA 91109-8099, USA
Mary G. Chapman, U.S. Geological Survey, 2255 N. Gemini Drive, Flagstaff, AZ 86001, USA
David A. Crown, Department of Geology and Planetary Science, 321 Engineering Hall, University of Pittsburgh, Pittsburgh, PA 15260-3332, USA
Larry Crumpler, Museum of Natural History and Science, 1801 Mountain Road NW, Albuquerque, NM 87104-1375, USA
Sarah A. Fagents, Department of Geology, Box 871404, Arizona State University, Tempe, AZ 85287-1404, USA
Ronald Greeley, Department of Geology, Box 871404, Arizona State University, Tempe, AZ 85287-1404, USA
Tracy K. P. Gregg, Department of Geology, 876 Natural Sciences and Math. Complex, State University of New York, Buffalo, NY 14260-3050, USA
Eric B. Grosfils, Geology Department, 609 N. College Ave., Pomona College, Claremont, CA 91711, USA
Magnus T. Gudmundsson, Science Institute, University of Iceland, Hofsvallagata 53, 107 Reykjavik, Iceland
Virginia C. Gulick, NASA Ames Research Center, MS 245-3, Moffett Field, CA 94035, USA
James W. Head III, Department of Geological Sciences, Box 1846, Brown University, Providence, RI 02912, USA
Harald Hiesinger, Department of Geological Sciences, Box 1846, Brown University, Providence, RI 02912, USA
Sveinn P. Jakobsson, Icelandic Institute of Natural History, Hiemmur 3, IS-105 Reykjavik, Iceland
Laszlo Keszthelyi, Lunar and Planetary Lab., University of Arizona, Tucson, AZ 85721, USA
Susan W. Kieffer, Kieffer & Woo, Inc., P.O. Box 130, Paigrave, Ontario, L0N 1P0, Canada
Rosaly Lopes-Gautier, Jet Propulsion Lab., 4800 Oak Grove Dr., Pasadena, CA 91109-8099, USA
Baerbel K. Lucchitta, U.S. Geological Survey, 2255 N. Gemini Drive, Flagstaff, AZ 86001, USA

Curtis R. Manley, Department of Geological Sciences, CB# 3315, University of North Carolina, Chapel Hill, NC 27599-3315, USA

Alfred S. McEwen, Lunar and Planetary Lab., University of Arizona, Tucson, AZ 85721, USA

Peter J. Mouginis-Mark, University of Hawaii, SOEST, 2025 Correa Road, Honolulu, HI 96822-2285, USA

Harry Pinkerton, Environmental Sciences Department, Institute of Environmental and Natural Sciences, Lancaster University, Lancaster, LA1 4YQ, United Kingdom

Louise Prockter, Department of Geological Sciences, Box 1846, Brown University, Providence, RI 02912, USA

Mark Robinson, Department of Geological Sciences, Northwestern University, Evanston, IL 60208, USA

Scott K. Rowland, University of Hawaii, SOEST, 2025 Correa Road, Honolulu, HI 96822-2285, USA

Susan Sakimoto, Geodynamics Branch, Code 921, NASA Goddard Space Flight Center, Greenbelt, MD 20771, USA

Paul Schenk, Lunar and Planetary Institute, 3600 Bay Area Blvd., Houston, TX 77058, USA

Ian P. Skilling, Department of Geology, Rhodes University, P.O. Box 94, 6140 Grahamstown, South Africa

Richard B. Waitt, U.S. Geological Survey, Cascades Volcano Observatory, 5400 MacArthur Blvd., Vancouver, WA 98661, USA

Catherine Weitz, Jet Propulsion Lab., MS 183-335, 4800 Oak Grove Dr., Pasadena, CA 91109-8099, USA

David A. Williams, Department of Geology, Box 871404, Arizona State University, Tempe, AZ 85287-1404, USA

Lionel Wilson, Environmental Sciences Department, Institute of Environmental and Natural Sciences, Lancaster University, Lancaster, LA1 4YQ, United Kingdom

Aileen Yingst, Department of Planetary Sciences, Lunar and Planetary Lab., University of Arizona, Tucson, AZ 85721, USA

James R. Zimbelman, CEPS MRC 315, National Air and Space Museum, Smithsonian Institution, Washington, DC 20560-0315, USA

Preface

The topic of volcanism in the solar system can imply many things. The ongoing eruptions on Io, the enormous shield volcanoes of Mars, and the low-profile volcanoes of Venus are but a few of the dramatic examples of planetary volcanism discovered in recent years. Yet each of these examples occurs within a distinctive environment. "Environment" is often viewed as merely the backdrop to the main events that created the impressive volcanic features evident throughout the solar system. However, in this book we hope to demonstrate that the environment into which a volcanic eruption takes place is a significant—and perhaps even a controlling—factor on the resulting eruption styles and products. This general subject is an outgrowth of special sessions on "Volcanic Eruptions in Diverse Environments" held at the 1997 annual meeting of the Geological Society of America, which met in Salt Lake City, Utah. The editors served as cochairs for these stimulating sessions, which led directly to the subject of this book. Much can be learned by comparing volcanic features observed in various locations throughout the solar system. The goal of the present compilation is to increase awareness of volcanic diversity on solid surfaces, paying special attention to the local planetary environment.

The book is intended to be a supplement to an advanced undergraduate or beginning graduate level course in either volcanology or planetary geology. It is organized to progress roughly from more accessible to less familiar subjects, with liberal cross-referencing to aid the reader in the comparison process. Volcanism in the solar system occurs on objects of varying sizes, with a range of physical conditions present (Chapter 1). The order of discussion proceeds from subaerial volcanism on Earth (Chapter 2), to extraterrestrial examples with similar—and increasingly different—ambient conditions. The interaction of lava with water and ice on Earth and Mars results in very unique products (Chapter 3). Mars (Chapter 4) displays a great range of volcanic features found under a thin (but nonzero) atmospheric pressure. In contrast, the elevated pressure environment found on both Earth's seafloor and the surface of Venus leads to interesting similarities, as well as some important differences, between volcanic landforms observed in both locations (Chapter 5). Eruption into a near-vacuum characterizes the volcanic features seen on the Moon and Mercury (Chapter 6) as well as Jupiter's moon Io (Chapter 7), but the ongoing tidal forcing on the latter makes it the most volcanically active object in the solar system other than Earth. Basaltic lavas are probably most common on the planets mentioned above, but the possibility of more rare "exotic" lavas illustrates that volcanic activity is not limited to common silicate lavas (Chapter 8). The book closes with a brief summary and a compilation of physical parameters important for modeling volcanic phenomena (Chapter 9).

Each chapter was reviewed by at least two reviewers. We wish to acknowledge the following for their many comments and suggestions on various chapters: Nathan Bridges, Mary Chapman, David Crown, Larry Crumpler, Sarah Fagents, Ronald Greeley, Tracy Gregg, Eric Grosfils, Jim Head, Rosaly Lopes-Gautier, Peter Mouginis-Mark, Harry Pinkerton,

Louise Prockter, Susan Sakimoto, Debbie Smith, Catherine Weitz, David Williams, and Jim Zimbelman. We are very grateful to Ken Howell, editor at Plenum (and subsequently Kluwer), and those working with him, who helped to put the manuscript into book format. Some chapters acknowledge grant support for individual contributing authors. We hope that the reader will find this compilation a helpful resource for placing volcanic eruptions into a solar-system-wide context.

James R. Zimbelman, *Washington, DC, USA*
Tracy K. P. Gregg, *Buffalo, NY, USA*

Contents

1. VOLCANIC DIVERSITY THROUGHOUT THE SOLAR SYSTEM   1
   James R. Zimbelman and Tracy K. P. Gregg

   1.1. Introduction: "A Little Ambiance"   1
   1.2. Environmental Effects on Volcanism   3
   1.3. Lava Composition   4
   1.4. Lava Rheology   5
   1.5. Emplacement Styles   6
   1.6. Environmental Impact of Volcanism   7
   1.7. Conclusion   7
   1.8. References   8

2. SUBAERIAL TERRESTRIAL VOLCANISM: ERUPTIONS IN OUR OWN BACKYARD   9
   James R. Zimbelman, Sarah A. Fagents, Tracy K. P. Gregg, Curtis R. Manley, and Scott K. Rowland

   2.1. Introduction   9
   2.2. Magma Generation and Rise Conditions   10
   2.3. Global Setting: Plate Tectonics and "Hot Spots"   12
   2.4. Effusive Eruptions   17
   2.5. Explosive Eruptions   25
   2.6. Large Igneous Provinces   28
   2.7. Discussion: Volcano/Environment Interactions   30
   2.8. Conclusion   32
   2.9. References   32

3. VOLCANISM AND ICE INTERACTIONS ON EARTH AND MARS   39
   Mary G. Chapman, Carlton C. Allen, Magnus T. Gudmundsson, Virginia C. Gulick, Sveinn P. Jakobsson, Baerbel K. Lucchitta, Ian P. Skilling, and Richard B. Waitt

   3.1. Introduction   39
   3.2. Nevado del Ruiz, Colombia: Eruption beneath Alpine Ice   41
   3.3. Gjálp, Vatnajökull: Eruption beneath an Ice Sheet   42
   3.4. Subglacial Volcanic Hyaloclastic Ridges   45
   3.5. Subglacial Volcanic Tuyas   50
   3.6. Lahars and Jökulhlaups   55
   3.7. Phreatic Craters on Earth and Mars   58

3.8. Hyaloclastic Ridges, Tuyas, Lahars, and Jökulhlaups on Mars	60
3.9. Volcano–Ground Water Interactions and Snow/Ice Perturbations on Mars	64
3.10. Conclusion	68
3.11. References	68

4. VOLCANISM ON THE RED PLANET: MARS — 75
Ronald Greeley, Nathan T. Bridges, David A. Crown, Larry Crumpler, Sarah A. Fagents, Peter J. Mouginis-Mark, and James R. Zimbelman

4.1. Introduction	75
4.2. Background	76
4.3. Mars Composition	78
4.4. Influence of Mars' Environment on Volcanism	84
4.5. Large Shield Volcanoes	87
4.6. Martian Highland Paterae	96
4.7. Small Volcanic Constructs	102
4.8. Volcanic Plains	102
4.9. Conclusion	104
4.10. References	107

5. VOLCANISM ON EARTH'S SEAFLOOR AND VENUS — 113
Eric B. Grosfils, Jayne Aubele, Larry Crumpler, Tracy K. P. Gregg, and Susan Sakimoto

5.1. Introduction	113
5.2. Insights Into Volcanic Style: Intrusions, Effusive Eruptions, and Explosive Events	125
5.3. Conclusion	137
5.4. References	137

6. MOON AND MERCURY: VOLCANISM IN EARLY PLANETARY HISTORY — 143
James W. Head III, Lionel Wilson, Mark Robinson, Harald Hiesinger, Catherine Weitz, and Aileen Yingst

6.1. Introduction	143
6.2. Lunar Volcanism	145
6.3. Morphology	147
6.4. Pyroclastic Volcanism	151
6.5. Nonmare Volcanism	154
6.6. Recognition and Assessment of Cryptomaria	155
6.7. Lava Ponds	156
6.8. Contiguous Maria	159
6.9. Effects of A Global Low-Density Crust	161
6.10. Lunar Eruption Conditions	162
6.11. Synthesis of Lunar Volcanism	163
6.12. Volcanism on Mercury	165
6.13. Conclusion: Environmental Influences, Synthesis, and Outstanding Problems	171
6.14. References	173

7. EXTREME VOLCANISM ON JUPITER'S MOON IO — 179
Alfred S. McEwen, Rosaly Lopes-Gautier, Laszlo Keszthelyi, and Susan W. Kieffer

- 7.1. Introduction and Background — 179
- 7.2. Volcanic Hot Spots — 185
- 7.3. Plumes — 191
- 7.4. Sulfur Flows and Lakes? — 195
- 7.5. Global Distribution of Volcanism — 196
- 7.6. Discussion — 197
- 7.7. Conclusion and Future Exploration — 202
- 7.8. References — 203

8. EXOTIC LAVA FLOWS — 207
Harry Pinkerton, Sarah A. Fagents, Louise Prockter, Paul Schenk, and David A. Williams,

- 8.1. Introduction — 207
- 8.2. Cool Lavas: Carbonatites — 208
- 8.3. Primitive Lava: Komatiites — 214
- 8.4. Ice As Lava: Cryovolcanism — 220
- 8.5. Discussion and Conclusion — 234
- 8.6. References — 234

9. VOLCANIC VESTIGES: PULLING IT TOGETHER — 243
Tracy K. P. Gregg and James R. Zimbelman

- 9.1. Introduction — 243
- 9.2. Information — 244
- 9.3. Models — 244
- 9.4. Magma Intrusion — 249
- 9.5. Summary and Conclusion — 249
- 9.6. References — 250

INDEX — 253

Environmental Effects on Volcanic Eruptions

From Deep Oceans to Deep Space

1

Volcanic Diversity throughout the Solar System

James R. Zimbelman and Tracy K. P. Gregg

1.1. INTRODUCTION: "A LITTLE AMBIANCE"

What comes to mind when one considers the subject of volcanism in the solar system? The ongoing eruptions on Io, the enormous shield volcanoes of Mars, the low-profile volcanoes of Venus, or the nitrogen-powered geysers of Triton are but a few of the dramatic examples of planetary volcanism discovered in recent years. Yet each of these examples occurs within a distinctive environment. We usually think of "environment" as merely the backdrop to main events that created the impressive volcanic features evident throughout the solar system. However, in this book we hope to demonstrate that the environment into which a volcanic eruption takes place is a significant—and perhaps even a controlling—factor on the resulting eruption styles and products. Much can be learned by comparing volcanic features observed in various locations throughout the solar system. The goal of the present compilation is to increase awareness of volcanic diversity on solid surfaces, paying special attention to the local planetary environment.

The book is organized to progress roughly from more accessible to less familiar subjects, with liberal cross-referencing to aid the reader in the comparison process. Volcanism in the solar system occurs on objects of varying sizes (Figure 1.1), and which encompass a range of basic physical conditions (Table 1.1). The order of discussion proceeds from subaerial volcanism on Earth (Chapter 2—the primary source of current information on volcanic processes), to extraterrestrial examples with similar—and increasingly different—ambient conditions. Chapter 3 compares the interaction of lava with water and ice on Earth and Mars. Mars (Chapter 4) displays a great range of volcanic features found under a thin (but nonzero) atmospheric pressure. In contrast, the elevated pressure environment found on both Earth's seafloor and the surface of Venus leads to interesting similarities, as well as some important differences, between the volcanic landforms observed in both locations (Chapter 5). Eruption into a near vacuum characterizes the volcanic features seen on the Moon and Mercury (Chapter 6) as well as Jupiter's moon Io (Chapter 7), but the ongoing tidal forcing on the latter makes it the most volcanically active object in the solar system other than Earth. Basaltic

Figure 1.1. Sizes of the principal volcanic bodies in the solar system. Actual spacecraft images are shown at the same relative scale. NASA Public Affairs image P-41468.

Table 1.1. Basic Physical Parameters of Volcanic Planets and Moons

Object[a]	Mass (10^{22} kg)	Eq. radius (10^6 m)	Density (10^3 kg m^{-3})	Gravity (m s^{-2})	Surface atm. pressure (10^5 N m^{-2})	Surface temp. (K)
Mercury	33.01	2.439	5.44	3.70	∼0	90–740
Venus	486.8	6.051	5.24	8.87	∼90	740
Earth	597.4	6.380	5.52	9.81	1	190–338
Moon	7.35	1.738	3.34	1.62	∼0	100–360
Mars	64.18	3.396	3.93	3.71	0.007	150–320
(Jupiter)	189900	71.492	1.33			
Io	8.93	1.821	3.53	1.80	∼0	80–160
Europa	4.80	1.560	3.04	1.31	∼0	80–120
Ganymede	14.8	2.630	1.94	1.42	∼0	80–150
(Callisto)	10.8	2.400	1.86	1.25	∼0	80–160
(Saturn)	56850	60.268	0.69			
(Titan)	13.5	2.575	1.86	1.36	1.44	∼92
Enceladus	0.0084	0.250	1.24	0.09	∼0	∼60–145
(Neptune)	10240	24.764	1.64			
Triton	2.14	1.350	2.06	0.78	0.000014	∼38

[a] Natural satellites are indented below primary planet. Names in parentheses are included for comparison only; they are not necessarily volcanic objects.

lavas are probably most common on the planets mentioned above, but the possibility of more rare "exotic" lavas illustrates that volcanic activity is not limited to common silicate lavas (Chapter 8). The book closes with a brief summary and a compilation of physical parameters important for modeling volcanic phenomena (Chapter 9). As an aid to understanding these diverse topics, we next briefly discuss several themes encountered in the subsequent text, highlighting aspects the reader should note while progressing through the book.

1.2. ENVIRONMENTAL EFFECTS ON VOLCANISM

Understanding the relation between the environment and volcanic processes and products is the primary objective of this book. It is important to be aware of the many ways in which the environment can exert a strong influence on several aspects of a volcanic eruption. The "environment" consists of a complex array of factors, each of which contributes differently to the physical condition encountered by erupted volcanic products. Some of the factors that are likely to be the most significant in terms of affecting the condition and state of lava or pyroclastic materials are highlighted below.

Temperature. The rate of cooling of a lava flow has a significant effect on both the morphology and the emplacement condition of lava. When the lava temperature is several times that of the surrounding environment, rapid quenching generates a glassy crust that provides important insulation to the fluid lava beneath the crust (see Chapters 2, 4, and 5). Cooling is dominated by radiation when the lava is in a thin to nonexistent atmosphere (Chapters 6 and 7), but a thick atmosphere or seawater may actually increase the cooling rate through advection or convection in the surrounding fluid (e.g., Wilson and Head, 1983; Chapter 5). If the surroundings are maintained at a high temperature, as on Venus, the elevated temperatures may reduce the effective strength of the rock both in a lava flow and in the underlying country rock (Chapter 5). Eruption temperatures may range considerably, depending on the lava composition (e.g., komatiites apparently had eruption temperatures more than 200°C greater than those of typical basalts; see Chapters 7 and 8).

Pressure. Vesiculation can be modified by external pressure. For example, pressure from the overburden of seawater (Chapter 5) or ice (Chapter 3) inhibits vesiculation in erupting lavas, and a similar phenomenon occurs on Venus (Chapter 5). High atmospheric density, such as exists on Venus, may also contribute to convective cooling, as mentioned above. Ambient pressure, in conjunction with temperature, might affect the stability of mineral phases, which in turn might affect the solidification of lava or pyroclastic materials.

Gravity. The acceleration of gravity is the basic source behind most body forces experienced by volcanic products once they have left the vent. The magnitude of this force will affect all of the rates associated with emplacement of either fluid lavas or pyroclastic particulates. This value ranges over two orders of magnitude for the objects summarized in Table 1.1, which must have consequences for the resulting volcanic landforms and deposits found on their surfaces (Chapters 2, 4–7).

Slope and Roughness. The relief and texture of the surface onto which an eruption takes place will influence the emplacement of the volcanic materials. Slope, in conjunction with the acceleration of gravity, vectorially resolves some fraction of the downward force of gravity into a force for movement over the surface. The roughness of the underlying surface, when comparable to the thickness of the flow, can similarly affect emplacement by retarding the effects of the downslope motive force. These effects will be most important for prolonged eruptions of relatively limited effusion rate, as opposed to large effusion rates that would rapidly emplace lavas irrespective of local relief (see Chapters 2 and 5).

Meteorology. Where an eruption takes place into an atmosphere, motion of that atmosphere can contribute to the emplacement of the products. This is most important for pyroclastic eruptions where dispersal patterns conform to any prevailing wind (Chapters 2 and 4). Eruptions into near-vacuum conditions clearly would not experience such effects (Chapters 6 and 7).

Composition. The composition of the surrounding bedrock and atmosphere, as well as that of the erupting magma, might influence the volcanic products. Differing equilibrium and nonequilibrium reactions become possible, depending on the specific materials involved. The presence and state of water is a major factor for subaqueous and subglacial eruptions (Chapters 3 and 5).

Other Factors. Certainly factors in addition to those mentioned above help to characterize various planetary environments. However, some of these additional environmental factors likely do not have a significant influence on volcanic eruptions. For example, here we will not deal with the presence or absence of electromagnetic fields and energetic particles, a situation that varies greatly throughout the solar system depending on the presence and strength of a planetary magnetic field. All such environmental factors that are unlikely to have an impact on volcanic phenomena are excluded from what we consider here to be the environment of an eruption.

1.3. LAVA COMPOSITION

Most of the volcanism discussed in this book involves basalt, the most common rock on Earth and, it is presumed, on many other planetary surfaces as well (Basaltic Volcanism Study Project, 1981). It is important to realize that only for Earth (Chapter 2) and the Moon (Chapter 6) do we "know" that the dominant volcanic material is basalt; these are the only two extraterrestrial objects from which documented samples of basalt have been collected and analyzed. There is a suite of meteorites whose trapped gases indicate an origin on Mars, and these meteorites include several examples of basalts (Chapter 4), but unfortunately we do not know where on Mars the meteorites came from. In spite of these limitations, it is still widely held that basalt is likely to be the dominant volcanic rock not only on Earth, but also throughout the inner solar system, and even out to at least one moon of Jupiter, Io (Chapter 7). From remote sensing studies we know that the reflected and emitted electromagnetic radiation from most planetary volcanic materials is at least consistent with a basaltic composition, even if we cannot as yet remotely determine the precise mineralogic makeup of planetary surfaces. The volcanism discussed in this book is primarily basaltic volcanism except where otherwise noted.

Many existing remote sensing techniques are more sensitive to surface textures than to compositional variations. Not all "dark" (low albedo) rocks are likely to be basalts. Even where we are confident that we are indeed looking at volcanic rocks, our compositional inferences are commonly related to textural details or gross topographic properties that are not particularly diagnostic of composition. The observed surface textural properties may be more closely related to the specific physical situation during emplacement than to compositional variations. A significant compositional unknown, even if we are dealing primarily with basaltic magmas, is the volatile content and constituents. The limited samples currently available from other planets do not provide strong constraints on the type and abundance of volatiles actually present in extraterrestrial lavas. Since volatile abundance and type may have a significant impact on eruption and emplacement mechanisms (Chapters 2, 4, and 8), the reader should be cognizant of the limitations to our present models and theories.

We conclude this brief discussion of lava composition by considering viable alternatives to the silicate magma presumed to dominate throughout much of the solar system. Carbon-rich volcanic products are a rare but still important part of the terrestrial volcanic record (Chapter 8), and such magmas might have played a role in some volcanic features present on Venus and Mars. Sulfur flows are also rare but documented on Earth, and were a favored alternative for some flows on Io (although recent temperature measurements reduce the likelihood that sulfur flows comprise a substantial part of the volcanic record on this moon—see Chapter 7). Water and even ammonia act as magma in the cold temperatures of the outer solar system (Chapter 8), and at the extreme edge of the solar system, vaporizing nitrogen powers geysers on the frigid (38 K) surface of Triton (McKinnon and Kirk, 1999). These examples should encourage us to keep an open mind with regard to both the prevalence of basaltic volcanism and how the conditions in the distal portions of the solar system fundamentally alter our concept of "magma" and "volatile."

1.4. LAVA RHEOLOGY

Similar to the tacit assumption that most extraterrestrial volcanism in the inner solar system is basaltic, we also assume that the majority of extraterrestrial lava flows have a rheology consistent with our current understanding of the behavior of basaltic magmas. It is challenging to obtain measurements of rheologic parameters in actively flowing lava (e.g., Pinkerton, 1993), whose high temperatures make it hard on both the instrumentation and the humans involved. Yet an accurate understanding of lava rheology is crucial to a complete description of lava emplacement, which in turn is the current limit on our ability to model the details of flowing lava. Laboratory measurements of melted volcanic rocks and simulants that mimic the bulk properties of lava are providing important insights about flowing lava (Figure 1.2), yet in the laboratory it is difficult, if not impossible, to re-create the actual volatile and crystal content of fresh lava, both of which contribute to its overall rheologic properties (e.g., Dragoni, 1993). There is also a great need for laboratory and field studies of nonsilicate lavas, some of which may behave quite different from "typical" basalt flows (see Chapter 8). Add to this the complete lack of information about the rheologic properties of extraterrestrial lava flows and you begin to see how serious is this deficiency in our current state of knowledge, since morphology is not as yet indicative of *unique* composition and rheology information.

In spite of these limitations, we can still appreciate some of the physical parameters that are important contributors to lava rheology. These include, but are not limited to, composition (with all of the caveats discussed in the previous section), temperature (both of the lava and of the surroundings, as well as the temperature gradient across the growing solidified crust), volatile content (both the abundance and the composition of the volatiles), crystal content (abundance, size, and size distribution), flow rate and slope (both of which contribute to the rate of shear experienced throughout the flow), and flow front characteristics (perhaps the least understood of all factors). Flow front dynamics are important for assessing how fresh liquid lava is supplied to the growing flow, which varies greatly between pahoehoe and a'a flows, for example. The reader should keep such factors in mind whenever lava rheology is discussed throughout the book.

Figure 1.2. A laboratory simulation of lava flow emplacement using carbowax in a sugar water solution that reproduces the essential rheologic properties of flowing basalt. Lines on the floor are 2 cm apart; vent is to the left. Solid wax is light; liquid wax is dark. Courtesy of Jonathan H. Fink, Arizona State University, and Tracy K. P. Gregg, SUNY-Buffalo.

1.5. EMPLACEMENT STYLES

The manner in which volcanic materials are emplaced can be as important as the nature of the volcanic materials themselves. Pyroclastic materials are by definition the result of explosive ejection that disrupts the rising magma into small clots and particles, but the size distribution of the resulting materials is closely related to the explosive intensity and magma supply rate to the vent. For lava flows, effusion rate remains a controversial but still fundamental factor to emplacement. Hawaiian subaerial flows have been documented with effusion rates of tens to hundreds of cubic meters per second, with pahoehoe flows comprising practically all flows <10 m^3 s^{-1} (Rowland and Walker, 1990; see Chapter 2). In spite of our familiarity with Hawaiian-style eruptions, they may not be typical of eruptions on a planetary scale. For example, volcanism along midocean ridges is likely the dominant way in which lava is emplaced on our planet, and yet no seafloor eruption or lava flow has been observed during emplacement (see Chapter 5). The largest effusion rate for a historic eruption is likely that of the Laki eruption in Iceland where fissures supplied $\sim 4 \times 10^4$ m^3 s^{-1} of lava at peak discharge (Thordarson and Self, 1993), which may be more indicative of midocean ridge volcanic activity. Effusion rates for extraterrestrial lava flows can only be inferred from final flow dimensions; such estimates range from $\sim 10^2$ to 10^7 m^3 s^{-1} but concentrate around $\sim 10^5$ m^3 s^{-1} (Zimbelman, 1998).

The question of effusion rate is directly related to the broader issue of whether the emplacement was "fast" or "slow" (see Chapter 2). Advocates of fast emplacement traditionally were required to invoke turbulent flow, although recently this requirement has been dropped because of chemical and field-related arguments (Reidel, 1998). Advocates of slow emplacement use the mechanism of flow inflation to raise the flow crust and increase its resultant thickness severalfold (Thordarson and Self, 1998). This debate has spread to include seafloor flows (Gregg and Chadwick, 1996), and there is no a priori reason to exclude either option while considering flows in diverse planetary environments. The role of lava tubes in the emplacement of large flows is beginning to undergo quantitative evaluation and it too requires careful consideration for planetary applications.

Eruption of lavas beneath an ice cover is a specific example where the environmental situation may strongly influence the details of emplacement. Subglacial eruptions have led to well-documented cases of the sudden release of large volumes of meltwater, called *jökulhlaups* in Iceland (see Chapter 3). Unfortunately, it is not possible to observe directly how eruptions progress beneath the ice, but when exposed by the subsequent removal of the ice cover, distinctive deposits of hyaloclastites (essentially pyroclastic material strongly influenced by interaction with the meltwater surrounding a subglacial eruption), underlain by pillow lavas, are the common result (see Chapter 3). The interaction of volcanics and ice is also possible on Mars (Chapter 3) and the ice-rich satellites in the outer solar system (Chapter 8).

1.6. ENVIRONMENTAL IMPACT OF VOLCANISM

Much of the above discussion has centered around how volcanic materials interact with their surroundings on leaving the volcanic vent. However, it has also become clear that volcanic eruptions can in themselves influence and alter their surroundings. This situation is most often cited for the consequences of very large subaerial volcanic eruptions that may result in a pronounced, if temporary, effect on regional and even global climate (see Chapter 2). Both acidic aerosol components and quantities of greenhouse gases released during a large eruption can contribute to climatic effects, some of which can be devastating to agriculture and its dependent human population (Thordarson and Self, 1993). Pyroclastic flows also may generate a local environment around the rapidly traveling cloud of heated gas and particulates (see Chapter 2). This is perhaps most dramatic for the long-lived large eruptions occurring on Io, where localized column collapse and pyroclastic flow generation may produce local environments substantially different from the normal Ioian conditions (see Chapter 7). There is great potential for synergy between studies of terrestrial and planetary eruptions in evaluating their potential impact on the regional and global climate; each area of investigation likely can learn much from the other, and hopefully this book can contribute to this process.

1.7. CONCLUSION

When the reader has reached the end of this book, we hope that he or she will go away with the thought that the environment may significantly modify compositional and rheologic constraints on lavas. Traditional geologic training emphasizes the dominant role of magma composition on volcanic mechanisms and resulting landforms, with physical variations often attributed to rheologic changes to the flowing lava. Here we do not contest this view, but we hope that the solar system perspective provides readers with a range of examples in which they can judge for themselves the degree to which magma properties may be affected by the environment into which the eruption takes place. The wealth of new information obtained from many planetary spacecraft, as well as continuing investigations of both subaerial and subaqueous volcanism on Earth, has greatly expanded our range of "known" examples of volcanism. We have learned a lot, as reported in this book. However, we also still have a long way to go to understand fully the mechanisms and processes that cause a volcanic feature to look the way it does (see Chapter 9).

1.8. REFERENCES

Basaltic Volcanism Study Project, *Basaltic Volcanism on the Terrestrial Planets*, 1286 pp., Pergamon Press, New York, 1981.

Dragoni, M., Modelling the rheology and cooling of lava flows, in *Active Lavas*, edited by C. R. J. Kilburn and G. Luongo, pp. 235–261, University College London Press, London, 1993.

Gregg, T. K. P., and W. W. Chadwick, Jr., Submarine lava-flow inflation: A model for the formation of lava pillars, *Geology*, 24, 981–984, 1996.

McKinnon, W. B., and R. L. Kirk, Triton, in *Encyclopedia of the Solar System*, edited by P. R. Weissman, L. McFadden, and T. V. Johnson, pp. 405–434, Academic Press, San Diego, 1999.

Pinkerton, H., Measuring the properties of flow lavas, in *Active Lavas*, edited by C. R. J. Kilburn and G. Luongo, pp. 175–192, University College London Press, London, 1993.

Reidel, S. P., Emplacement of Columbia River flood basalt, *J. Geophys. Res.*, 103, B11, 27393–27410, 1998.

Rowland, S. K., and G. P. L. Walker, Pahoehoe and a'a in Hawaii: Volumetric flow rate controls the lava structure, *Bull. Volcanol.*, 52, 615–628, 1990.

Thordarson, T., and S. Self, The Laki (Skaftár Fires) and Grimsvötn eruptions, 1783–1785, *Bull. Volcanol.*, 55, 233–263, 1993.

Thordarson, T., and S. Self, The Roza member, Columbia River basalt group: A gigantic pahoehoe lava flow field formed by endogenous processes? *J. Geophys. Res.*, 103, B11, 27411–27446, 1998.

Wilson, L., and J. W. Head, A comparison of volcanic eruption processes on Earth, Moon, Mars, Io and Venus, *Nature*, 302, 663–669, 1983.

Zimbelman, J. R., Emplacement of long lava flows on planetary surfaces, *J. Geophys. Res.*, 103, B11, 27503–27516, 1998.

2

Subaerial Terrestrial Volcanism

Eruptions in Our Own Backyard

James R. Zimbelman, Sarah A. Fagents, Tracy K. P. Gregg,
Curtis R. Manley, and Scott K. Rowland

2.1. INTRODUCTION

Current understanding of volcanic eruptions is the result of millennia of written observations refined by decades of scientific research. There is still much to learn about the details of how individual volcanoes work, but the existing body of literature about subaerial volcanism on Earth represents the basis against which all other volcanic eruptions are compared. Excellent books summarize the processes and products of subaerial volcanism (e.g., Macdonald, 1972; Williams and McBirney, 1979; Cas and Wright, 1987; Cattermole, 1989; Francis, 1994), and it would be impossible to condense all of this information into one chapter. However, it is important to provide a concise compilation of the most salient aspects of subaerial volcanism to which all other examples of volcanism can be compared.

The purpose of this chapter is to review the basic eruptive styles, landforms, and products that result from volcanism on Earth's surface, where observations and samples are both readily obtainable. At present, documented samples of extraterrestrial lavas have been collected and returned only from the Moon, and these materials are all basaltic in composition (Heiken *et al.*, 1991). However, to these Apollo and Luna samples can be added a suite of basaltic meteorites now thought to come from Mars (McSween, 1994; see Chapter 4), plus various remote sensing data sets that indicate a preponderance of basaltic materials on rocky surfaces throughout the solar system (Basaltic Volcanism Study Project, 1981). Thus, we focus primarily on basaltic volcanism, but information on other compositional types is also included both to illustrate the diversity of volcanism on Earth, as well as to cover the strong likelihood that our present information substantially underrepresents the full volcanic diversity throughout the solar system (see Chapter 8).

This chapter begins by following volcanic magma at depth as it rises toward the surface. The location and style of volcanic eruptions on Earth are strongly influenced by both the plate tectonic setting and the presence of localized, deep-seated hot spots, which guides our understanding of most terrestrial volcanic centers. Effusive eruptions result in a plethora of products, but there are some basic forms and textures that transcend the location or type of individual volcanoes. Explosive eruptions occur when escaping magma is violently disrupted; for basaltic lavas, this tends to involve interaction with ground or surface water, but more evolved magma chemistries fragment through the degassing and violent explosion of magmatic volatiles. Some parts of Earth's surface were the sites of massive outpourings of lava, resulting in localized Large Igneous Provinces (LIPs), apparently special cases of how magma reaches Earth's surface. The chapter closes with a brief discussion of how lava and ash interact with Earth's subaerial environment, including how that environment is altered through magmatic input.

2.2. MAGMA GENERATION AND RISE CONDITIONS

2.2.1. Magma Generation

Many of the processes involved in the genesis of materials that will eventually erupt at the Earth's surface take place at such great depths that little is known about their details. However, the synthesis of many lines of investigation (e.g., geophysical, geochemical, and theoretical) permits formulation of some general models. Basaltic magmas are thought to originate at depths of 50–170 km as a consequence of partial melting of mantle materials. The source rocks most likely are garnet peridotites or lherzolites, which are crystalline aggregates of olivine, orthopyroxene, and clinopyroxene with other constituents (e.g., garnet) (Yoder, 1976; Wright, 1984). Energy to melt the parent rock is derived from a wide variety of sources. The most significant sources for basalt include: (1) pressure-release melting induced by diapiric rise (Ramburg, 1972) or solid-state convection (Verhoogen, 1954) in the mantle; (2) friction related to viscous strain (Shaw, 1973), shear strain along propagating cracks (Griggs and Baker, 1969), or tidal dissipation (Shaw, 1970); (3) reduction in melting temperature by the addition of volatiles (Yoder and Tilley, 1962); (4) conductive heat trapping as a result of changes in thermal conductivity with temperature and pressure (McBirney, 1963); and (5) internal radiogenic heat production. The temperatures at which partial melts form are a function of source depth and composition, but because of the release of latent heat the melting process should be approximately isothermal for any given case. For example, the formation of Hawaiian tholeiite magma at a depth of 60 km will take place at $\sim 1350–1400°C$ (Decker, 1987). Melting commences along boundaries between mineral grains, initially forming thin films (Yoder, 1976). As melting progresses the films increase in volume until segregation into discrete melt pockets takes place, a process governed by the rheology of the liquid-rock mush and the balance between melt buoyancy, local stresses, and viscous forces (Spera, 1980).

2.2.2. Rise Mechanisms

The accumulation of the melt pockets to form larger magma bodies will eventually cause net vertical motion as a result of increased buoyancy forces. In the asthenosphere, magma may rise as diapirs by deforming the ductile country rock, the ascent rate being a function of diapir size, density contrast, and rheologic properties. However, ascent through the lithosphere must overcome the rigidity of the cooler rock, and magma may stall at the asthenosphere–lithosphere boundary (Pitcher, 1979) before sufficient stresses accumulate to initiate brittle fracture and continue motion through the resultant dikes (Sleep, 1988). Buoyant ascent may also be halted in the shallow crust as the country rock density decreases as a result of the decreasing compaction related to lithostatic overburden. Such levels of neutral buoyancy may therefore define storage zones where trapped batches of melt accumulate to form a magma reservoir at depths of a few kilometers (Rubin and Pollard, 1987; Ryan, 1987). Further dike propagation is precluded until some driving mechanism overcomes the barrier imposed by the density trap. Possible examples include pressurization of the magma reservoir related to continued magma input from below, or redevelopment of buoyancy as a result of volatile exsolution and generation of a zone of low-density bubble-rich magma (Jaupart and Vergniolle, 1989), and/or loss of dense mineral components related to fractional crystallization (Sparks et al., 1980). An increase in reservoir pressure might lead to lateral dike propagation within the edifice, whereas buoyancy-driven motion would be more nearly vertical. The velocity of magma in the dike and consequent eruption rate are governed by the magnitude of the driving forces (pressure or buoyancy), dike width, and magma rheologic properties (Wilson and Head, 1981). Eruption volume may be linked to the size of the subsurface reservoir, where one exists (Blake, 1981). However, not all magmas will erupt: Velocities must be sufficient to avoid excessive cooling and solidification, and pressure forces must be sufficient to drive the magma all the way to the surface.

2.2.3. Vesiculation and Fragmentation

During the final stages of magma ascent, the behavior of magmatic volatiles such as H_2O, CO_2, and SO_2 can critically influence the manifestation of the eruption. Volatile solubility is a function of magma composition and pressure (i.e., depth) (Mysen, 1977). CO_2 and SO_2 are less soluble than H_2O and usually they will exsolve at much greater depths. Consequently, these may be lost during transport or residence in a magma chamber. As the confining pressure decreases during ascent, the volatiles exsolve and gas bubbles nucleate and grow, initially by diffusion of the volatiles through the melt and into bubbles; later, as the pressure drops (i.e., approaching the external ambient pressure), bubble size increases and decompression dominates the growth (Sparks, 1978). If the near-surface magma rise speed is relatively low, bubble rise velocity may exceed that of the melt because the buoyancy forces increase with bubble size. Runaway bubble collision and coalescence may take place to form very large bubbles ascending much faster than the magma, producing intermittent, strombolian-type eruptions (see Section 2.5.2) (Blackburn et al., 1976). Alternatively, such gas pockets might form in shallow reservoirs by coalescence of a bubbly foam. On reaching the surface of the magma column the bubbles burst, ejecting a spray of small magma clots and fragmentation products formed from disrupted bubbles. At magma rise speeds of ~ 0.5 to 1 m s^{-1} the bubbles remain effectively locked to the magma during ascent. Once the gas occupies a critical fraction (~ 75 vol%) of the total magma volume (such that the bubbles are close-packed and further expansion is prohibited) which typically occurs at depths of a few tens to a few hundred meters, disruption of the magma into a collection of magma clots entrained in a gas stream is almost inevitable (Sparks, 1978). These conditions favor the formation of sustained, Hawaiian-type lava fountains or plinian columns (Wilson and Head,

1981). However, if enough gas is lost during either storage or transport, lava will simply effuse passively from the vent.

A number of interrelated factors, associated mainly with magma composition, influence the "explosivity" of an eruption. The low viscosity of basalt means that viscous opposition to bubble expansion is negligible, allowing bubbles to grow relatively large and precluding significant excess pressures in the bubble (Sparks, 1978). Although fragmented to some degree, the resulting eruptive products are coarse-grained. Bubbles in magmas with higher silica contents and viscosities can develop significant excess pressures and undergo sudden, violent decompression on reaching the fragmentation level, disrupting the magma into fine particles that are able to transfer heat efficiently to the surrounding atmosphere. The heated atmosphere is then entrained to drive buoyantly a vigorous eruptive plume (Wilson, 1976). By comparison, basaltic eruptions involve low gas pressures, large pyroclasts and, together with the low total volatile contents, these factors contribute to the low fountains (and lack of a tall convecting plume) that characterize typical explosive basaltic eruptions (Head and Wilson, 1989). More vigorous explosive basaltic eruptions are possible as a result of interaction with external water sources (e.g., Williams, 1983; Walker *et al.*, 1984). The above scenario can be compared to magma rise on other planetary bodies (see Chapters 4, 6, and 8).

2.3. GLOBAL SETTING: PLATE TECTONICS AND "HOT SPOTS"

2.3.1. The Unifying Theory of Plate Tectonics

For many the ears the location and distribution of volcanoes on Earth could not be easily reconciled with existing theories. During the 1960s, geophysical and geologic data led to the "Plate Tectonic Revolution" that was finally successful in presenting a unified theory that accounted for the global distribution of volcanic and seismic phenomena (see Press and Siever, 1974; Summerfield, 1991; Francis, 1994). The strong correlation of active volcanism and seismicity with the margins of crustal plates (Figure 2.1) provided a rationale for why

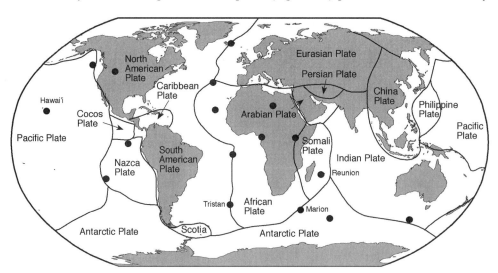

Figure 2.1. Plate margins superposed on the outline of the continents. Selected prominent hot spots are also indicated by dots. Redrawn from Simkin *et al.* (1994).

most volcanic eruptions occurred where they do. Concentrations of volcanoes are found at island arcs and continent-margin subduction zones, mid-ocean ridges, and continental rift valleys, representing either divergent or convergent boundaries.

2.3.2. Divergent Margins: Generation of New Crust

Plates of Earth's crust move away from each other at divergent margins. The spreading apart of two plates causes passive upwelling in the underlying asthenosphere, resulting in partial melting of the asthenosphere and subsequent volcanism. The gross morphologic and volcanologic characteristics of divergent margins vary widely, and depend on whether the rifting is occurring on land (such as the East African Rift) or underwater (such as the Mid-Atlantic Ridge). The vast majority of Earth's divergent margins are found on the ocean floor, where they form a submarine volcanic chain approximately 75,000 km long called the mid-ocean ridge (MOR). Spreading rate has a strong control on morphologic and volcanologic characteristics. For discussion, MORs are classified on the basis of spreading rate: slow ($\lesssim$30 mm a^{-1}, full spreading rate), intermediate ($\sim$60–90 mm a^{-1}), and fast (>90 mm a^{-1}). The northern Mid-Atlantic Ridge, excluding Iceland (the only subaerial portion of a MOR), typifies a slow-spreading center while the Juan de Fuca Ridge and northern East Pacific Rise are type examples of intermediate- and fast-spreading margins, respectively.

At a MOR, the forces of volcanism, magmatism, and tectonism interact to generate the overall ridge morphology, which is reflective of the relative dominance of these processes over the scale of a million years. However, all spreading ridges are characterized by a broad topographic rise capped by a central valley, graben, or collapse trough (Chadwick and Embley, 1998; Fornari et al., 1998). The width and depth of the summit depression depend on the long-term magma supply. For example, the slow magma supply to the Mid-Atlantic Ridge results in a central valley as wide as 40 km that is flanked by faulted walls reaching as high as 2 km above the valley floor (Mutter and Karson, 1992). The floor of the axial valley contains large volcanic constructs termed *axial volcanic ridges* that can be up to several kilometers long and as high as 2000 m (Smith and Cann, 1996). In contrast, the magmatically robust East Pacific Rise near 9°50'N is characterized by a narrow ($\sim$40 m wide), shallow (5–8 m) summit collapse trough formed by repeated near-surface collapse of lava flow crusts (Haymon et al., 1991; Fornari et al., 1998). The southern East Pacific Rise near 17°30'S does not display a summit trough, but instead appears to be covered with a relatively young lava flow that could have obliterated any previous structure on the axial summit (Auzende et al., 1996). These trough-filling eruptions may take place approximately every 5–10 years per 10 km of ridge (Fornari et al., 1998).

The precise frequency and duration of volcanic eruptions at MORs are unknown because technologic and financial constraints preclude constant monitoring of the global MOR system (see Chapter 5). The Juan de Fuca Ridge has been monitored with a SOund SUrveillance System (SOSUS) since 1993, and since that time, four eruptions have been detected (Fox et al., 1995; Chadwick et al., 1998). These events lasted up to 14 days, and emplaced $\sim 10^5$–10^6 m^3 of lava, giving an average effusion rate of $\sim$1 to 100 m^3 s^{-1} (Gregg and Fink, 1995; Fox, 1998; Chadwick et al., 1999). The site of an eruption between 9°46' and 52'N on the East Pacific Rise was visited shortly after activity ceased (Haymon et al., 1993). Results of recent modeling suggest that here $\sim 10^6$ m^3 of lava was emplaced in <2 h, giving an average effusion rate of 10^4–10^5 m^3 s^{-1} (Gregg et al., 1996). Based on mapping as well as numerical and physical modeling of submarine lava flow behavior, it appears that eruptions at fast-spreading centers are more frequent, of shorter duration, and have smaller volumes than those at slow-spreading centers (Smith *et al.*, 1995; Gregg *et al.*, 1996; Chadwick *et al.*, 1998).

Iceland, located directly on the Mid-Atlantic Ridge, provides the unique opportunity to study MOR volcanism and tectonism in a subaerial environment. The volcanism on Iceland consists primarily of basalt and icelandite (a term recently applied to possible basaltic andesite compositions on Mars, see Chapter 4). The Mid-Atlantic Ridge is visible on Iceland as a series of fissures across which the spreading of the rift has been closely monitored (e.g., Thorarinsson, 1967). Historic fissure eruptions in Iceland have provided valuable insight into divergent margin volcanism; these include the Laki eruption of 1783 (Thorarinsson, 1969; Thordarson and Self, 1993) and a fissure eruption within the Askja caldera, which was closely observed by geologists in 1961 (Thorarinsson and Sigvaldason, 1962). The manifestation of Icelandic (and MOR) volcanism is generally very distinct from that associated with rifting within continents, such as along the East African Rift, where exotic lavas like carbonatite (see Chapter 8) accompany the basalts.

2.3.3. Subduction: Destruction of Old Crust

Lithospheric material descends back into Earth's interior along subduction zones; as the name implies, lithosphere is forced under either oceanic or continental crustal materials. Subduction zones are compressive tectonic regimes, with driving forces resulting primarily from the pulling force of the descending slab, along with the relative motion of the plates on either side of the trench (see Summerfield, 1991, pp. 48–54; Bebout *et al.*, 1996).

Subduction zones formed between two oceanic plates are characterized by a topographic trench along the line where the subducted slab disappears, with a line of volcanic islands paralleling the trench on the overriding slab. Because of the geometry of a spherical planet, volcanoes associated with oceanic subduction zones tend to form broadly curving archipelagos ("arcs") of volcanic islands. The subducted lithosphere consists of both basalts and peridotites, plus a liberal coating of oceanic sediments and trapped seawater. As the slab descends into the hot mantle, water and other volatiles are "boiled" out of the slab. The volatiles ascending from the slab alter the properties of the asthenospheric wedge between the downgoing and overlying plates, leading to extensive partial melting (see Stein and Stein, 1996). These melts rise through the overlying oceanic plate to produce island arc volcanoes, typically found between 70 and 90 km from the associated subduction trench. Erupted materials include basalts, broadly similar to those found at ocean ridges but with a slightly different chemistry distinctive of island arc basalts, as well as minor amounts of silica-enriched basaltic andesites and andesites.

Where an oceanic plate collides with a continent, the resulting subduction zone is broadly similar to that of an oceanic subduction zone described above, but with the added complication of continental crustal materials. The ascending magma can now cause partial melting of the lower continental crust, providing new (usually silica-rich) materials for incorporation into the magma as it rises through this region. Continental margin volcanism, such as that found in the Andes along the western margin of South America, tends to be dominated by more evolved materials such as basaltic andesite, copious quantities of andesite and dacite, but very little basalt, most of which is underplated at the base of the crust (Francis, 1994, p. 38). Fractional crystallization can eventually lead to very silica-rich magma such as rhyolite.

Both island arc and continental margin volcanoes contribute to make one of the largest subaerial volcanic regions on Earth. The margin of the Pacific Ocean, comprised of many hundreds of active or recently active volcanoes, has been termed the "Ring of Fire"; it is the result of subduction processes at the convergent boundaries around the Pacific plate. In comparison with subduction zones, the magmatic sources for intraplate volcanism tend to be much deeper within the Earth.

2.3.4. Hot Spots: Deep Mantle Sources

The deep interior of the Earth accumulates heat at a rate faster than it is lost by the combined processes of conduction to the Earth's surface and of volcanism at oceanic spreading centers and subduction-related volcanoes. This causes large diapirs of hot mantle material known as *mantle plumes* (Figure 2.2) to rise toward the surface, presumably from the deep mantle (see Francis, 1994, pp. 38–47). As the large plume head rises, it remains connected to its source by a much narrower tail, through which additional material also rises.

When the plume head flattens itself against the base of the oceanic or continental lithosphere, it causes doming of the crust via both physical uplift and thermal expansion of the crustal rock (Phipps Morgan *et al.*, 1995), as well as radiating dike swarms (see Chapter 5). The material in the 800- to 1000-km-diameter plume head begins to melt primarily as a result of the decrease in lithostatic pressure. The volume of such a diapiric plume head is great, so the amount of magma it can produce is tremendous. Huge areas of flood basalts are the result, and they have been found across the globe (see Section 2.6). The Columbia River Plateau basalt in the USA's Pacific Northwest is a young and well-studied example; its total area is 164,000 km^2 and total volume is approximately 174,000 km^3, with individual basalt lava flows reaching up to 2000 km^3; 90% of its volume erupted over a period of 1.5 Myr (Tolan *et al.*, 1989; Reidel and Hooper, 1989). Other flood basalt provinces include the Deccan Traps in India and the Paraná Basin in Brazil, both on continental crust. Not all of the

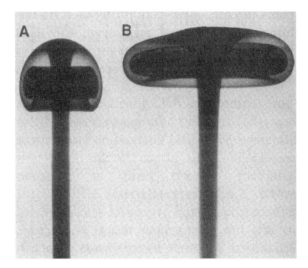

Figure 2.2. Plumes of material rising from the base of the mantle are thought to have a similar morphology and behavior to these plumes of hot glucose syrup (dark) rising diapirically through cooler syrup (light) in a lab experiment at the Australian National University. (A) Rising plume entraining cooler material. (B) Impingement of the plume head against a horizontal barrier (i.e., the base of the lithosphere). After the head becomes stagnant, hot material continues to rise through the tail of the plume. Modified from Hill *et al.* (1992).

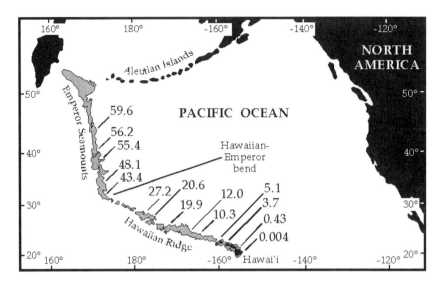

Figure 2.3. The Hawaiian-Emperor chain of seamounts and islands, marking the motion of the Pacific plate over the Hawaiian hot spot. Numbers show the oldest reliable ages (in Ma) of selected basaltic volcanoes. The bend in the chain marks a change in the vector of motion of the Pacific plate about 40 Ma. Redrawn from Calgue and Dalrymple (1987).

magma generated by the plume head is erupted; much may be intruded as dikes and sills within the crust, adding heat to the crust and changing its bulk composition. The plume head eventually equilibrates thermally with the upper mantle and ceases producing magma. However, hot mantle material may continue to rise through the tail for many millions of years. This represents a "hot spot" beneath the crust, so magmatism continues, though at much reduced rates (Richards *et al.*, 1989).

Hot spots are thought to be fairly stationary relative to the mobile tectonic plates that make up the Earth's surface. Thus, as the lithospheric plates move over hot spots, tracks of volcanic centers are formed on the crust, in a manner similar to the way burn holes form in a sheet of paper being moved horizontally just above a candle flame. In the ocean basins, hot spot traces are expressed as linear chains of islands and seamounts, each of which is composed of several basaltic shield volcanoes. The Hawaiian-Emperor island and seamount chain (Figure 2.3), active for at least 80 Myr, is perhaps the most extensive and best-known example (Clague and Dalrymple, 1987).

On the continents, hot spots generally leave tracks of silicic volcanic centers instead of basaltic shields. The Snake River Plain (White *et al.*, in press), in southern Idaho, USA, is believed to be primarily the track of the tail of the hot spot presently under Yellowstone (Figure 2.4). The head of this same plume may also be responsible for the Columbia River basalts. Volcanic activity along the Snake River Plain (SRP) began about 16 Ma in southeastern Oregon and southwestern Idaho and progressed to the Yellowstone Plateau, which has been volcanically active for the past 2 Myr. Both discrete and less well-defined silicic eruptive centers, up to at least 100 km in diameter, were formed sequentially as the North American tectonic plate moved southwestward over the plume. Basaltic magma generated from the material in the plume tail provides the heat source for SRP plume track volcanism, leading to abundant basalt and silicic products in close proximity. Fractional crystallization of the basaltic magma, plus partial melting of basalts intruded into the deep crust, generated the silicic magmas; melting of silicic crustal rocks can also occur. Hot spot magmatism tends

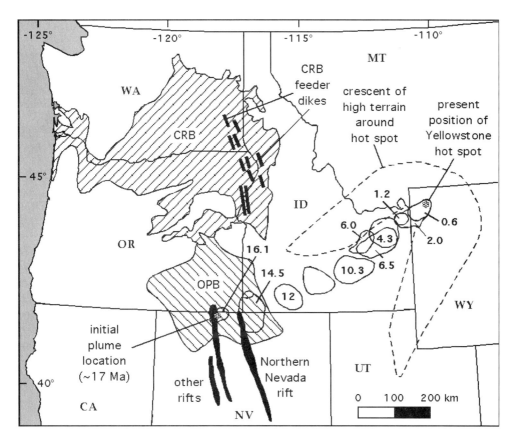

Figure 2.4. The Columbia River Basalt (CRB) and related volcanics in the northwestern United States. Impingement of the head of the Yellowstone plume at ~17 Ma is commonly thought to have led to rifting in northern Nevada, and possibly to the eruption of the CRB and the Oregon Plateau Basalts (OPB), both ~17–13 Ma. After magmatism associated with the plume head waned, the plume tail created a track of silicic eruptive centers (labeled with ages in Ma) as the North American plate moved southwest, relative to the hot spot. The hot spot is still active and is presently located beneath Yellowstone National Park. Redrawn from Pierce and Morgan (1992).

toward silicic magmas with somewhat lower water contents compared with silicic magmas from non-hot-spot-related centers. These lower water contents manifest themselves in extremely voluminous silicic lava flows, which are common features of hot spot centers (Bonnichsen, 1982; Manley, 1995). Elsewhere on the continents are large, long-lived, silicic centers similar to those along hot spot tracks, but which do not seem associated with any hot spot.

2.4. EFFUSIVE ERUPTIONS

2.4.1. Lava Flows

The flow of lava is the primary mechanism through which effusive eruptions build edifices and invade the surrounding terrain. Flow properties are inherently associated with a host of parameters such as viscosity, density, temperature, and crystal content (see Chapter 9), but remarkably most lava flow textures fall within a few broad categories. The following discussion focuses primarily on basaltic lava (see Williams and McBirney, 1979, pp. 106–112), but includes brief references to other significant compositional types. Also note that transitional textural types can be found between these basic types.

A'a. The Hawaiian word *a'a* refers to flows covered with jumbles of rough, clinkery, and spinose fragments, ranging from small chips to blocks meters across (Figure 2.5a). Flows with a'a texture range from one-half to tens of meters in thickness, and are usually comprised of distinct lobes fed by a central leveed channel. Once solidified, a'a lava typically displays rubbly upper and lower surfaces, with a massive central unit where the liquid core of the flow solidified under the insulating effects of the rubble zones. Basalts, basaltic andesites, and andesites all commonly display a'a textures.

Pahoehoe. The Hawaiian word *pahoehoe* refers to flows with smooth crusts that can occur in a bewildering array of shapes, ranging from twisted ropes (Figure 2.5b) to shelly blisters. Pahoehoe textures result from fluid lava usually emplaced at relatively low effusion rates (see Section 2.4.3); pahoehoe flow fields are composed of many thousands of individual flow units or "toes." These toes are usually <50 cm thick, up to 1 m wide, and up to 10 m long. Individual flows can transition from pahoehoe to a'a, particularly if the localized flow rate increases such as on a steep slope. Adjectives such as *scaly, shelly, slabby,* and *massive* have been used in field descriptions of pahoehoe flows, along with tumuli and "squeeze-ups" for pahoehoe that is locally extruded along cracks into the flow crust. A glassy pahoehoe surface is readily weathered in temperate environments, leaving vesicular blocks lacking the intricate pahoehoe surface textures. Near-vent basalts commonly have pahoehoe textures, but pahoehoe is usually absent from more silica-rich lavas. Recently portions of the Columbia River Basalt (CRB) sequence have been interpreted to be inflated pahoehoe flows (Self *et al.*, 1996, 1997).

Block Flows. Block flows are sometimes considered synonymous with a'a flows, but the term should be restricted to flows made up largely of detached, polyhedral blocks with planar to curved faces (Figure 2.5c). Block basalt flows are much less common than a'a flows, and they are best developed in intermediate to siliceous flows that formed thick, glass-rich crusts that subsequently fractured into angular blocks. Andesite, dacite, and rhyolite flows display blocky characteristics that grade with increasing volume into domes, discussed in Section 2.4.4 below.

Pillow Lavas. Pillow lavas may be the most abundant volcanic rocks on Earth, since they form where lava is slowly erupted under water (see Chapters 3 and 5); they are essentially the underwater equivalent of pahoehoe. Because this chapter concentrates on subaerial eruptions, we will not discuss them further here. However, the distinctive budded pillow morphology occurs in flows ranging from ultrabasic to rhyolitic in composition; more viscous lavas tend to generate larger pillows.

2.4.2. Constructs

When a centralized volcanic vent is active for a prolonged period of time, substantial constructs are formed through the accumulation of lava flows and pyroclastic material. As with lava flows, volcanic constructs take on a myriad of shapes and forms. Fortunately, two broad types of constructs encompass the majority of volcanic landforms encountered throughout the world.

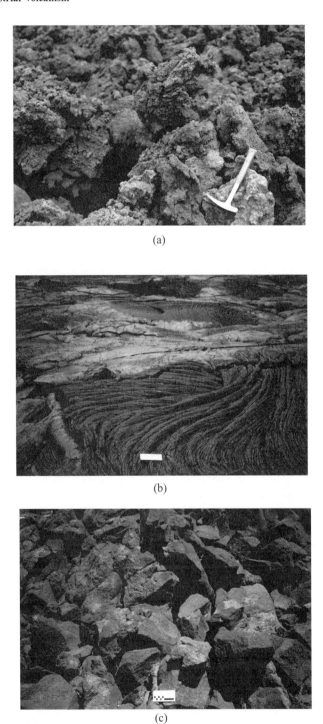

Figure 2.5. Lava flow textures. (a) A'a flow. Episode 5 of the Pu'u O'o eruptions of Kilauea, Hawaii. (Photo by J. Zimbelman, Jan. 1984.) (b) Pahoehoe flow. 1969–1972 flows from Mauna Ulu, Hawaii. (Photo by J. Zimbelman, Sept. 1983.) (c) Block flow. SP flow near Flagstaff, Arizona. (Photo by J. Zimbelman, June 1998.)

Shield Volcanoes. Repeated eruption of fluid lava from a central vent produces a broad convex-upward volcanic edifice called a *shield volcano* (Figure 2.6a), based on its resemblance to the rounded shields of early Germanic warriors (Macdonald, 1972, p. 271). Shield volcanoes typically have flanks with slopes of 4° to 5°, culminating in a collapse caldera or caldera complex near the summit. The edifice is comprised of countless individual lava flows stacked adjacent to and on top of each other, with only a very small amount of included pyroclastics. Shield volcanoes are predominantly comprised of basaltic lavas, although basaltic andesites are not uncommon. Well-known examples include Mauna Loa and other volcanoes in Hawaii, the smaller but very symmetrical Icelandic shields like Skjalbreiður, and the Galapagos archipelago where some volcanoes have flank slopes >20° (Williams and McBirney, 1979, pp. 197–205). Huge shield volcanoes occur on other planets, including both Mars (see Chapter 4) and Venus (see Chapter 5).

Composite Volcanoes. If effusive flows (generally more viscous than basalt) are intermixed with substantial pyroclastic deposits, a composite volcano is formed (Figure 2.6b), the classic conical form that even nongeologists associate with a volcano. The term *composite* is preferable to the earlier term *stratovolcano* because shields and many domes can also be considered to be stratified (Williams and McBirney, 1979, p. 179). Rarely is the composite structure built of regularly alternating effusive and pyroclastic layers, but the relatively common mixture of flows and pyroclastics (Figure 2.6c) lets the volcano flanks attain slopes of 10° to 12°, more than twice the average slope of a shield volcano. Composite volcanoes are predominantly comprised of andesite and dacite materials, which contributes to the abundance of pyroclastics, although once again basaltic andesites are not uncommon. Composite volcanoes are common throughout the world, but Fujiyama in Japan, several of the Cascade peaks in the western United States, and preeruption Mt. Pinatubo in the Philippines are among the better-known examples.

Other Volcanic Centers. Many volcanoes do not fall into either the shield or composite category. Pyroclastic (cinder) cones are monogenetic constructs composed of scoriaceous ejecta (Figure 2.6d). When a cinder cone eruption stops, it rarely erupts again at the same vent; instead, a new cinder cone forms in the general vicinity of the previous eruption, resulting in local concentrations of tens to hundreds of cinder cones. Fissure eruptions likely produce the largest volumes of effusive materials (see Section 2.7), but they rarely result in significant landforms other than low spatter cones or ramparts along the active segment of the fissure. Volcanic domes represent an important volcanic construct distinct from other effusive or explosive volcanoes (see Section 2.4.4).

2.4.3. Flow Effusion Rates

The emplacement of lava flows is governed by the lava rheology (a complex function of composition, temperature, volatile content, and so on), preflow topography (slope and roughness), and effusion rate, perhaps the single most important parameter in characterizing an eruption. An evaluation of documented Hawaiian eruptions led Rowland and Walker (1990) to conclude that effusion at a rate <10 m^3 s^{-1} results in almost exclusively tube-fed pahoehoe flows whereas eruptions at >20 m^3 s^{-1} are almost exclusively a'a flows. The average effusion rate for current eruptions at Kilauea is $\sim$5 m^3 s^{-1} (Wolfe *et al.*, 1989). From 1983 to mid-1986 the Kilauea eruption was characterized by distinct episodes at high effusion rates that produced high fountains and fast-moving a'a flows. Since mid-1986 the eruption has been more or less continuous, producing pahoehoe flows and little or no pyroclastic activity. In contrast, the 1984 Mauna Loa eruption emplaced a 27-km-long a'a flow over 21

Subaerial Terrestrial Volcanism 21

(a)

(b)

(c)

Figure 2.6. Volcanic constructs. (a) Shield volcano. Mauna Loa, Hawaii. (Photo by J. Zimbelman, Sept. 1983.) (b) Composite volcano. Villarrica, Chile. (Photo by J. Zimbelman, Dec. 1985.) (c) Ash interbedded with basaltic andesite lava, on the northern flank of Villarrica volcano, Chile. (Photo by J. Zimbelman, Dec. 1985.)

Figure 2.6. Volcanic constructs. (d) Cinder cone. Sunset Crater, north of Flagstaff, Arizona. (Photo by J. Zimbelman, July 1992.)

days at an average eruption rate of ~ 110 m^3 s^{-1} (Lipman and Banks, 1987). The largest effusion rate inferred for a historical flow comes from the Laki eruption of 1783–1785 in Iceland, with a peak effusion rate of 4600 m^3 s^{-1}, equivalent to an effusion rate of ~ 6 m^3 s^{-1} per meter of active fissure (Thordarson and Self, 1993). Such observations are important considerations when examining the emplacement of large lava flows on other planets (Zimbelman, 1998).

Lava tubes can significantly enhance the delivery of relatively unaltered lava to an active flow front well removed from the vent (e.g., Greeley, 1987). Tubes develop primarily within fluid basaltic lavas, at least in some cases through the roofing over of an active channel (Greeley, 1971). Complex networks of tubes can feed the emplacement of many large flow fields. Some drained lava tubes can be very large; tube sections in the Undara volcanic field in northeastern Australia attain widths of up to 20 m (Atkinson *et al.*, 1975). Calculations of the thermal efficiency of a well-developed tube indicate that lava can be transported many hundreds of kilometers without losing enough heat to alter the lava rheology (Keszthelyi, 1995; Sakimoto *et al.*, 1997). Tubes may play an important role in the emplacement of enormous volumes of basaltic lava found around the world (see Section 2.6).

2.4.4. Volcanic Domes

In the simplest definition, a lava dome is an extrusion of lava with a thickness comparable to its diameter. Lava domes form most often from lavas with high viscosities—andesite, dacite, rhyolite, and trachyte. Several broad types of domes have been distinguished, based largely on gross morphology, but with differences in lava rheology (primarily the yield strength of the lava) as the underlying factor. Domes can display a variety of surface textures (e.g., Anderson and Fink, 1990), but the various emplacement mechanisms lead to the following basic types.

Upheaved Plug Domes. When the lava solidifies in the vent and becomes very strong before reaching the surface, it may be forced slowly upward in a steep-sided, coherent mass of rock that can barely spread under its own weight. Lassen Peak, in northern California, USA, and the Chain of Puys in France are well-known examples.

Peléean Domes. Lava that reaches the surface before solidifying, but which still has an appreciable strength, will form a dome similar to that extruded during 1902–1903 at Mont Pelée, Martinique. Spines (small versions of upheaved plugs) and craggy piles of rubble protrude from the summits of Peléean domes, and their lower portions are usually completely engulfed by aprons of rubbly talus. The spines themselves are transitory, as they often crumble and collapse as they cool. Episodic growth over many years, punctuated by avalanches and explosions, is common with these domes. It was a block-and-ash flow from the Mont Pelée dome that destroyed the town of St. Pierre on Martinique in 1902 (see Section 2.5.2).

Coulees. Lava that would form a circular dome on a flat surface will tend to flow slowly down a slope, forming a coulee (a short but very thick lava flow). Most coulees are covered by a carapace of broken pumiceous or scoriaceous blocks. The Mono Domes in eastern California, USA, contain several classic examples. If a large volume of magma is available to erupt, a coulee may continue to increase its lateral extent, usually with only a minor increase in height; a dome may thus lose its domical aspect and become a lava flow.

Low Domes. Lava sufficiently low in viscosity so as to spread slowly away from the vent on a flat surface will form a generally circular, smooth-profile dome called a *low lava dome* by Blake (1990). Although their upper surfaces can be covered with loose rubble, they lack the craggy relief typical of a Peléean dome. Avalanches and explosions are uncommon. A well-studied example is the 1979 Soufrière dome on the island of St. Vincent (Huppert *et al.*, 1982).

Cryptodomes (Intrusive Domes). In some cases, the magma does not reach the surface but intrudes at shallow depths, doming or lifting up the overlying rock and soil. These intrusive domes have been termed *cryptodomes*, for the magma itself remains hidden.

The Dome Spectrum. A complete spectrum exists from plug domes through low domes to coulees and lava flows (Figure 2.7), because the controlling factors of viscosity and erupted volume also vary smoothly. For example, the dome that grew in the crater of Mount St. Helens from 1980 to 1986 showed aspects of both low and Peléean-type domes. Other domes change character abruptly during their growth. In the Inyo Domes of eastern California, USA, the last lava that erupted was more viscous and had a higher yield strength than the first, so it piled up high over the vent, forming a craggy Peléean-type dome surrounded by a low-dome aureole (Figure 2.8) only half as thick (Sampson, 1987).

Figure 2.7. South Deadman dome, in the Inyo volcanic chain, eastern California, USA. The dashed line separates the earlier-erupted, low-yield-strength, crystal-poor lava, which makes up a flat-topped aureole that surrounds the tall (Peléean-type) center, comprised of later-erupted, high-yield-strength, crystal-rich lava. Photo courtesy of Allen F. Glazner.

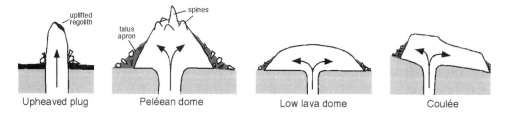

Figure 2.8. The four types of extrusive lava domes. Their morphologies are primarily controlled by the viscosity and yield strength of the lava. Redrawn from Blake (1990) and Francis (1994).

Viscosity and Size. Although the high viscosity of most silicic magmas might seem to limit the size of lava domes and flows, in fact it simply slows the rate of advance of the lava across the Earth's surface. The distance the lava may move from the vent then becomes dependent on the balance between the lava's cooling and crystallization on the one hand, and its rate of supply on the other (Manley, 1992). Voluminous silicic lava flows have now been mapped in many locations on Earth, but the great majority are associated with the trails of mantle plume hot spots across the continents, such as the Snake River Plain hot-spot track across southern Idaho, USA (Bonnichsen, 1982), and the Paraná continental flood basalt (and rhyolite) province of southern Brazil (Garland et al., 1995).

2.4.5. Interactions with Meteoric Water

When a subaerial lava flow advances into wet areas such as a lake bed or a marshy depression, water can become trapped beneath the lava and flash into steam, bursting explosively through the lava to form a pseudocrater or rootless vent (Francis, 1994, pp. 151–152). One of the best-known locations for pseudocraters is around Lake Myvatn in Iceland where dozens of pseudocraters are clustered. Pseudocraters are discussed in more detail in Chapter 3, because lava–ice interactions can also contribute to steam explosions. If ground water within a shallow aquifer is brought into contact with ascending magma, steam explosions can excavate into the preexisting substrate and produce a broad, shallow depression termed a *maar*, after lake-filled craters in Germany generated in this fashion (Francis, 1994, pp. 341–342). As the water–magma ratio increases, phreatomagmatic eruptions can achieve significantly increased mechanical energy release (governed by fuel–coolant reactions) that result in tuff cones or tuff rings as the energy of emplacement increases (Wohletz and Sheridan, 1983). If the water–magma ratio continues to increase beyond the tuff cone/ring stages, eventually pillow lavas result from the rapid chilling of lava lobes in the surrounding water. When lava enters standing water >1500 m deep, the hydrostatic pressure of the water is sufficient to inhibit the formation of steam, leading to the benign generation of pillow lavas (see Chapter 5).

Interaction of basaltic volcanic ash with water shortly after emplacement can lead to alteration of the glassy shards by hydration to form palagonite. The orange-brown color of palagonite is distinctive of this wet alteration process; these materials have reflectance spectra very similar to those of the bright dusty regions on Mars. If palagonite is a significant component of the Martian fines, this would strengthen the likelihood that both basaltic products and access to a hydrating agent such as water were readily present at some point in Martian history (see Chapter 4).

2.5. EXPLOSIVE ERUPTIONS

2.5.1. Pyroclastics

Pyroclastic rocks (Greek, "fire-broken") consist of material that is fragmented and ballistically ejected by expanding gases, and which subsequently accumulates into a deposit (Macdonald, 1972, p. 23). Pyroclastics can encompass other descriptive terms like *pumice* or *ash*, although the latter term also indicates a specific size range. What mechanisms contribute to this breaking of rocks by expanding gases? By far, the most important mechanism is related to the initial fragmentation of magma, either during ascent or at the surface (see Section 2.2). Fragment sizes relate directly to basic properties of the magma itself, such as viscosity and volatile content. The greater the gas content and viscosity, the higher the gas pressure can build up within the magma; the stronger the explosions, the smaller the resulting fragments. After eruption, additional processes may further reduce the size of the erupted products. Some fragmentation results from the tensile stress exerted when the particle cools from a molten state, and abrasion between fragments during transportation can both round the particles as well as generate abundant fine abrasion products. If conditions are right, some pyroclastic particles actually grow in size after eruption, such as accretionary lapilli formed when moist ash particles adhere together, but this requires abundant meteoric water to form.

Pyroclastic materials can be broadly subdivided by the readily observable quantity of particle size. Pyroclastic deposits are termed *ash* when individual particles are <2 mm across, *lapilli* (Latin for "little stones") when between 2 and 64 mm, and *bombs* when >64 mm (Fisher, 1961). If particle size is not part of the intended description, the more generic Greek term *tephra* is preferred (Francis, 1994, p. 185). Deposit thickness measurements can be combined with the areal extent of a deposit; if measurements are sufficiently distributed, this provides a reasonable estimate of the deposit volume. Other important quantities either measured in the field or determined from later laboratory measurements include the maximum clast size in a given deposit at a given location (related to the energy of emplacement), the degree of sorting, the vesicularity of the clasts, and the presence or abundance of either crystal or lithic fragments.

Pyroclastic materials do not in themselves imply any particular composition. Magmas from basaltic to rhyolitic compositions can all produce pyroclastic materials if the eruption is sufficiently vigorous to break apart the erupting liquid. In general, the areal extent of mafic pyroclastics tends to be more restricted than that of either andesitic or especially rhyolitic magmas, which typically produce increasingly explosive eruptions.

2.5.2. Modes of Emplacement

The depositional products from explosive eruptions can be broadly divided between falls and flows where, as the names suggest, the former descend either ballistically or roughly vertically out of the atmosphere while the latter are emplaced by movement across the surface. There is no way to cover the diversity found within such deposits within one summary chapter; the interested reader is referred to standard texts for more detailed treatment (e.g., Williams and McBirney, 1979; Fisher and Schmincke, 1984; Cas and Wright, 1987; Francis, 1994).

Falls. The degree of explosivity of an eruption is a useful way to classify eruptive styles and the pyroclastic products that result from them. Early on these distinctions were assigned to type volcanoes displaying specific attributes, but more recently the Volcanic Explosivity

Index (VEI) has provided a way to quantify some of these eruptive types based on the size and dispersal of pyroclasts (Newhall and Self, 1982). At the low end of the explosive sequence are Hawaiian eruptions, which are primarily effusive but which can be accompanied by fire-fountaining that deposits a wide variety of glassy particles (e.g., spatter, clots, "Pélé's tears," reticulite). Strombolian eruptions launch magmatic cobble-sized vesicular cinders (scoria) that collect around the vent to produce the classic cinder cone; rare larger blocks are termed *volcanic bombs*. Vulcanian eruptions incorporate a large fraction of nonmagmatic lithic fragments derived from the initial "throat-clearing" and subsequent erosion of the vent wall, with products ranging in size from ash to bombs. Strombolian and Vulcanian eruptions discussed thus far deposit materials almost exclusively in the immediate vicinity of the vent, with very limited dispersal, in marked contrast to the very explosive Plinian eruption.

Plinian eruptions derive their name from both the famous Roman naturalist, Pliny the Elder, who died in the catastrophic outburst of Vesuvius in 79 AD, and Pliny the Younger, who meticulously described the eruption (Macdonald, 1972, p. 231). Plinian eruption columns can enter the stratosphere in the most dramatic cases. In Plinian and sub-Plinian explosive eruptions, magma (commonly with preeruptive water contents of 4 to 6 wt%; Lowenstern, 1995) vesiculates to the point of explosive comminution as it rises in the conduit toward the Earth's surface. This produces tephra with a large proportion of fine ash particles, so the eruptive column above the vent becomes a roiling, turbulent mass of gas and dust. Entrainment of air into the column completely cools the fine-grained tephra and carries much of it high into the atmosphere and far from the vent. The prodigious ash produced by these eruptions may travel large distances, blanketing the surroundings with deep layers of bedded deposits. Both nonmagmatic lithics and highly vesiculated magmatic pumice fragments abound; the sizes of these can provide information on the variability of intensity during the course of an eruption (Sigurdsson *et al.*, 1985). A Plinian deposit represents the most explosive product from a single vent, surpassed only by massive caldera collapse eruptions that lead to ignimbrite deposits, discussed below.

Flows. Pyroclastic flow products range from low-density, vesiculated pumice to dense, unvesiculated lava clasts; the following discussion utilizes the broad division of pyroclastic flows derived from this distinction (Fisher and Schmincke, 1984).

Ignimbrites. In broad terms, an ignimbrite is the deposit left behind by a pyroclastic flow, consisting of poorly sorted mixtures of pumice blocks and lapilli in a matrix of fine ash (Williams and McBirney, 1979, pp. 161–165). Ignimbrites are predominantly rhyolitic in composition, although rhyodacites and dacites are also common. An ignimbrite can result from one or more individual pyroclastic flows, but if multiple flows are emplaced rapidly enough, they can result in a single compound cooling unit. The pyroclastic flows that produce an ignimbrite result from many causes, but the most widely held model involves flow by collapse of a Plinian column when the local cloud density can no longer be supported by buoyancy (Sparks *et al.*, 1997, pp. 144–178). Ignimbrite materials do not conduct heat very efficiently, so that the heat remaining after emplacement can partially to completely weld the pumice and ash into a central zone of dense obsidianlike glass (Fisher and Schmincke, 1984). The top of an individual flow within an ignimbrite sometimes consists of fine co-ignimbrite ash, which settles from the cloud of hot gas and ash that rises above an active pyroclastic flow (Sparks *et al.*, 1997, pp. 180–208). The 1815 eruption of Tambora produced the largest accumulation of ignimbrite and co-ignimbrite deposits yet identified for a historic eruption (Sigurdsson and Carey, 1989).

Block-and-Ash Flows. Observers of the eruptions of Mt. Pelée in 1902 introduced the term *nuées ardentes* ("glowing clouds") to describe the pyroclastic block-and-ash flows that periodically swept down the volcano, one of which wiped out the town of St. Pierre. For our purposes, nuées ardentes can be considered pyroclastic flows where the magmatic component is dense rock rather than the vesiculated pumice found in ignimbrites (Francis, 1994, p. 247). The mechanism of generation of such flows can range from explosive disruption of a volcanic dome, as occurred on Mt. Pelée (Boudon and Gourgaud, 1989), to hot avalanches caused by the nonexplosive gravitational collapse of a dome, as at Merapi, Indonesia, and Unzen, Japan (Francis, 1994, pp. 250–253), to eruption column collapse incorporating hot juvenile blocks, such as the 1968 eruption of Mayon (Moore and Melson, 1969). The deposits from all of these pyroclastic flows are generally very poorly sorted, chaotic, and unstratified; these deposits are now termed *block-and-ash flows* because of the distinct particle sizes involved.

Surges. The precise origin of pyroclastic surges remains rather controversial, but their distinctive characteristics include reduced bulk density (relative to ignimbrites and block-and-ash flows) and evidence of turbulent rather than laminar emplacement regimes (Francis, 1994, p. 236) Surges are usually associated with hydromagmatic eruptions, where the interplay of nonjuvenile water and magma is dominant and which leads to a high degree of fragmentation (Sheridan and Wohletz, 1983; Wohletz and Heiken, 1992, pp. 26–37). Surges are also identified with the initiation of many ignimbrites, and they may form crucial parts of the column collapses involved in the AD 79 eruption of Vesuvius (Sheridan et al., 1981; Sigurdsson et al., 1985). Surge deposits often include sections with sand waves and climbing dunes, indicative of emplacement by saltation (bouncing along the ground) rather than full suspension (Sigurdsson et al., 1987), which is directly related to the intensity of energy release associated with a hydromagmatic eruption (Sheridan and Wohletz, 1983).

2.5.3. Fountain-Fed Rhyolitic Eruptions

Fire-fountaining is commonly considered an eruptive mode confined to basalts and other mafic magma compositions with relatively low viscosities, but an increasing number of rhyolitic units are interpreted to have formed by fountaining. They range from small welded tephra rings through moderately sized secondary lava flows to widespread welded airfall tuffs that blanket any preeruptive topography (Christiansen and Blank, 1975; Duffield, 1990; Manley, 1994; Christiansen, in press; Manley and McIntosh, in press), and are known primarily from Yellowstone National Park in Wyoming, southwestern New Mexico, and southwestern Idaho, all in the United States. The fountaining mode of eruption in these units is facilitated not by low viscosities caused by high temperatures, but rather by the behavior of the erupting magma before and during disruption in the conduit. These units all seem to have preeruptive water contents lower than those erupted in a Plinian manner, and many have high fluorine contents. Rhyolitic magma with a preeruptive water content of less than about 3 wt% (Manley, 1994, 1996) becomes less comminuted and produces a much smaller proportion of ash-sized tephra than those with more water. This results in a low-bulk-density eruptive column; entrainment of air is minimized, as is heat loss, and the clasts are emplaced ballistically, as in a basaltic fire-fountain. Because more of the tephra lands nearer the vent and remains hot enough to weld, a welded airfall deposit forms. High magmatic fluorine contents of up to 3 wt% (Webster and Duffield, 1991) may play an important role by decreasing the viscosity (Dingwell et al., 1985) of the rather dry magma sufficiently that it is still able to rise through the conduit and erupt.

2.5.4. Modification of Ash Deposits

Ash deposits can generate distinctive geomorphic terrains. Large ignimbrite deposits fill in preexisting topographic lows to produce remarkably flat units that can bury valleys and embay the surrounding hills, often with only a gradual slope extending away from the vent area. Individual pyroclastic flows can have a planform similar to that of lava flows but which can demonstrate remarkable fluidity in detail in skirting around topographic obstacles. If remobilized by water, ash can produce enormous mudflows (identified by the Philippine word *lahar*) capable of doing considerable geomorphic work and moving at rapid speeds when traveling over steep slopes. Remobilized ash is often difficult to distinguish from a traditional mudflow without close examination of the deposits.

The erosional characteristics of ash deposits also are not uniquely diagnostic of a pyroclastic origin, but they do highlight variations in consolidation that may be difficult to duplicate with other processes. Thick ignimbrites usually show a vertical progression of alteration and competency that results from the slow cooling of such deposits (see Section 2.4.2). A densely welded zone can form near the middle of the deposit, which can make a prominent cliff-forming unit as compared with the overlying alteration zone (where vapor-phase interactions are important) and the underlying weakly consolidated ash, both of which rapidly weather. Large ignimbrites in the Andes are locally sculpted into immense yardang fields where wind and sand supply are sufficient, as along the elevated Altiplano in Bolivia and Chile (de Silva and Francis, 1991). However, yardangs themselves are not diagnostic of ignimbrites, only of a material strong enough to hold steep slopes but weak enough to be scoured by aeolian processes (see Chapter 4 for a discussion of yardang fields associated with hypothesized ignimbrite deposits on Mars).

2.6. LARGE IGNEOUS PROVINCES

2.6.1. Distribution

LIPs represent some of the largest volcanic features on Earth, both in terms of areal coverage and total volume of lava (Table 2.1). LIPs occur across the entire globe (Figure 2.9); several are located on the ocean floor, including the Ontong Java Plateau, which at $\sim 3.6 \times 10^7$ km^3 is the largest LIP on Earth (Coffin and Eldholm, 1993). The most intensely studied LIPs (to date) are found on continents, such as the CRB group in the Pacific Northwest, USA, and the Deccan Traps in western India. The diverse settings for LIPs result in considerable variability in what is known about a given site, yet to date an impressive global data set has been compiled for various LIPs (Macdougall, 1988; Mahoney and Coffin, 1997).

Why do such voluminous outpourings of lava occur in these locations? There are several hypotheses, numbering almost as many as there are identified LIPs. The most widely cited explanation ties together both a mantle plume and the initiation of plate rifting. White and McKenzie (1989) relate the Paraná basalts in southeastern South America (see Renne *et al.*, 1992) and the Etendeka basalts in southwestern Africa (see Renne *et al.*, 1996) to continental rifting and the opening of the south Atlantic. Rifting was associated with the buoyant rise of the mantle into the thinned lithosphere, which led to decompression melting that generated copious quantities of basalt magma. This process may have contributed to the opening of the south Atlantic as well as the formation of a stable mantle plume whose volcanic products can be traced from the South American and African basalts to the Tristan island group, which

Subaerial Terrestrial Volcanism

Table 2.1. Large Igneous Provinces[a] Emplaced in the Past 300 Myr

Name	Province	Age (Ma)	Area (10^6 km^2)
Snake River	Western United States (SRP)	0.002–15	0.3
Columbia River	Western United States (CRB)	6–17	0.2
Iceland	North Atlantic Tertiary Basalts	7–20	0.04
Traps	Ethiopia	20–40	0.8
Britain/Greenland	North Atlantic Tertiary Basalts	45–65	0.7
Deccan	India	60–67	0.5
Caribbean	Caribbean	70–93	1.0
Ontong Java	Pacific Ocean	90 & 122	1.5
Kerguelen	Southern Indian Ocean	90–115	0.7
Paraná	Brazil	115–135	1.2
Manihiki	Pacific Ocean	115–125	0.7
Karoo	South Africa	110–200	1.4
Patagonia	Argentina	190–210	0.7
Siberia	Russia	250–280	1.5

[a] Information from Coffin and Eldholm (1993) and Mursky (1996).

represents the recent output from the plume source (Ewart *et al.*, 1998) (see also Section 2.3.4). Similar conclusions can be drawn for other LIPs such as those around the north Atlantic, but with volcanic tracks less obvious than the Paraná–Tristan–Etendeka volcanics. Interested readers are referred to Mahoney and Coffin (1997) for a recent compilation of both observations and hypotheses for many LIPs.

2.6.2. Emplacement

There is considerable debate concerning how LIPs were emplaced. Favored mechanisms can be separated into diametrically opposed camps: rapid emplacement of fast-moving flows and slow emplacement with progressive inflation of the individual flows. The rapid

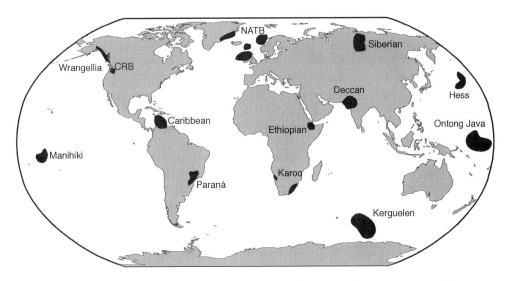

Figure 2.9. Large Igneous Provinces (black), present on both continental (gray) and oceanic (white) crust. See Table 2.1 for information on selected examples.

emplacement scenario was pioneered by Shaw and Swanson (1970), who made the first quantitative estimates of CRB emplacement by relating steady effusion along feeder dikes to turbulently flowing lava powered by the hydraulic head resulting from the gentle slope of the lava surface. Their calculations led to emplacement times for individual CRB flows of days to weeks, an interpretation favored by some more recent investigations (Swanson et al., 1989, pp. 21–26; Tolan et al., 1989; Reidel and Tolan, 1992). In contrast, recent observations of inflated basaltic flows in Hawaii provide a mechanism for slow emplacement of what eventually becomes a thick flow (Walker, 1991; Hon et al., 1994). Inflation features recently have been tentatively identified in CRB flows at several locations, suggesting that individual CRB flows may have been emplaced over years to decades (Self et al., 1996, 1997). An advantage of the slow emplacement mechanism is that required effusion rates are comparable to documented effusion at Hawaii and Laki (see Section 2.4.3), rather than invoking effusion rates that are orders of magnitude larger than any documented eruption. One recent model attempts to incorporate elements of both fast and slow emplacement to explain the CRB (Reidel, 1998).

The inflation mechanism has applications to flows other than just those of basalts or LIPs. Ultramafic komatiite flows have compositions suggestive of very fluid lava (see Chapter 8), but komatiites typically are quite old and preserved evidence of flow emplacement is often difficult to find. Inflation has recently been proposed as a way to reconcile known komatiite characteristics with models for the generation and eruption of ultramafic lavas (Cas and Self, 1998). Clearly all investigators need to look for details that might test the inflation hypothesis, not only in basaltic terrains, but also with more exotic lavas.

2.7. DISCUSSION: VOLCANO/ENVIRONMENT INTERACTIONS

2.7.1. Physical Interaction between Volcanic Products and Earth's Environment

As volcanic materials leave the vent, they will immediately interact with whatever environment is surrounding the volcano. This interaction can take the form of generation of a solid crust, mechanical degradation of solidified materials, or even agradation of fine particles (e.g., accretionary lapilli). Here we will not explore all of the variety of ways this interaction manifests itself, because that is one of the main goals of this book. However, it may be helpful to discuss briefly some of the physical consequences of fresh volcanic materials erupted subaerially on Earth (subaqueous eruptions are discussed in Chapter 5).

The tensile strength of a geologic material is a fundamental physical property that affects not only possible tectonic deformation but also how the material responds to sudden environmental change. Without question, the abrupt transition from conditions within the vent to those at the Earth's surface will be the most dramatic change the magma experiences. Physical properties have been measured for numerous igneous and volcanic rocks (e.g., Shaw et al., 1968; Murase and McBirney, 1973; Robertson and Peck, 1974; Horai, 1991), but the important issue here is the total magnitude of the temperature contrast experienced by a volcanic product and the rate at which the thermal change is imposed. In both effusive and explosive eruptions, the temperature contrast will be very large (up to 1000°C) and rapidly applied (seconds). This combination exerts a substantial mechanical stress on the cooling material.

For effusive eruptions, the temperature contrast and the thermal conductivity of the lava allow crustal growth to be modeled (Crisp and Baloga, 1990; Kilburn and Lopes, 1991;

Manley, 1992; Kilburn, 1993). Crusts on slow-moving pahoehoe flows allow the discrete "toes" (~10 to 30 cm in width) typical of pahoehoe to form and grow at a rate comparable to the average rate of flow advance. This newly formed glassy crust remains sufficiently heated (by the still-molten lava within the flow) to deform plastically, so that intense fracturing of the pahoehoe surface typically does not occur during emplacement. Faster-moving a'a flows disrupt the surface of the active lava channel into centimeter-scale clinkers that aid somewhat in insulating the flowing lava beneath them. However, gaps between solidified portions of an active flow surface, produced either by thermal stress or by the shearing action of the flow, allow incandescent lava to radiate energy more efficiently than does the cooled crust, which can significantly affect the cooling history of the flow (Crisp and Baloga, 1990). When a flow section comes to rest, thermal stresses can still contribute to pervasive fracturing of the outer sections of the flow.

Which is more important: the physical conditions of the magma at eruption, or the external environment into which the eruption takes place? It is unlikely that any one answer will apply to all volcanic eruptions, especially when the discussion extends to the plethora of environmental conditions found throughout the solar system. Instead, the combination of lava properties and the environmental conditions will be playing against each other to constrain the condition of the final volcanic product. In each individual situation, it is important to consider this dynamic interplay of material properties and environment when evaluating volcanic features (see Chapter 9).

2.7.2. Volcanic Effects on the Atmosphere and Climate

Historic eruptions provide the opportunity to look for potential links between volcanism and environmental change (Francis, 1994, pp. 368–379). Probably the first documented case of making the connection between a volcanic eruption and unusual atmospheric conditions is a report by Benjamin Franklin (1784), describing several weather anomalies from summer through winter that he conjectured may have been related to an eruption reported in Iceland. Although Franklin did not know specifics about the eruption, it was in fact the 8-month-long Laki fissure eruption whose consequences were devastating for Iceland and very serious elsewhere in Europe. Recent analysis indicates this eruption released massive quantities of SO_2 and other caustic gases (Thordarson and Self, 1993), which turned into an acid rain and haze in Scandinavia and damaged much of the fall harvest. The consequences were worst in Iceland, where the acid haze demolished the summer crops, resulting in famine that led to the death of 75% of Iceland's livestock and 25% of the human population (Thordarson and Self, 1993).

The massive Krakatau eruption of 1883 also resulted in enormous loss of life (primarily through tsunamis). The estimated 20 km^3 of erupted pyroclastic material, in association with stratospheric droplets condensed from volcanic gases, decreased the intensity of sunlight reaching Earth's surface and resulted in worldwide unusual optical effects (e.g., dramatic sunsets and solar rings). Available evidence supports a cooling in the Northern Hemisphere of 0.25°C for 1 to 2 years (Rampino and Self, 1982). The even larger 1815 eruption of Tambora released >50 km^3 of magma in a series of explosions from April 5 to 10 that produced large Plinian columns, massive pyroclastic flows, and co-ignimbrite ash (Sigurdsson and Carey, 1989). Unusual optical effects also accompanied this eruption. More importantly it produced an observable decrease in both sunlight and starlight around the world, which when combined with documented weather changes (e.g., the "year without a summer" in the Northern

Hemisphere) make a strong case for volcanically induced climate change of more than 2 years' duration (Stommel and Stommel, 1983; Stothers, 1984).

A better understanding of the mechanism behind these volcanic environmental consequences resulted from two more recent eruptions in Mexico and the Philippines. Several Plinian columns and associated pyroclastic surges erupted from El Chichón volcano in Mexico between March 29 and April 4, 1982, releasing ~ 1 km^3 of magma with severe consequences to nearby villages (Sigurdsson *et al.*, 1987). The climatic significance of this eruption was that the tephra contained up to 2 wt% sulfates (including anhydrite crystals), resulting in an eruption cloud unusually rich in sulfuric acid aerosols that quickly encircled the globe, as monitored by orbiting satellites (Rampino and Self, 1984). The importance of sulfur-rich eruptions was highlighted by the larger eruption of Mt. Pinatubo in the Philippines on June 15–16, 1991, which released 3 km^3 of magma (Wolfe and Hoblitt, 1996), and which also contained abundant anhydrite and sulfur-rich gases (Bluth *et al.*, 1992; Hattori, 1996). Sulfur-rich gas, once in the stratosphere, condenses into sulfuric acid droplets that are the primary long-term agents of global cooling. In this situation, volcanic eruptions that place a sufficient volume of the right materials into the stratosphere might affect the environment more than the eruption itself is affected by its surrounding environment.

2.8. CONCLUSION

In this chapter we have reviewed the considerable diversity of volcanic products emplaced within a subaerial environment on Earth. Field observations and theoretical developments have provided a good indication of what happens to magma during its ascent to the Earth's surface. Plate tectonics provides the unifying theory through which the wide range of volcanic styles and locations can be understood. Volcanic materials are derived from both effusive and explosive eruptions, resulting in a wide range of volcanic constructs and deposits. Massive outpourings of basaltic lavas occur in LIPs distributed throughout the world, but most are likely associated with mantle plumes and the initiation of plate rifting. All of the volcanic materials described here have interacted with the subaerial environment found at Earth's surface, and in some cases the environment itself may have been influenced by the eruptions. This information represents the basis against which we can evaluate volcanic eruptions and associated products encountered throughout the rest of the solar system, as discussed in the remainder of this book.

2.9. REFERENCES

Anderson, S. W., and J. H. Fink, The development and distribution of surface textures at the Mount St. Helens dome, in *Lava Flows and Domes*, edited by J. H. Fink, pp. 25–46, Springer-Verlag, Berlin, 1990.

Atkinson, A., T. J. Griffin, and P. J. Stephenson, A major lava tube system from Undara volcano, North Queensland, *Bull. Volcanol.*, 39, 266–293, 1975.

Auzende, J.-M., V. Ballu, R. Batiza, D. Bideau, J.-L. Charlou, M.-H. Cormier, Y. Pouquest, P. Geistodoerfer, Y. Lagabrielle, J. Sinton, and P. Spadea, Recent tectonic, magmatic, and hydrothermal activity on the East Pacific Rise between 17°S and 19°S: Submersible observations, *J. Geophys. Res.*, 101, 17995–18010, 1996.

Bailey, R. A., G. B. Dalrymple, and M. A. Lanphere, Volcanism, structure, and geochronology of Long Valley Caldera, Mono County, California, *J. Geophys. Res.*, 81, 725–744, 1976.

Basaltic Volcanism Study Project, *Basaltic Volcanism on the Terrestrial Planets*, 1286 pp., Pergamon Press, New York, 1981.

Bebout, G. E., D. W. Scholl, S. H. Kirby, and J. P. Platt (Eds.), *Subduction: Top to Bottom*, 384 pp., *Geophys. Monogr. 96*, American Geophysical Union, Washington, DC, 1996.

Blackburn, E. A., L. Wilson, and R. S. J. Sparks, Mechanisms and dynamics of strombolian activity, *J. Geol. Soc. London, 132*, 429–440, 1976.

Blake, S., Volcanism and dynamics of open magma chambers, *Nature, 289*, 783–785, 1981.

Blake, S., Viscoplastic models of lava domes, in *Lava Flows and Domes*, edited by J. H. Fink, pp. 88–126, Springer-Verlag, Berlin, 1990.

Bluth, G. J. S., S. D. Doiron, C. C. Schnetzler, A. J. Krueger, and L. S. Walter, Global track of the SO_2 clouds from the June, 1991 Mount Pinatubo eruptions, *Geophys. Res. Lett., 19*(2), 151–154, 1992.

Bonnichsen, B., Rhyolite lava flows in the Bruneau–Jarbidge eruptive center, southwestern Idaho, in *Cenozoic Geology of Idaho*, edited by B. Bonnichsen and R. M. Breckenridge, *Idaho Bureau Mines Geol. Bull., 26*, 283–320, 1982.

Boudon, G., and A. Gourgaud (Eds.), Mount Pelée, 200 pp., *J. Volcanol. Geotherm. Res., 38*, 1989.

Cas, R. A. F., and S. Self, Flow front behavior of komatiite lavas, *IAVCEI Int. Vol. Congress Abstracts*, University of Cape Town, South Africa, 11, 1998.

Cas, R. A. F., and J. V. Wright, *Volcanic Successions: Modern and Ancient*, 487 pp., Allen & Unwin, Boston, 1987.

Cattermole, P. J., *Planetary Volcanism: A Study of Volcanic Activity in the Solar System*, 443 pp., Wiley, New York, 1989.

Chadwick, W. W., Jr., and W. W. Embley, Graben formation associated with recent dike intrusions and volcanic eruptions on the mid-ocean ridge, *J. Geophys. Res., 103*, 9807–9825, (1998).

Chadwick, W. W., Jr., R. W. Embley, and T. M. Shank, The 1996 Gorda Ridge eruption: Geologic mapping, sidescan sonar, and SeaBeam comparison results, *Deep Sea Res., 45*, 2547–2570, 1998.

Christiansen, R. L., The Quaternary and Pliocene Yellowstone Plateau Volcanic Field of Wyoming, Idaho, and Montana, *U.S. Geol. Surv. Prof. Pap., 729-G*, in press.

Christiansen, R. L., and H. R. Blank, Jr., Geologic map of the Canyon Village quadrangle, Yellowstone National Park, Wyoming, *U.S. Geol. Surv. Map, GQ-1192*, scale 1:62,500, 1975.

Clague, D. A., and G. B. Dalrymple, The Hawaiian-Emperor volcanic chain. Part 1. Geologic evolution, *U.S. Geol. Surv. Prof. Pap., 1350*, 5–54, 1987.

Coffin, M. F., and O. Eldholm, Scratching the surface: Estimating dimensions of large igneous provinces, *Geology, 21*, 515–518, 1993.

Crisp, J. A., and S. M. Baloga, A model for lava flows with two thermal components, *J. Geophys. Res. 95*, 1255–1270, 1990.

Decker, R. W., Dynamics of Hawaiian volcanoes: An overview, *U.S. Geol. Surv. Prof. Pap., 1350*, 997–1018, 1987.

de Silva, S. L., and P. W. Francis, *Volcanoes of the Central Andes*, 216 pp., Springer-Verlag, Berlin, 1991.

Dingwell, D. B., C. M. Scarfe, and D. J. Cronin, The effect of fluorine on viscosities in the system $Na_2O-Al_2O_3-SiO_2$: Implications for phonolites, trachytes and rhyolites, *Am. Mineral., 70*, 80–87, 1985.

Duffield, W., Eruptive fountains of silicic magma and their possible effects on the tin content of fountain-fed lavas, *Geol. Soc. Am. Spec. Pap., 246*, 251–261, 1990.

Ewart, A., S. C. Milner, R. A. Armstrong, and A. R. Duncan, Etendeka volcanism of the Goboseb mountains and Messum igneous complex, Namibia. Part 1. Geochemical evidence of early Cretaceous Tristan plume melts and the role of crustal contamination in the Paraná–Etendeka CFB, *J. Petrol., 39*(2), 191–225, 1998.

Fisher, R. V., Proposed classification of volcaniclastic sediments and rocks, *Geol. Soc. Am. Bull., 72*, 1409–1414, 1961.

Fisher, R. V., and H. U. Schmincke, *Pyroclastic Rocks*, 472 pp., Springer-Verlag, Berlin, 1984.

Fornari, D. J., R. M. Haymon, M. R. Perfit, T. K. P. Gregg, and M. H. Edwards, Axial summit trough of the East Pacific Rise 9°–10°N: Geological characteristics and evolution of the axial zone on fast spreading mid-ocean ridges, *J. Geophys. Res., 103*, 9827–9855, 1998.

Fox, C. G., In situ ground deformation measurements from the summit of Axial Volcano during the 1998 Volcanic Episode, *Trans. Am. Geophys. Union, 79*, F921, 1998.

Fox, C. G., W. E. Radford, R. P. Dziak, T.-K. Lau, H. Matsumoto, and A. E. Schreiner, Acoustic detection of a seafloor spreading episode on the Juan de Fuca Ridge using military hydrophone arrays, *Geophys. Res. Lett., 22*, 131–134, 1995.

Francis, P., *Volcanoes: A Planetary Perspective*, 443 pp., Oxford University Press (Clarendon), London, 1994.

Franklin, B., The meteorological imaginations and conceptions, *Mem. Lit. Philos. Soc. Manchester, 3*, 373–377, 1784.

Garland, F., C. J. Hawkesworth, and M. S. M. Mantovani, Description and petrogenesis of the Paraná rhyolites, southern Brazil, *J. Petrol., 36*, 1193–1227, 1995.

Greeley, R., Observations of actively forming lava tubes and associated structures, Hawai'i, *Mod. Geol.*, *2*, 207–223, 1971.

Greeley, R., The role of lava tubes in Hawaiian volcanoes, *U.S. Geol. Surv. Prof. Pap.*, *1350*, 1584–1602, 1987.

Gregg, T. K. P., and J. H. Fink, Quantification of submarine lava-flow morphology through analog experiments, *Geology*, *23*, 73–76, 1995.

Gregg, T. K. P., D. J. Fornari, M. R. Perfit, R. M. Haymon, and J. H. Fink, Rapid emplacement of a mid-ocean ridge lava flow on the East Pacific Rise at 9°46′–51′N, *Earth Planet. Sci. Lett.*, *144*, E1–E7, 1996.

Griggs, D. T., and D. W. Baker, The origin of deep-focus earthquakes, in *Properties of Matter Under Unusual Conditions*, ed. by H. Mark and S. Fernbach, Interscience Publishers, New York, p. 23–42, 1969.

Hattori, K., Occurrence and origin of sulfide and sulfate in the 1991 Mount Pinatubo eruption products, in *Fire and Mud: Eruptions and Lahars of Mount Pinatubo, Philippines*, edited by C. G. Newhall and R. S. Punongbayan, pp. 807–824, University of Washington Press, Seattle, 1996.

Haymon, R. M., D. J. Fornari, M. H. Edwards, S. Carbotte, D. Wright, and K. C. Macdonald, Hydrothermal vent distribution along the East Pacific Rise crest (9°09′–54′N), *Earth Planet. Sci. Lett.*, *104*, 513–534, 1991.

Haymon, R. M., D. J. Fornari, K. L. Von Damm, M. D. Lilley, M. R. Perfit, J. M. Edmond, W. C. Shanks, III, R. A. Lutz, J. M. Grebmeier, S. Carbotte, D. Wright, E. McLaughlin, M. Smith, N. Beedle and E. Olson, Volcanic eruption of the mid-ocean ridge along the East Pacific Rise at 9°45′–52′N: I. Direct submersible observation of seafloor phenomena associated with an eruption event in April, 1991, *Earth Planet. Sci. Lett.*, *119*, 85–101, 1993.

Head, J. W., and L. Wilson, Basaltic pyroclastic eruptions: Influence of gas-release patterns and volume fluxes on fountain structure, and the formation of cinder cones, spatter cones, rootless flows, lava ponds and lava flows, *J. Volcanol. Geotherm. Res.*, *37*, 261–271, 1989.

Heiken, G., D. Vaniman, and B. M. French, *Lunar Sourcebook: A User's Guide to the Moon*, 735 pp., Cambridge University Press, London, 1991.

Hill, R. I., I. H. Campbell, G. F. Davies, and R. W. Griffiths, Mantle plumes and continental tectonics, *Science*, *256*, 186–193, 1992.

Hon, K., J. Kauahikaua, R. Denlinger, and K. Mackay, Emplacement and inflation of pahoehoe sheet flows: Observations and measurements of active lava flows on Kilauea volcano, Hawai'i, *Geol. Soc. Am. Bull.*, *106*, 351–370, 1994.

Horai, K., Thermal conductivity of Hawaiian basalt: A new interpretation of Robertson and Peck's data, *J. Geophys. Res.*, *96*, 4125–4132, 1991.

Huppert, H. E., J. B. Shepherd, H. Sigurdsson, and R. S. J. Sparks, On lava dome growth, with application to the 1979 lava extrusion of the Soufrière of St. Vincent, *J. Volcanol. Geotherm. Res.*, *14*, 199–222, 1982.

Jaupart, C., and S. Vergniolle, The generation and collapse of a foam at the roof of a basaltic magma chamber, *J. Fluid Mech.*, *203*, 347–390, 1989.

Keszthelyi, L., A preliminary thermal budget for lava tubes on the Earth and planets, *J. Geophys. Res.*, *100*, 20411–20420, 1995.

Kilburn, C. R. J., Lava crusts, aa flow lengthening and the pahoehoe–aa transition, in *Active Lavas*, edited by C. R. J. Kilburn and G. Luongo, pp. 263–280, University College London Press, London, 1993.

Kilburn, C. R. J., and R. M. C. Lopes, General patterns of flow field growth: Aa and blocky lavas, *J. Geophys. Res.*, *96*, 19721–19732, 1991.

Lipman, P. W., and N. G. Banks, Aa flow dynamics, Mauna Loa 1984, *U.S. Geol. Surv. Prof. Pap.*, *1350*, 1527–1567, 1987.

Lowenstern, J. B., Applications of silicate melt inclusions to the study of magmatic volatiles, in *Magmas, Fluids and Ore Deposits*, edited by J. F. H. Thompson, pp. 71–99, *Mineral Assoc. Canada Short Course*, *23*, 1995.

Macdonald, G. A., *Volcanoes*, 510 pp., Prentice–Hall, Englewood Cliffs, NJ, 1972.

Macdougall, J. D. (Ed.), *Continental Flood Basalts*, 341 pp., Kluwer Academic, Norwell, MA, 1988.

Mahoney, J. J., and M. F. Coffin (Eds.), *Large Igneous Provinces*, 520 pp., *Geophys. Monogr.*, *100*, American Geophysical Union, Washington, DC, 1997.

Manley, C. R., Extended cooling and viscous flow of large, hot rhyolite lavas: Implications of numerical modeling results, *J. Volcanol. Geotherm. Res.*, *53*, 27–46, 1992.

Manley, C. R., Rhyolitic fire-fountaining on the Owyhee Plateau, SW Idaho: Very low-H_2O magmas from higher-H_2O magmatic systems (abstract), *Trans. Am. Geophys. Union*, *75*, 751, 1994.

Manley, C. R., How voluminous rhyolite lavas mimic rheomorphic ignimbrites: Eruptive style, emplacement conditions, and formation of tuff-like textures, *Geology*, *23*, 349–352, 1995.

Manley, C. R., Rhyolitic fire-fountains, low pre-eruptive volatile contents, and petrogenesis by wall-rock melting (abstract), *Trans. Am. Geophys. Union*, *77*, 818, 1996.

Manley, C. R., and W. C. McIntosh, The Juniper Mtn. volcanic center, Owyhee County, southwestern Idaho: Age relations and physical volcanology, in *Tectonic and Magmatic Evolution of the Snake River Plain Volcanic Province*, edited by C. White, M. McCurry, and B. Bonnichsen, *Idaho Geol. Surv. Bull.*, in press.

McBirney, A. R., Conductivity variations and terrestrial heat-flow distribution, *J. Geophys. Res.*, *68*, 6323–6329, 1963.

McSween, H. Y., What we have learned about Mars from SNC meteorites, *Meteoritics*, *29*, 757–779, 1994.

Moore, J. G., and W. G. Melson, Nuées ardentes of the 1968 eruption of Mayon volcano, Philippines, *Bull. Volcanol.*, *33*, 600–620, 1969.

Murase, T., and A. R. McBirney, Properties of some common igneous rocks and their melts at this temperature, *Geol. Soc. Am. Bull.*, *84*, 3563–3592, 1973.

Mursky, G., *Introduction to Planetary Volcanism*, 293 pp., Prentice–Hall, Englewood Cliffs, NJ, 1996.

Mutter, J. C., and J. A. Karson, Structural processes at slow-spreading centers, *Science*, *257*, 627–634, 1992.

Mysen, B. O., The solubility of H_2O and CO_2 under predicted magma genesis conditions and some petrological and geophysical implications, *J. Geophys. Res.*, *15*, 351–361, 1977.

Newhall, C. G., and S. Self, The Volcanic Explosivity Index (VEI): An estimate of explosive magnitude for historical volcanism, *J. Geophys. Res.*, *87*, 1231–1238, 1982.

Phipps Morgan, J., W. J. Morgan, and E. Price, Hotspot melting generates both hotspot volcanism and a hotspot swell? *J. Geophys. Res.*, *100*, 8045–8062, 1995.

Pierce, K. L., and L. A. Morgan, The track of the Yellowstone hot spot: Volcanism, faulting, and uplift, in *Regional Geology of Eastern Idaho and Western Wyoming*, edited by P. K. Link, M. A. Kuntz, and L. B. Platt, pp. 1–53, *Geol. Soc. Am. Mem.*, *179*, 1992.

Pitcher, W. S., The nature, ascent and emplacement of granitic magmas. *J. Geol. Soc. London*, *136*, 627–662, 1979.

Press, F., and R. Siever, *Earth*, 945 pp., Freeman, San Francisco, 1974.

Ramburg, H., Mantle diapirism and its tectonic and magmatic consequences, *Phys. Earth Planet. Inter.*, *5*, 45–60, 1972.

Rampino, M. R., and S. Self, Historic eruptions of Tambora (1815), Krakatua (1883) and Agung (1963), their stratospheric aerosols and climatic impact, *Quat. Res.*, *18*, 127–143, 1982.

Rampino, M. R., and S. Self, The atmospheric effects of El Chichón, *Sci. Am.*, *250*, 48–57, 1984.

Reidel, S. P., Emplacement of the Columbia River flood basalt, *J. Geophys. Res.*, *103*, B11, 27393–27410, 1998.

Reidel, S. P., and P. R. Hooper (Eds.), *Volcanism and Tectonism in the Columbia River Basalt Province*, 386 pp., *Geol. Soc. Am. Spec. Pap.*, *239*, 1989.

Reidel, S. P., and T. L. Tolan, Eruption and emplacement of flood basalt: An example from the large-volume Teepee Butte member, Columbia River basalt group, *Geol. Soc. Am. Bull.*, *104*, 1650–1671, 1992.

Renne, P. R., M. Ernesto, I. G. Pacca, R. S. Coe, J. M. Glen, J. M. Prévot, and M. Perrin, The age of Paraná flood volcanism, rifting of Gondwanaland, and the Jurassic–Cretaceous boundary, *Science*, *258*, 975–979, 1992.

Renne, P. R., J. M. Glen, S. C. Milner, and A. R. Duncan, Age of the Etendeka flood volcanism and associated intrusions in southwestern Africa, *Geology*, *24*(7), 659–662, 1996.

Richards, M. A., R. A. Duncan, and V. E. Courtillot, Flood basalts and hot-spot tracks: Plume heads and tails, *Science*, *246*, 103–107, 1989.

Robertson, E. C., and D. L. Peck, Thermal conductivity of vesicular basalt from Hawai'i, *J. Geophys. Res.*, *79*, 4875–4888, 1974.

Rowland, S. K., and G. P. L. Walker, Pahoehoe and aa in Hawai'i: Volumetric flow rate controls the lava structure, *Bull. Volcanol.*, *52*, 615–628, 1990.

Rubin, A. M., and D. D. Pollard, Origins of blade-like dikes in volcanic rift zones, *U.S. Geol. Surv. Prof. Pap.*, *1350*, 1449–1470, 1987.

Ryan, M. P., Neutral buoyancy and the mechanical evolution of magmatic systems, in *Magmatic Processes: Physicochemical Principles*, edited by B. O. Mysen, pp. 259–287, Geochemical Society, Special Publication No. 1, University Park, PA, 1987.

Sakimoto, S. E. H., J. Crisp, and S. M. Baloga, Eruption constraints on tube-fed planetary lava flows, *J. Geophys. Res.*, *102*, 6597–6613, 1997.

Sampson, D. E., Textural heterogeneities and vent area structures in the 600-year-old lavas of the Inyo volcanic chain, eastern California, *Geol. Soc. Am. Spec. Pap.*, *212*, 89–101, 1987.

Self, S., T. Thordarson, L. Keszthelyi, G. P. L. Walker, K. Hon, M. T. Murphy, P. Long, and S. Finnemore, A new model for the emplacement of Columbia River basalts as large, inflated pahoehoe lava flow fields, *Geophys. Res. Lett.*, *23*(19), 2689–2692, 1996.

Self, S., T. Thordarson, and L. Keszthelyi, Emplacement of continental flood basalt lava flows, in *Large Igneous Provinces*, edited by J. J. Mahoney and M. F. Coffin, pp. 381–410, *Geophys. Monogr. 100*, American Geophysical Union, Washington, DC, 1997.

Shaw, H. R., Earth tides, global heat flow, and tectonics, *Science, 168*, 1084–1087, 1970.

Shaw, H. R., Mantle convection and volcanic periodicity in the Pacific: Evidence from Hawai'i, *Geol. Soc. Am. Bull., 84*, 1505–1526, 1973.

Shaw, H. R., and D. A. Swanson, Eruption and flow rates of flood basalts, in *Proceedings of the Second Columbia River Basalt Symposium*, edited by E. H. Gilmour and D. Stradling, pp. 271–299, East Washington State College Press, Cheney, 1970.

Shaw, H. R., T. L. Wright, D. L. Peck, and R. Okamura, The viscosity of basaltic magma: An analysis of field measurements in Makaopuhi lava lake, Hawai'i, *Am. J. Sci., 261*, 255–264, 1968.

Sheridan, M. F., and K. H. Wohletz, Hydrovolcanism: Basic considerations and review, *J. Volcanol. Geotherm. Res., 17*, 1–29, 1983.

Sheridan, M. F., F. Barberi, M. Rose, and R. Santacroce, A model for Plinian eruptions of Vesuvius, *Nature, 289*, 282–285, 1981.

Sigurdsson, H., and S. Carey, Plinian and co-ignimbrite tephra from the 1815 eruption of Tambora volcano, *Bull. Volcanol., 51*, 243–270, 1989.

Sigurdsson, H., S. Carey, W. Cornell, and T. Pescatore, The eruption of Vesuvius in AD 79, *Nat. Geogr. Res., 1*(3), 332–387, 1985.

Sigurdsson, H., S. N. Carey, and R. V. Fisher, The 1982 eruption of El Chichón volcano, Mexico (3): Physical properties of pyroclastic surges, *Bull. Volcanol., 49*, 467–488, 1987.

Simkin, T., J. D. Unger, R. I. Tilling, P. R. Vogt, and H. Spall, This dynamic planet: World map of volcanoes, earthquakes, impact craters, and plate tectonics, *U.S. Geol. Surv.*, scale 1 : 30,000,000, 1994.

Sleep, N. H., Tapping of melt by veins and dikes, *J. Geophys. Res., 93*, 10255–10272, 1988.

Smith, D. K., J. R. Cann, M. E. Dougherty, J. Lin, S. Spencer, C. MacLeod, J. Keeton, E. McAllister, B. Brooks, R. Pascoe, and W. Robertson, Mid-Atlantic Ridge volcanism from deep-towed side-scan sonar images, 25°–29°N, *J. Volcanol. Geotherm. Res., 67*, 233–262, 1995.

Sparks, R. S. J., The dynamics of bubble generation and growth in magmas: A review and analysis, *J. Volcanol. Geotherm. Res., 3*, 1–37, 1978.

Sparks, R. S. J., P. Meyer, and H. Sigurdsson, Density variation amongst mid-ocean ridge basalts: Implications for magma mixing and the scarcity of primitive lavas, *Earth Planet. Sci. Lett., 46*, 419–430, 1980.

Sparks, R. S. J., M. I. Bursik, S. N. Carey, J. S. Gilbert, L. S. Glaze, H. Sigurdsson, and A. W. Woods, *Volcanic Plumes*, 574 pp., Wiley, New York, 1997.

Spera, F. J., Aspects of magma transport, in *Physics of Magmatic Processes*, edited by R. B. Hargraves, pp. 265–323, Princeton University Press, Princeton, NJ, 1980.

Stein, S., and C. A. Stein, Thermo-mechanical evolution of oceanic lithosphere: Implications for the subduction process and deep earthquakes, in *Subduction: Top to Bottom*, edited by G. E. Bebout, D. W. Scholl, S. H. Kirby, and J. P. Platt, pp. 1–17, *Geophys. Monogr. 96*, American Geophysical Union, Washington, DC, 1996.

Stommel, H., and E. Stommel, in *Volcano Weather*, 177 pp., Seven Seas, Newport, RI, 1983.

Stothers, R. B., The great Tambora eruption in 1815 and its aftermath, *Science, 224*, 1191–1198, 1984.

Summerfield, M. A., *Global Geomorphology*, 537 pp., Wiley, New York, 1991.

Swanson, D. A., K. A. Cameron, R. C. Evarts, P. T. Pringle, and J. A. Vance, Cenozoic volcanism in the Cascade range and Columbia plateau, southern Washington and northernmost Oregon, *Field Trip Guidebook Vol. T106*, 60 pp., American Geophysical Union, Washington, DC, 1989.

Thorarinsson, S., Some problems of volcanism in Iceland, *Geol. Rundsch., 57*, 1–20, 1967.

Thorarinsson, S., The Lakagigar eruption of 1783, *Bull. Volcanol., 33*, 910–927, 1969.

Thorarinsson, S., and G. E. Sigvaldason, The eruption of Askja 1961, *Am. J. Sci., 260*, 641–651, 1962.

Thordarson, T., and S. Self, The Laki (Skaftár Fires) and Grímsvötn eruptions, 1783–1785, *Bull. Volcanol., 55*, 233–263, 1993.

Tolan, T. L., S. P. Reidel, M. H. Beeson, J. L. Anderson, K. R. Fecht, and D. A. Swanson, Revisions to the estimates of the areal extent and volume of the Columbia River Basalt Group, in *Volcanism and Tectonism in the Columbia River Basalt Province*, edited by S. P. Reidel and P. R. Hooper, pp. 1–20, *Spec. Pap. Geol. Soc. Am., 239*, 1989.

Verhoogen, J., Petrological evidence on temperature distribution in the mantle of the Earth, *Trans. Am. Geophys. Union, 35*, 85–92, 1954.

Walker, G. P. L., Structure, and origin by injection under surface crust, of tumuli, "lava rises", "lava rise pits", and "lava inflation clefts" in Hawai'i, *Bull. Volcanol., 53*, 546–558, 1991.

Walker, G. P. L., S. Self, and L. Wilson, Tarawera 1886, New Zealand—A basaltic Plinian fissure eruption, *J. Volcanol. Geotherm. Res.*, *21*, 61–78, 1984.

Webster, J. D., and W. A. Duffield, Volatiles and lithophile elements in Taylor Creek rhyolite: Constraints from glass inclusion analysis, *Am. Mineral.*, *76*, 1628–1645, 1991.

White, C., M. McCurry, and B. Bonnichsen (Eds.), Tectonic and Magmatic Evolution of the Snake River Plain Volcanic Province, *Idaho Geol. Surv. Bull.*, in press.

White, R., and D. McKenzie, Magmatism at rift zones: The generation of volcanic continental margins and flood basalts, *J. Geophys. Res.*, *94*, 7688–7729, 1989.

Williams, H., and A. R. McBirney, *Volcanology*, 397 pp., Freeman and Cooper, San Francisco, 1979.

Williams, S. N., Plinian airfall deposits of basaltic composition, *Geology*, *11*, 211–215, 1983.

Wilson, L., Explosive volcanic eruptions—III. Plinian eruption columns, *Geophys. J. R. Astron. Soc.*, *45*, 543–556, 1976.

Wilson, L., and J. W. Head, Ascent and eruption of basaltic magma on the Earth and Moon, *J. Geophys. Res.*, *86*, 2971–3001, 1981.

Wohletz, K., and G. Heiken, *Volcanology and Geothermal Energy*, 432 pp., University of California Press, Berkeley, 1992.

Wohletz, K. H., and M. F. Sheridan, Hydrovolcanic explosions II: Evolution of basaltic tuff rings and tuff cones, *Am. J. Sci.*, *283*, 385–413, 1983.

Wolfe, E. W., and R. P. Hoblitt, Overview of the eruptions, in *Fire and Mud: Eruptions and Lahars of Mount Pinatubo, Philippines*, edited by C. G. Newhall and R. S. Punongbayan, pp. 3–20, University of Washington Press, Seattle, 1996.

Wolfe, E. W., C. A. Neal, N. G. Banks, and T. J. Duggan, Geologic observations and chronology of eruptive events, *U.S. Geol. Surv. Prof. Pap.*, *1463*, 1–97, 1989.

Wright, T. L., Origin of Hawaiian tholeiite: A metasomatic model, *J. Geophys. Res.*, *89*, 3233–3252, 1984.

Yoder, H. S., *Generation of Basaltic Magma*, 265 pp., National Academy of Sciences, Washington, DC, 1976.

Yoder, H. S., and E. C. Tilley, The origin of basalt magmas: An experimental study of natural and synthetic rock systems, *J. Petrol.*, *3*, 342–532, 1962.

Zimbelman, J. R., Emplacement of long lava flows on planetary surfaces, *J. Geophys. Res.*, *103*, B11, 27503–27515, 1998.

3

Volcanism and Ice Interactions on Earth and Mars

Mary G. Chapman, Carlton C. Allen, Magnus T. Gudmundsson,
Virginia C. Gulick, Sveinn P. Jakobsson, Baerbel K. Lucchitta,
Ian P. Skilling, and Richard B. Waitt

3.1. INTRODUCTION

When volcanoes and ice interact, many unique types of eruptions and geomorphic features result. Volcanism appears to occur on all planetary bodies, but of the inner solar system planets, ice is limited to Earth and Mars. Earth, the water planet, is covered by ice wherever the temperature is cold enough to freeze water for extended periods of time. Ice is found in sheets covering the Antarctic continent and Greenland, as glacial caps in high mountainous regions, as glaciers in polar and temperate regions, and as sea ice in the northern and southern oceans. With changes in climate, landmass, solar radiation, and Earth orbit, ice masses can contract or expand over great distances, as occurred during the Pleistocene Ice Age. The Earth is still in an ice age—at the beginning of the Cenozoic, 65 million years ago, our planet was ice-free. In fact, there is now so much ice that about 70% of the world's total fresh water is contained within the Antarctic ice sheet.

It is less well known that Mars also has a significant amount of ice, including polar caps somewhat like (those on) Earth. Its north polar cap consists of mostly water ice, its south polar cap is surfaced by CO_2 ice (Kieffer, 1979). The most important reservoir of ice, however, is retained in the ground as pore fillings or segregated masses. The average mean temperature on Mars is $-80°C$ overall, $-60°C$ at the equator (Fanale *et al.*, 1992), resulting in a permafrost layer (ice rich?) ranging from about 1 to 2 km thick at the equator to 3 to 6 km at the poles (Rossbacher and Judson, 1981; Clifford, 1993); a common estimate for midlatitude permafrost thickness is 2 km (Squyres *et al.*, 1992). However, in near-surface rocks or soil between about 50° north and south latitude (Mellon and Jakosky, 1995), ice may not be retained, because in the Martian thin atmosphere water vapor migrates out of the ground in the warmer equatorial areas. On the other hand, complexities of the surface expressed in varying albedo (relative brightness), thermal inertia (ability to absorb and release heat), and orbital obliquity (angle between equatorial plane and orbit) may allow for the existence of near-surface ice even in this region (Paige, 1992; Mellon and Jakosky, 1995). Geologic observations attest to a large reservoir of ground ice, near the surface or at depth, almost everywhere on Mars, leading Carr (1995) to suggest an ancient water layer several hundred meters thick (if evenly distributed). These observations include flow lobes on crater ejecta, features possibly attributed to thermokarst such as chaotically collapsed ground and irregular depressions, flow lobes on scarps that resemble rock glaciers, and numerous fluvial valleys and outflow channels. Water from channels appears to have pooled in the northern lowlands, forming a lake or ocean (Lucchitta, 1981; Witbeck and Underwood, 1983; Lucchitta et al., 1986; Parker et al., 1989, 1993; Jöns, 1990; Baker et al., 1991; Scott et al., 1992; Lucchitta, 1993; Kargel et al., 1995). This ocean, though blanketed by dust, may remain as a vast reservoir of subsurface ice.

Our knowledge of planetary volcano and ice interactions begins on Earth. Many terrestrial composite volcanoes have considerable ice covers on their slopes and some volcanic regions are covered by large ice sheets and icecaps. These different settings give rise to two types of eruptions: (1) those in alpine situations beneath mountain snow or summit and valley glaciers and (2) those beneath broad continental-scale glaciers or ice sheets. A third type are not true eruptions but result from explosions caused by lava flowing over water/ice-saturated ground and form phreatic craters. Terrestrial volcano/ice interactions of eruption types 1 and 2 both involve large volumes of meltwater and therefore usually generate floods, ranging from water floods to mudflows or lahars (an Indonesian term) composed chiefly of volcaniclastic materials. *Jökulhlaup* is an Icelandic term for glacier outburst flood.

On Earth, sub-ice-sheet eruptions (type 2) from fissures or point sources produce pillow lavas and hyaloclastites (vitroclastic tephra produced by interaction of water and lava). Overlain by meltwater and ice, such eruptions from fissure sources form hyaloclastic ridges, while those from point sources form mounds. If the eruption penetrates the ice, subaerially extruded layered lava may cap these volcanoes, resulting in a tuya. This feature is often commonly and perhaps incorrectly called a *table mountain*, since a table mountain in a geomorphic sense refers to a flat-topped mountain of any origin and translates directly to mesa.

Volcanic accumulations of hyaloclastites, combined with pillow lavas, irregular intrusions, and occasionally subaerial lavas, are widely exposed in the volcanic zones of Iceland. *Möberg*, a generic Icelandic term, literally means "rock that looks like peat" and is used to describe all brownish, rich in palagonite (hydrated volcanic glass), basaltic, hyaloclastic rocks of Iceland, including all subglacial volcanic materials that formed during the upper Pleistocene but also during recent times. Therefore the use of the term *möberg ridge*, when discussing a hyaloclastic ridge, is a misnomer.

The origin of the hyaloclastite volcanoes was for a considerable time a matter of dispute. Peacock (1926) was first to present a "volcano-glacial" origin for hyaloclastites. Noe-Nygaard (1940) gave a description of a subglacial volcanic eruption in Iceland. Independently in the 1940s, Kjartansson (1943) in Iceland and Mathews (1947) in British Columbia presented the hypothesis that the Pleistocene hyaloclastite mountains are piles formed in subglacial eruptions. The work of Van Bemmelen and Rutten (1955) in the northern part of Iceland established the theory. Moore and Calk (1991) provided definitive geochemical studies of Icelandic tuyas. Because of the late acknowledgment of subglacial volcanism as a process, it is a relatively young science and geologists are unfamiliar with the subject. This chapter seeks to help alleviate this problem with detailed terrestrial examples of recent volcano/ice eruptions, lithologic details of hyaloclastic ridges and a tuya deposit, and descriptions of recent lahars and jökulhlaups.

Our planetary discussion of volcano/ice interactions is limited to Mars, whose surface contains ice and a variety of volcanic landforms, including volcanic plains, ancient highland volcanoes, and two large volcanic regions, Tharsis and Elysium (see Chapter 4). Crater density data suggest that volcanism continued throughout Mars's geologic history (Tanaka et al., 1992). Because extensive morphologic evidence exists for ground water and ground ice throughout the planet's history, ground ice and frozen waters likely interacted with subice volcanoes. In fact, there are Martian geomorphic features that have been interpreted to be analogous to those terrestrial edifices that form from all three types of subice eruptions. Therefore, in addition to the discussion of terrestrial volcano/ice interactions, other sections in this chapter note these examples of Martian analogues.

Volcano/ice interactions on Earth produce catastrophic lahars and distinctive edifices. On Mars these types of interactions seem to produce geomorphic change on two time scales. The emplacement of the volcano itself can produce large-scale geomorphic signatures (e.g., cones, tuyas). On a more gradual time scale the subsurface heat source could heat and cycle ground water to the near-surface environment over periods of thousands to millions of years, depending on the size of the intrusion. The resulting hydrothermal systems may also provide hospitable environments for the evolution of life on Mars. Therefore, the important implications of ground water and snow/ice perturbations from hydrothermal systems will also be discussed in this chapter.

3.2. NEVADO DEL RUIZ, COLOMBIA: ERUPTION BENEATH ALPINE ICE

Although Baker et al. (1991) have suggested the possibility of local snowfall on elevated areas surrounding the hypothetical paleo-ocean of Mars, certain volcano/ice interactions may be unique to the Earth, where precipitation occurs regularly. This limitation may be the case for eruptions beneath ice-capped volcanoes. A recent, relatively small example of this type of eruption occurred on November 13, 1985, at Nevado del Ruiz, Colombia. Nevado del Ruiz is 5400 m high, the northernmost-active volcano of the Andes, and a typical subduction zone andesite (Williams et al., 1986) volcano, that produced a horrible catastrophe, killing more than 23,000 people living 72 km away, in the city of Armero. The 1985 eruption is only the most recent event in a series of 12 eruptive stages that have occurred in the past 11,000 years at the 0.2-Myr-old volcano (Thouret et al., 1990).

About 25 to 30% of the ice on the icecap of Nevado del Ruiz was fractured and destabilized by earthquake and explosion; nearly 16% (4.2 km^2) of the surface area of the ice and snow (about 0.06 km^3 or 9% of the total ice and snow volume) was lost (Thouret, 1990). Large volumes of meltwater were produced (20 million m^3 and combined peak discharge of 80,000 m^3 s^{-1}) that flowed downslope, liquefied some of the new volcanic deposits, and generated avalanches of saturated snow, ice, and rock debris within minutes of the eruption (Pierson et al., 1990; Pierson, 1995). Lahar extents varied from 7.2 to 102.6 km from the vent. Thouret (1990) estimates that the meltwater volume released during the 1985 eruption was two to four times that which was incorporated into the lahars. This noncontributing water was: (1) included in snow avalanches or bound up in pyroclastic surge deposits, (2) incorporated in the phreatic explosive products (see Chapter 2), (3) sublimated into steam, or (4) stored within the current icecap (Thouret, 1990).

The key lessons of the Nevado del Ruiz eruption are: (1) catastrophic lahars (Section 3.6) can be generated on ice- and snow-capped volcanoes by relatively small eruptions, (2) the surface area of an icecap can be more critical than total ice volume when considering lahar potential because of mobilizing processes acting predominantly at the surface, (3) placement of hot rock debris on snow is insufficient to generate lahars—the two materials must be mechanically mixed together (in pyroclastic flows, surges, and mixed avalanches) for sufficiently rapid heat transfer, (4) lahars can increase their volumes significantly by entrainment of water and eroded sediment, and (5) valley-confined lahars can maintain relatively high velocities and can have catastrophic impacts as far as 100 km downstream (Pierson et al., 1990). Based on these lessons, one might expect much larger lahar volumes and extents on the comparatively larger volcanoes of Mars.

Many factors continue to render the Ruiz domes unstable, including deeply dissected troughs, strongly hydrothermally altered summit flanks, currently glaciated high cliffs, and steep unstable icecap margins (Thouret et al., 1990). The presence of a large volume of hydrothermally altered rock within the volcano not only increases the potential for edifice collapse because the altered rock is less competent, but also increases its potential to become a saturated, liquefied debris flow because the rock can hold more water (Carrasco-Núñez et al., 1993). Alteration gases and waters are emitted from fumaroles and thermal springs on the flanks of the volcano and are likely to be derived from a two-phase vapor-brine envelope surrounding the magmatic system (Giggenbach et al., 1990).

Hydrothermal alteration leading to unstable flanks may have also occurred on Mars. Olympus Mons and Apollinaris Patera are bound by steep scarps, indicating failure of flanks, and Olympus Mons appears to be surrounded by some type of debris lobes (aureole; see Section 3.9.3). Furthermore, hydrothermal discharge from fumaroles and springs may have formed valleys on the flanks of Martian volcanoes (see Section 3.9).

3.3. GJÁLP, VATNAJÖKULL: ERUPTION BENEATH AN ICE SHEET

Because outflow channels may have accumulated water in ponded situations, subglacial eruptions under broad ice sheets may be a more likely occurrence on Mars. Subglacial eruptions are frequent in Iceland, and the majority occur within the Vatnajökull icecap (Figure 3.1A; Thorarinsson, 1967, 1974; Björnsson and Einarsson, 1990; Voight, 1990; Williams, 1990; Larsen et al., 1998). However, owing to low activity in recent decades, the first such eruption to be monitored in any detail was the 13-day-long fissure eruption in Gjálp, Vatnajökull, in October 1996 (Gudmundsson et al., 1997). The progress of the eruption is shown schematically in Figure 3.1B. The eruption melted its way through 500 m of ice in 30 hr, forming a subaerial crater (Figures 3.2 and 3.3). The eruption produced basaltic andesite with a SiO_2 content of 54% (Gudmundsson et al., 1997). A thin layer of tephra was dispersed over about 10,000 km^2 but its volume is no more than 1–2% of the total erupted. The eruption formed a 6-km-long ridge with three peaks in the bedrock (Figure 3.1B), with an estimated volume of 0.7 km^3. The bedrock relief increased by 200–350 m; the highest peak is about 150 m below the initial ice surface. The new ridge is partly built on a ridge formed in a similar eruption in 1938.

Data on melted volumes allowed calorimetry to be used to infer roughly the eruption rate with time and provide an estimate of the total volume erupted (Gudmundsson et al., 1997). Magma discharge was highest in the first 4 days (melting rate 0.5 km^3 day^{-1}) and by the end of the eruption melted ice had produced 3 km^3 of water. After 100 days the melted volume

(a)

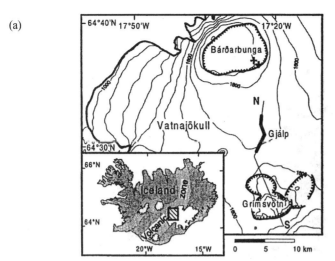

(b)

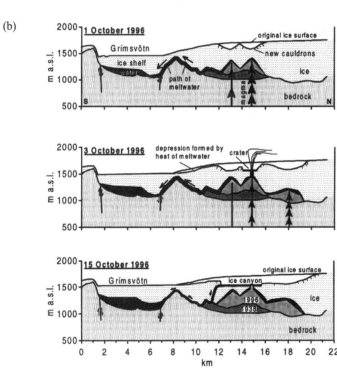

Figure 3.1. (a) Distribution of Upper Pleistocene (<0.78 Ma) and Recent hyaloclastites and associated lithofacies in Iceland. Modified after Johannesson and Saemundsson (1998). The hyaloclastite units shown cover an area of about 10,600 km^2, those hidden by present icecaps may bring the total area to some 15,000 km^2. The total volume is unknown, but by rough estimate, 60–70% of the exposed units are primary or redeposited hyaloclastites, 15–25% are pillow lavas, with the remainder being subaerial lavas and intrusions. Basalts are 90% of the volume, probably <5% are intermediate rocks and ~5% silicic rocks. Insert shows the location of Gjálp and Grímsvötn. During the latter part of the eruption an ice canyon formed in the ice surface, where part of the meltwater flowed before disappearing down into the glacier (after Gudmundsson et al., 1997). (b) Sections showing schematically the development of the subglacial ridge in the Gjalp eruption and the accumulation of meltwater in the subglacial lake of the Grímsvötn caldera. Five weeks after the eruption started, the subglacial lake was drained in a very swift jökulhlaup.

Figure 3.2. The eruption on October 3, 1996. The crater in the ice surface is about 500 m in diameter. The ice surface is covered with tephra. From Gudmundsson *et al.* (1997).

was 4 km³ and the total reached 4.7 km³ in January 1998. The high rate of heat transfer during the eruption suggests that fragmentation of magma into glass (hydroclasts or vitriclasts) was the dominant form of activity, not eruption of pillow lava. About 20% of the melting during the eruption occurred over the subglacial path of the water, indicating that its temperature was 15–20°C when it left the eruption site. This high temperature must have greatly enhanced the opening of subglacial drainage channels.

The meltwater from the eruption accumulated in the Grímsvötn caldera lake for 5 weeks before it was released in a very swift jökulhlaup. An estimated volume of 3.5 km³ was drained in 2 days, flooding the Skeidarársandur outwash plain to the south of Vatnajökull, destroying bridges and roads.

In June 1997, inspection of the exposed uppermost 40 m of the edifice showed that it was mainly made of hydrothermally altered clastic rocks with ash- to lapillus-sized glass particles; only a small fraction was crystalline rock (maximum diameter of fragments ∼20 cm). The temperature at a depth of 0.5 m was 60–70°C. By June 1998, ice flow had covered the edifice. Monitoring of heat flow and other observations may show to what extent the ridge becomes palagonitized and consolidated.

Significant storage of meltwater at the eruption site did not occur at Gjálp, as explained by applying the theory of water flow under temperate glaciers (Paterson, 1994; Björnsson, 1988). The direction of flow is down the slope of a potential function determined by (1) the slope of the bedrock and (2) the slope of the overlying ice surface. The ice surface slope is nine times more influential and controls the flow direction except in areas of very rugged bedrock (flow may be uphill, see Figure 3.1A). Thus, meltwater was drained away from the beginning, as if it were flowing on the ice surface. (This has potential implications for Mars,

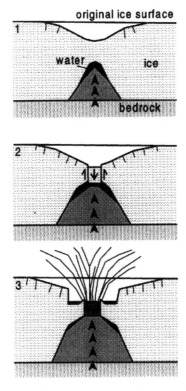

Figure 3.3. Schematic sections showing the penetration of the icecap by the eruption. The central piston was rapidly melted as it collapsed onto the underlying vent. From Gudmundsson *et al.* (1997).

where overlying, nonpolar surface ice has long since sublimated away and paths of meltwater are assumed to follow bedrock slopes.) After depressions had formed, subglacial drainage continued, since melting related to the heat of the water prevented closure of the subglacial conduits.

The opening formed in the ice (Figure 3.3) remained only 200–300 m in diameter throughout the eruption; wall melting was apparently compensated by ice flow. The water level at the eruption site was 150–200 m below the original ice surface; the icecap in the surrounding area is 600–700 m thick, hence, the height of the water column was 70–80% of the ice thickness. Outside the subsidence cauldrons, no signs of enhanced ice flow have been detected, indicating that for temperate glaciers the effects of subglacial eruptions on ice flow are localized.

3.4. SUBGLACIAL VOLCANIC HYALOCLASTIC RIDGES

Subglacial hyaloclastic ridges are unique indicators of volcano/ice interaction and appear to have only been described in Iceland. However, some formations on the volcanic island Jan Mayen in the Norwegian Sea (Imsland, 1978) may be classified as short ridges. Although unreported, hyaloclastite ridges probably occur in other volcanic subarctic or arctic regions.

A distinction is made between three main phases in the generation of a subglacial volcano: pillow lava, hyaloclastites, and subaerial lava. A pile of pillow lava forms if the hydrostatic pressure exceeds the gas pressure of the magma (see Chapter 5). Typically, only a minor amount of hyaloclastites form between pillows. Pillow lavas are often found at the base of hyaloclastite mounds, ridges, and tuyas. The thickness of basal basaltic pillow lava piles often exceeds 60–80 m, and a 300-m-thick section has been reported (Jones, 1970).

The depth at which effusive eruption of pillow lava transitions to fragmentation of hyaloclastites is not well known. The transition depth appears to be controlled by two factors: (1) the volatile content of the magma and (2) the hydrostatic pressure at the vent. Eruption rate may also be an influencing factor. On the basis of Late Pleistocene formations, Jones (1970) suggested basaltic phreatomagmatic explosions produce hydroclastites at a water depth less than approximately 100–200 m. Recent field studies in Iceland, however, suggest that a somewhat greater depth may be typical for basalts (Kokelaar, 1986).

Hyaloclastites of Icelandic deposits can be subdivided into two main types. Type one is poorly bedded, unsorted hyaloclastite (Jakobsson and Moore, 1982) that probably was quenched and rapidly accumulated below sea level without penetrating the surface. The coarse-grained, often massive core of many of the larger tuyas and ridges may have formed in this way. Type two is more fine grained than the lower, more massive part and forms at shallow water depths (approximately 20–30 m in the submarine Surtsey eruption; Kjartansson, 1966). At these shallow depths, the phreatic eruptions penetrate the water surface, and bedded hyaloclastite (tephra) is deposited by air fall or base surge on ice or land and sediment gravity flows into ponded water (e.g., in the Surtsey and Gjálp eruptions). In some cases, hyaloclastites form flows, but their extent is not known and may be limited to southeast and south Iceland (Walker and Blake, 1966). Walker and Blake (1966) have described such an Early Pleistocene mass of hyaloclastite with associated pillows and columnar basalt in Dalsheidi in southeastern Iceland. It is interpreted as a flow beneath a valley glacier in a channel previously excavated by englacial floods. It is 300 m thick and may originally have been 35 km long and about 0.5 km wide.

Lava caps the hyaloclastite section of the tuyas and possibly about 10–15% of the Icelandic hyaloclastite ridges. This lava is comparable to other subaerial lava; sheet lava is most common but massive, simple flows also occur. The lava grades into foreset breccia and sometimes into gently dipping, degassed pillow lava. Besides regular feeder dikes, apophyses and small irregular intrusions are common.

Moore and Calk (1991) studied the geochemistry of four Icelandic tuyas and one hyaloclastite ridge. A systematic upward change in rock composition was found in most cases, exemplified by decrease of the MgO content. The upward decrease in sulfur content is thought to be a measure of the change in depth of water (or ice) cover at the vent during eruption.

Evidence for gravitational sliding (Furnes and Fridleifsson, 1979) and collapse in the walls of the subglacial volcanoes is commonly seen. Evidently the walls became unstable when the confining ice retracted. Prior to consolidation the hyaloclastites are easily eroded by glaciers and running water, both during and after the eruption.

The bulk of the hyaloclastites are consolidated and altered. The volcanic glass is thermodynamically unstable and the basaltic glass alters easily to palagonite (Fisher and Schmincke, 1984). Palagonite is a complex substance and variable in composition; the overall process of alteration is not well understood. Alteration primarily occurs under two different types of conditions: in short-lived, mild, hydrothermal systems and in weathering conditions (Jakobsson, 1978). Intrusions in the hyaloclastite provide the heat for the hydrothermal system in the Surtsey tuya (Jakobsson and Moore, 1986). The basal pillow lava pile possibly serves as a heat source in subglacial and subaquatic volcanoes (Sigvaldason, 1968). The identification of bacteria in near-surface subaerially formed palagonite has led Thorseth et al. (1992) to suggest that microbes have played an important role in the development of certain textures in porous palagonite.

The distribution of upper Pleistocene and Recent Icelandic eruption units with significant hyaloclastite is shown in Figure 3.1A. Most of these units belong to the Moberg Formation (Kjartansson, 1960), formed during the upper Pleistocene (0.01–0.78 Ma), and are predominantly of subglacial origin, though a smaller part is formed in lacustrine and marine environments. Tuff and tuff breccia are common along with pillow lava. Moraines and fluvioglacial deposits are often intercalated with the hyaloclastites. Intraglacial and interglacial lavas are also an important constituent. Subglacial eruptions within the icecaps of Vatnajökull and Myrdalsjökull have continued up to the present.

Subglacial hyaloclastite ridges and mounds are distributed throughout the volcanic zones of Iceland. The number of exposed units is not known but may easily exceed 1000–1200. A recent study (unpublished) in the northern part of the Western Volcanic Zones (Figure 3.1A) has identified 83 englacial volcanoes, of which 59 are hyaloclastite ridges or mounds and 24 are tuyas. The most common lithofacies is poorly bedded tuff breccia, often with pods and lenses of pillow lava, and finely bedded tuff. Basal pillow lava is found in 30 mountains, 8 are solely made of pillows. The length of these ridges ranges from 1 to 18 km and the width from 0.5 to 2.2 km. Maximum height is commonly 200–400 m.

Two examples of medium-sized subglacial hyaloclastite ridges occur in the Western Volcanic Zone (Figure 3.4A, B). Pillow lava is seen at the base of both ridges. The distinction between the ridges and tuyas sometimes is not clear, as demonstrated by the mountain Stóra-Björnsfell (Figure 3.4C). The basal pillow lava appears to have erupted from a fissure. As the eruption progressed, activity concentrated at one vent and eventually produced the subaerial lava cap. Nine ridges in the Western Zone have remnants of subaerial lava on the top; therefore, these units, like the tuyas, have penetrated the ice sheet.

Jones (1969, 1970) published a structural analysis of the subglacial ridge Kalfstindar in the Western Volcanic Zone. Kalfstindar (Figure 3.5) is not a typical ridge in that it is unusually large, has an exceptionally thick basal pillow pile, and may be composed of a few smaller eruption units.

The area between Vatnajökull and Myrdalsjökull is dominated by hyaloclastite ridges (Figure 3.6 and Figure 3.17A); no basaltic tuyas are found in the area. The ridges are larger than in the Western Zone, with the largest eruption unit 44 km long (Vilmundardottir and Snorrason, 1997). The ridges in this area commonly have relief of 300–400 m. A number of ridges have been found with radio-echo soundings under the western part of Vatnajökull (Björnsson, 1988).

Silicic subglacial hyaloclastites are found within Pleistocene central volcanoes in Iceland, although comparatively rare (Furnes et al., 1980). Typical rhyolitic formations are ridges 2 to 2.5 km long and 0.7 to 1 km wide. Well-known examples are Bláhnúkur in the Eastern Volcanic Zone (Saemundsson, 1972) and Hlidarfjall in the Northern Zone (Jonasson, 1994). A few examples of rhyolitic tuyas have been briefly described [e.g., Kirkjufell in the Eastern Zone (Saemundsson, 1972) and Höttur in the Western Zone (Grönvold, 1972)].

The reason hyaloclastite ridges form in some cases and tuyas in other cases remains unclear. Analyses of rocks from the northern half of the Western Volcanic Zone have shown that the distinction is not governed by magma chemistry. Tectonic control may be important since in an eruption on a typical rifting-related fissure, the available magma will be quickly erupted along the length of the fissure. The hyaloclastite ridges most likely form in such

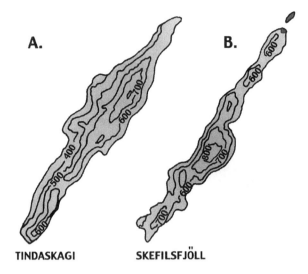
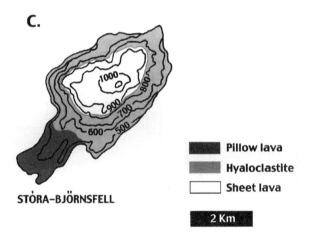

Figure 3.4. Morphology and main structural units of three Upper Pleistocene subglacial volcanoes in the region southwest of Langjökull, the Western Volcanic Zone in Iceland. (A) Tindaskagi. (B) Skefilsfjöll, two typical medium-sized hyaloclastite ridges. Only minor pillow lava is exposed. The ridges are partly buried by Recent lava flows. (C) Stóra-Björnsfell, an elongated medium-sized tuya. The basal pillow lava appears originally to have erupted from a fissure.

eruptions. In contrast, the tuyas in Iceland appear to form in eruptions along short tectonic fissures producing large volumes of magma per unit length (Figure 3.7). Possibly, the underlying tectonics in tuya-forming eruptions are vertical crustal movements related to loading/unloading when overlying ice sheet advanced/retreated in response to climatic fluctuations.

The height of a tuya/ridge subaerial lava cap must have some relation to the ice thickness, but the relationship is not simple. The level of a meltwater lake determines the height where subaerial flow of lava starts. In temperate icecaps, the water level of subglacial

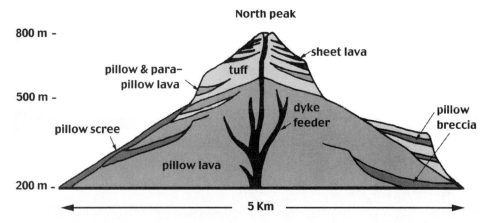

Figure 3.5. Schematic sections through the large Kalfstindar hyaloclastite ridge in the Western Volcanic Zone in Iceland. From Jones (1970).

lakes rarely rises to 90% of the ice thickness, the level required to float the ice and such lakes are drained in jökulhlaups at levels several tens of meters below the floating level (Björnsson, 1988). Thus, within temperate glaciers, semistable water levels higher than 80% of the ice thickness appear unlikely (see Section 3.3). Cold-based glaciers are frozen to their beds, and if water is to flow subglacially, brittle failure of the ice–bedrock contact must take place. However, because of the strength of the ice–bedrock contact, water flow along the ice surface may be dominant in eruptions within thin cold-based glaciers. No observations exist, but in

Figure 3.6. An oblique aerial photo of subglacial hyaloclastite ridges southwest of Vatnajökull, Iceland. The ridges are 20–30 km long and have a relief of 300–500 m. (Photo by Oddur Sigurdsson.)

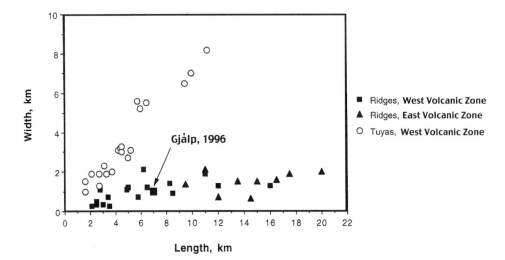

Figure 3.7. A diagram showing the relationship between what various authors have designated basaltic subglacial hyaloclastite ridges and tuyas, as regards length and maximum width. Data from the Western and Eastern Volcanic Zones in Iceland. The estimated position of the basalt–andesitic hyaloclastite ridge formed in the 1996 Vatnajökull eruption is indicated.

this case, the water level of an meltwater lake may be comparable to the ice thickness. At the peak of the last glaciation in Iceland, the ice sheet would have had a mean thickness of about 1.0 km, double the typical height of ridges and tuyas. This may suggest that the tuyas and lava-capped ridges were formed during intervals with less extensive glacial cover.

3.5. SUBGLACIAL VOLCANIC TUYAS

Hyaloclastic ridges capped by subaerial lavas are called *tuyas*. Basaltic hydrovolcanic sequences with tuya or table-mountain form, which have been interpreted as the products of subglacial and englacial eruptions, have been recorded at several localities on Earth. All of the localities are in present-day high-latitude areas, reflecting the low preservation potential of such sequences and/or problems of interpretation in areas situated far from present-day glaciers or ice sheets. Sequences have been recorded from Iceland (Saemundsson, 1967; Sigvaldason, 1968; Jones, 1969, 1970; Allen *et al.*, 1982), British Columbia (Mathews, 1947; Allen *et al.*, 1982; Hickson and Souther, 1984), and Alaska (Hoare and Coonrad, 1978).

Brown Bluff volcano (Smellie and Skilling, 1994; Skilling, 1994) is a 1-Myr-old (Rex, 1976) tuya, located at the northern tip of the Antarctic Peninsula (Figure 3.8). A number of factors suggest Brown Bluff was initially erupted subglacially and later within an englacial lake. Field evidence suggests that there were drainage and refilling episodes of the lake, with drawdown and rise of at least 100 m, during construction of the volcano (Figure 3.9). This is characteristic of englacial lakes, which are typically unstable, especially in active volcanic regions (Björnsson, 1988). Changes in sea level or a nonglacial lake level of this magnitude are unlikely to have occurred during construction of the volcano. There is also no likely paleotopography in this area that could have caused a lake to pond.

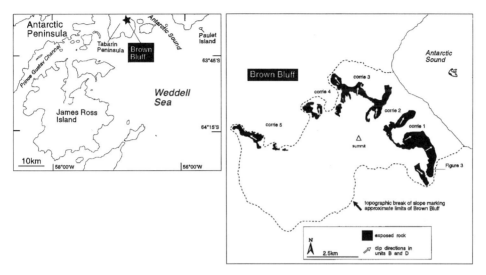

Figure 3.8. Maps illustrating location of Brown Bluff volcano. In B, the dips of hyalotuff cone deposits (Units B and D) radiate approximately from the present summit region, indicating a source region in that area. Adapted from Skilling (1994).

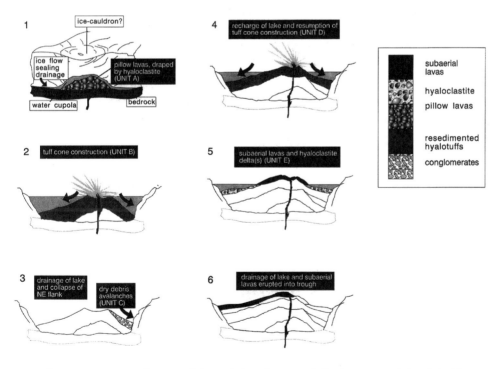

Figure 3.9. Summary diagrams of the evolution of Brown Bluff volcano. 1: entirely subglacial (pillow volcano stage); 2: hyalotuff cone stage, terminated by lake drainage; 3: slumping within tuff ring; 4: lake refilling and continuation of tuff cone construction; 5: subaerial lava flows erupted and formation of hyaloclastite deltas; 6: second episode of lake drainage and partial infilling of trough surrounding the cone by subaerial lava flows. Adapted from Skilling (1994).

The total thickness of >400 m for the Brown Bluff succession implies that the ice thickness was at least of a similar order. Skilling (1994) suggested that the original diameter of the volcano was about 12–15 km, probably with a single central vent. The rock is exposed in five corries (cirques) in the northern half of the volcano, labeled corries 1 to 5 (Figure 3.8). The southern half of the volcano is obscured by ice. Brown Bluff conforms to the simple fourfold structural and lithologic subdivision of tuyas, described by Jones (1969, 1970) from Laugarvatn, Iceland. This sequence comprises basal pillow lavas overlain by lapillus-sized (2–64 mm) vitric tuffs (hyalotuffs) and is capped by hyaloclastite delta deposits and overlying subaerial lava flows. The following is an account of the processes and products that gave rise to this sequence at Brown Bluff. A more detailed account of the sedimentology is given in Skilling (1994).

Brown Bluff is subdivided into four stages—pillow volcano, tuff cone, slope failure, and hyaloclastite delta/subaerial—and into five structural units (A to E, Table 3.1). Unit A is the basal and oldest unit, termed the *pillow volcano stage*, and is composed dominantly of pillow lavas draped by massive hyaloclastite. Units B and D comprise the tuff cone stage and consist dominantly of vitric lapillus tuffs. These units are separated in the northeast of Brown Bluff by deposits of Unit C. This unit represents the slope failure stage, and consists dominantly of debris avalanche deposits. Unit E is the uppermost hyaloclastite delta/subaerial stage, and is composed of hyaloclastite delta deposits and overlying subaerial lava flows. Figure 3.9 summarizes the evolution of Brown Bluff, including periods of lake drainage and recharge.

The major lithofacies and their relative abundances within each unit are illustrated in Table 3.1. More detailed information on lithofacies is given in Skilling (1994). Most of the facies are subdivided into "hyaloclastite" and "hyalotuff." These are defined on the basis of an estimate of the vesicularity of the lapillus (or coarser) clasts (<25%: hyaloclastite; >25%: hyalotuff). No particular fragmentation is implied by these terms. The sedimentary term *conglomerate* is used where lithification prior to resedimentation had taken place. Sedimentary grain sizes are used throughout.

Table 3.1. Description and Interpretation of Lithofacies of Brown Bluff Volcano, a Tuya in Antarctica

Stage	Units	Lithofacies
Hyaloclastite delta and subaerial	E	Subaerial: pahoehoe lavas > water-cooled lavas (near base)
		Hyaloclastite delta: stratified cobble and gravel hyaloclastite > massive cobble hyaloclastite > massive sandy hyaloclastite > normally graded sandy hyaloclastite > complete pillow-rich hyaloclastite > sandy conglomerates
Slope failure (interbedded between Units B and D; present in corries 1 and 2 only)	C	Sandy conglomerates > gravel hyalotuffs > stratified cobble and gravel hyaloclastite
Tuff cone	B and D	Gravel hyalotuffs > massive gravel hyalotuffs > water-cooled lavas > sandy conglomerates > normally graded sandy hyalotuffs > massive cobble hyaloclastite > complete pillow-rich hyaloclastite > pillow lavas
Pillow volcano	A	Pillow lavas > stratified cobble and gravel hyaloclastite > massive gravel hyalotuffs > complete pillow-rich hyaloclastite > massive sandy hyaloclastite > stratified gravel hyalotuffs

3.5.1. Unit A

Unit A has similarities to pillow volcano or seamount sequences, which often consist of a core of pillow lavas draped by massive and steeply bedded hyaloclastites (Lonsdale and Batiza, 1980; Staudigel and Schmincke, 1984). The presence of pillow lavas at the base of the Brown Bluff succession and at the base of several other subglacial tuyas (Jones, 1969, 1970; Wörner and Viereck, 1987) suggests suppressed explosivity related to the overlying water/ice hydrostatic head. Unit A was constructed by both extrusive and intrusive processes with intrusion of pillows and pillow-margined bodies continuing at depth while tuff cone construction had already commenced (Skilling, 1994). The massive hyaloclastites that drape the pillow lavas probably represent slumped deposits produced by failure of a steep-sided pillow volcano, or may represent slumped hyaloclastite delta deposits, from an early stage of emergence, developed without an intervening tuff cone.

3.5.2. Units B and D

Units B and D represent the deposits of a Surtseyan-type tuff cone. In the summit region, the presence of some inward-dipping beds, more pervasive hydrothermal alteration, and radial dips of the outer cone slope deposits away from this area, imply the proximity of a vent. A steep inward-dipping contact in this region is interpreted as a crater margin (Figure 7 in Skilling, 1994), and the facies as the crater-fill. Tuff cone deposits were probably generated initially by subaqueous and subaerial wet tephra fallout (and surge?), with redeposition into the crater by mass flows and slumps. The occurrence of massive tephra in this area is interpreted as a vent slurry (Figure 3.10). Curvicolumnar-jointed lavas in the summit area are interpreted as water-cooled lavas ponded within the crater. The interbedding of air and water-cooled lavas in the uppermost parts of Unit D implies periodic water flooding of the crater floor.

The deposits of the outer cone slope were interpreted by Skilling (1994) as density-modified grain flows that had transformed to turbidites on the lower slopes. The occurrence of turbidite deposits, the absence of fluidal gravity flow deposits, and the lack of reworking of the tephra, imply that the tuff cone was constructed within a ponded water environment (Figure 3.10). The lateral impersistence and amalgamation of most of the beds suggests that the primary emplacement was from low-volume, discrete, simultaneous eruptions, interpreted as Surtseyan "tephra jets" (Figure 3.10). There is no evidence of direct deposition by pyroclastic surges or airfall, implying either that all of the tephra was resedimented downslope or back into the crater, or that the subaerial upper part of the tuff cone has been lost to erosion.

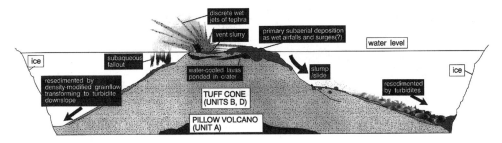

Figure 3.10. Diagram illustrating the eruptive and depositional processes that operated during the hyalotuff cone stage. Adapted from Skilling (1994).

3.5.3. Unit C

The conglomerates of this unit record the failure of the northeast flank of the volcano, during tuff cone construction. Jigsaw-brecciated clasts, very poor sorting and poor mixing of clast populations with steep internal contacts, imply that this unit comprises dry debris avalanche deposits. This means that the ponded water necessary for turbidite deposition in Unit B had drained away, but that the lake must have refilled to allow continued depostion of Unit D turbidites (Figure 3.9). The origin of the collapse is unclear, but the abundance of strongly hydrothermally altered clasts suggests that pervasive alteration of tephra to clay minerals in the vent region was a contributing factor. Drainage of the lake may also have contributed to collapse.

3.5.4. Unit E

The wedge-shaped morphology, steep to asymptotic bedding, and the capping of lava flows with lobes that extend down into hyaloclastite (Skilling, 1994), suggest that Unit E represents hyaloclastite delta deposits (Figure 3.11). The angle of deposition and the presence of inverse grading implies that the dominant process of deposition was density-modified grain flow. The occurrence of several conglomerate facies on the upper parts of the delta slope, and their juxtaposition with facies at the delta brink point from which they were derived, suggests that they originated by oversteepening and collapse in this area (Figure 3.11). Skilling (1994) suggested that a variable input rate and volume of lava streams into the lake influenced the type of deposits that ponded at the brink point, and consequently the type of flows that are generated by their collapse. Narrow lava streams generated density-modified grain flows of cobble hyaloclastite, while wider streams formed pillow lavas that ponded at the brink point, and subsequently collapsed down the delta front. The widest streams of lava ponded at the brink point as large masses of curvicolumnar-jointed lavas, which also collapsed downslope.

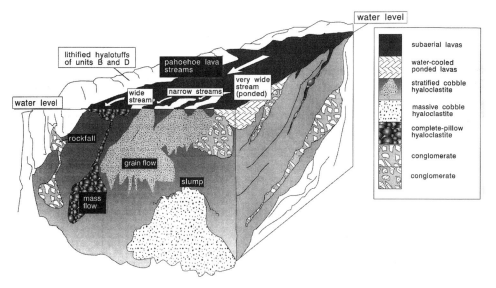

Figure 3.11. Diagram illustrating the processes and products generated during hyaloclastite delta construction (Unit E). Adapted from Skilling (1994).

Rockfalls from the surrounding lithified tuff cone deposits also collapsed onto the delta. The juxtaposition of subaerial lava flows and hyalotuffs in corrie 4 without any intervening hyaloclastite (Figure 7 in Smellie and Skilling, 1994), implies that a second drainage of the lake took place, allowing subaerial lava flows to fill a drained trough around the tuff cone (Figure 3.9).

3.6. LAHARS AND JÖKULHLAUPS

As we have learned in the previous sections, hyaloclastic ridges and tuyas are a vent phenomenon and are restricted to eruptions beneath ice sheets or water (Surtsey is a marine tuya); therefore they are a definitive indicator of volcano/ice interaction. Floods and resulting mass flows are triggered by numerous causes including volcano/ice interaction, which form lahars and jökulhlaups (mass flows consisting of volcanic material). Although mass flows are not definitive indicators of volcano/ice interaction, they are usually the most widespread, voluminous, and destructive products of subice eruptions.

Mass flows are gravity-driven, viscous non-Newtonian fluids. Features shared by most mass-flow deposits are poor sorting (Rodine and Johnson, 1976), lack of internal stratification (Crandell and Waldron, 1956), support of large clasts in a finer-grained matrix (Johnson, 1970), sharp contacts (Hooke, 1967), and somewhat uniform thickness over large expansive flats (Bull, 1972) like the Martian lowlands. Morphologically, they may have steeply dipping lobate snouts, elevated margins, and central plugs of coarse clasts (Johnson, 1970). Lahars have been known to abraid underlying bedrock, contain well-rounded cobbles and boulders with increased distance downstream, and deposit boulder clasts at flow obstructions (Pierson et al., 1990). The following discussion provides details of terrestrial lahars and jökulhlaups.

3.6.1. Alpine Situations

Snowmelt by Pyroclasts. Explosive eruptions and subglacial melting at snowclad volcanoes swiftly generate in several ways (Major and Newhall, 1989) huge floods that are hazardous and can alter landscape. A common process is turbulent hot flows entraining snow to produce diverse "mixed avalanches" consisting of debris-bearing snow avalanches, snowflows, and slushflows that may transform downvalley into great debris-laden floods (Pierson and Janda, 1994; Pierson, 1995).

During the first 2 minutes of the May 18, 1980, eruption of Mount St. Helens, a great turbulent pyroclastic surge melted snowpack, firn, and the surfaces of small glaciers. Thousands of small slushflows mixed with surge debris and accumulated unstably on steep slopes, coalescing downslope into slushy floods. Because constituent snow and debris retarded turbulence, the flows accelerated to 100 km hr^{-1} and more (Waitt, 1989) into valleys, which channeled the floods for tens of kilometers (Pierson, 1985; Scott, 1988). During eruptions at Mount St. Helens between 1982 and 1991, small pyroclastic flows and surges repeatedly melted snowpack and formed various mixed pyroclast–snow-grain flows and floods (Figure 3.12; Waitt et al., 1983; Waitt and MacLeod, 1987; Pierson and Waitt, 1997).

Alaska's snowy volcanoes often generate floods during eruption. Pyroclastic-flow deposits of eruptions of Augustine volcano in 1976 and 1986 grade into bouldery flood deposits (Waitt and Begét, 1996). During each of three eruptions of Crater Peak (Mount Spurr) volcano in 1992, hot pyroclasts melted snowpack to form diverse debris-rich snow avalanches and flows intermediate between pyroclastic flows and lahars (Waitt, 1995).

Figure 3.12. Snowy flood from crater of Mount St. Helens induced by small explosive eruption in March 1982. From outer margin to center is the succession: snowflow, slushflow, watery flow.

Influence of Glaciers. Like Nevado del Ruiz (Section 3.2), other historical summit eruptions have triggered remarkable lahars. In winter 1989–1990, dome-collapse pyroclastic flows at Mt. Redoubt, Alaska, stripped snowpack and firn to form a 3- to 20-m-thick icy diamict of snow grains and coarse ash freighted with huge blocks of andesite, glacier ice, and snow (Waitt *et al.*, 1994). Some icy flows reached 14 km laterally over an altitude drop of 2.3 km. Erupting hot andesite probably mixed turbulently with snow, triggering avalanches that rapidly entrained firn and ice blocks from the heavily crevassed glaciers. Successive eruptions melted snow and glacier ice, incising and eventually beheading Drift glacier (Trabant *et al.*, 1994). Resultant floods swept down Drift River valley for 35 km to the sea with discharges as high as 10,000 to 60,000 m^3 s^{-1} (Dorava and Meyer, 1994).

Glaciated Mount Rainier volcano has shed at least 60 sizable lahars in postglacial time, many of them clearly originating by eruption but a few very large ones originating as landslides, perhaps noneruptively (Crandell, 1971; Scott *et al.*, 1995). Though Rainier's eruptions tend to be far less explosive than Mount St. Helens, meltwater from pyroclastic eruption or geothermal heating near Rainier's summit must flow at least 5 km across or beneath steep, snowclad, much-crevassed glacier(s), a situation ripe for generating lahars and floods.

3.6.2. Icecaps and Ice Sheets

Many volcanoes, especially in middle to high latitudes, are crowned by icecaps or ice sheets. Besides the swift melting of snow and ice during explosive eruption, these volcanoes and their geothermal systems can slowly melt ice and store water at the glacier bed (Björnsson, 1975; Figure 3b in Björnsson, 1988). Such unstable icebound water may then break out as a subglacial jökulhlaup.

Iceland's Vatnajökull (icecap) overlies several hydraulic basins (Plate 8 of Björnsson, 1988), some of which are repeatedly swept by jökulhlaups. Archetypical jökulhlaups issue from the iceshelved caldera of Grímsvötn and flow down Skeidarársandur (Figure 3.1A), typically without eruption. Geothermal heating gradually melts ice and lifts the lake level in Grímsvötn until hydraulic stability is exceeded and the lake suddenly drains subglacially (Björnsson, 1974; Nye, 1976). Before 1940 the lake drained every 7–10 years and released large volumes (4.5 km^3) of water in large-discharge (>25,000 m^3 s^{-1}) floods; in 1938 a flood was thought to have peaked at about 30,000 m^3 s^{-1} (Björnsson, 1992; Gudmundsson et al., 1995) but was recently recalculated to 40,000 m^3 s^{-1}, although the peak value is very sensitive to the estimated duration of the flood (Arni Snorrason, personal communication, 1998). Since 1940 jökulhlaups recur every 4–5 years, have typically smaller volume (1 to 2.5 km^3), and have lower peak discharges of 600 to 11,000 m^3 s^{-1}. The unusually large 1996 Skeidará jökulhlaup of 52,000 m^3 s^{-1} (Jónsson et al., 1998a,b) was caused not just by passive geothermal melting but by a subglacial Gjálp eruption (Section 3.3) that rapidly melted glacier ice.

Skafá River on southwest Vatnajökull has flooded suddenly many times. Some 40 km above the glacier terminus where the flood emerges, the glacier surface sags as two deep cauldrons, where Björnsson (1977) concludes geothermal heating melts the glacier bed to form cupola lakes. Sudden draining of either of these unstable lakes causes the ice surface directly above it to subside. Glacier flow then gradually infills the cauldrons.

In southeast Vatnajökull, icecapped Öræfajökull—Iceland's largest volcano—erupted explosively in 1362, sending huge debris-laden floods down outlet glaciers and erasing farmsteads (Rórarinsson, 1958). Bouldery diamict hummocks 2–4 m high stand well out in front of the glacial moraines. Voluminous snow and glacier-surface ice may have been melted swiftly as at Mount St. Helens, but the floods were large enough to indicate release of subglacially stored water, as in 1918 at Katla (see below).

Jökulsá á Fjöllum, a large glacial river, drains from Vatnajökull icecap 190 km northward to the sea. A scabland-carving flood had swept down lower Jökulsá á Fjöllum about 8000 years ago. About 2000 years ago Kverkfjöll caldera generated a gigantic flood that carved recessional cataracts and other scabland topography and laid gravel bars along 150 km of Jökulsá valley (Figure 3.13). Smaller floods from Kverkfjöll in the fifteenth to eighteenth centuries destroyed farms in Axarfjördur. Many erosional and depositional features along this path resemble parts of Washington's Channeled Scabland, carved by one of Earth's largest Pleistocene floods. The late Holocene Jökulsá flood probably peaked between 0.4 and 1 million m^3 s^{-1} (Tómasson, 1973; Waitt, 1998). Subglacial Kverkfjöll caldera is a flood source, but farther west a subglacial fissure system and Bárdarbunga caldera are also potential sources. Had the 1996 Gjálp eruption (Section 3.3) been farther north along the subglacial

Figure 3.13. Scabland with relief of 20 m carved about 2000 years ago by huge jökulhlaup along Jökulsá á Fjöllum, north Iceland.

rift, its meltwater would have drained north to Jökulsá á Fjöllum rather than as it did south to Grímsvötn and Skeidará.

In south Iceland Myrdalsjökull (icecap) overlies Katla caldera, whose eruptions triggered enormous floods in 1660, 1725, 1755, and 1918 (Jónsson, 1982). For the great 1918 Katla flood, Tómasson (1996) recalculates peak discharge at about 300,000 m^3 s^{-1}—one of Earth's largest historic floods. It seems impossible that the enormous Katla jökulhlaups could have originated solely from melted snow and ice during brief eruption. Quick release of such volumes implies breakout of dammed water, probably geothermally melted water stored in a cupola at the base of the ice. The eighteenth-century jökulhlaups of Mÿrsdalssandur transported megaclasts, which Jónsson (1982) interprets to be deposits of dense debris flows. The poor sorting and weak stratification indeed resemble deposits of debris flows and hyperconcentrated flows from Mount St. Helens during its May 18, 1980, eruptions (Scott, 1988).

3.7. PHREATIC CRATERS ON EARTH AND MARS

Another type of volcano/ice interaction occurs not as a true volcanic eruption, but as a result of steam explosions when lava flows over ground saturated with water or ice (see Section 2.4.5). Many of the Earth's volcanic regions are dotted with craters formed by steam explosions. Eruptions near the ocean or lava flowing into the sea commonly produce tuff rings or littoral cones. The Hawaiian and Galápagos islands contain many such features. The encounter of magma with ground water at depth can yield spectacular maars and explosion-collapse depressions such as the Pinacate craters of Sonora, Mexico.

Tuff rings and maars are found in Iceland, but a third type of steam explosion crater seems unique to that country. This is the phreatic crater, or pseudocrater, a landform resulting from explosions caused by lava flowing over water-saturated ground (Thorarinsson, 1960). Phreatic craters are cones, generally resembling small cinder cones, which are rootless. That is, they are not formed above a vent.

The type locality for these pseudocraters is Lake Myvatn in northern Iceland, where basaltic lava flowed into the shallow lake basin around 2500 years ago. The southern shore and islands of Myvatn are dotted with over 1000 of these small cones (Figure 8 of Allen, 1979b). They can be as high as 30 m and have basal diameters as large as 320 m. The average Myvatn phreatic crater rises no more than 5 to 10 m and is approximately 50 to 100 m across. They were formed in a pahoehoe flow that ranged from 3 to 5 m thick.

Though the phreatic craters at Myvatn are the best known, similar features occur in many parts of Iceland. At least eight fields of these craters have been mapped in other parts of the country. Phreatic craters are diverse in terms of composition, size, and morphology. A broad central pit is common, often approximately one-half of the basal diameter across. Not even the crater form is requisite, however. Some of the features in southern Iceland have no summit pits, but are tumuli several meters in height. Most phreatic craters are composed primarily of spatter, indicating that the explosions occurred while the bulk of the lava was still molten or at least semiplastic. In some cases coherent basalt layers over a meter thick were uplifted and contorted without breaking (Allen, 1979b).

Individual phreatic craters are similar in many respects to small cinder and spatter cones, though their origin is significantly different (see Chapter 2). Morphologic evidence for an origin by steam explosion includes small size, large crater diameter/basal diameter ratio, lack of alignment among craters, and confinement to a single flow.

The formation of phreatic craters by steam explosion is well accepted, on the basis of field observations. Fluid basalt can flow rapidly enough to trap water, which flashes to superheated steam. The calculated steam pressure is around 100 times the pressure exerted by the overlying lava (Allen, 1979a). Violent explosions should be expected in such an environment.

What if, instead of water, lava flows over ice? The situations are actually very similar. The heat needed to melt ice and create superheated steam is only around 7% more than the energy required to make the steam by heating water (Allen, 1980). If the ice is buried under a few meters of crushed rock, the heat of a lava flow is still sufficient to produce a steam explosion (Allen, 1979a,b). This simplified description of the near surface may apply to some areas of Mars. With surface gravity less than 40% of that on Earth and a much more tenuous atmosphere, phreatic craters several times larger than their terrestrial counterparts might be formed (McGetchin et al., 1974).

Groups of small cones, suggested to be of phreatic origin, have been recognized by several independent investigators in the Chryse Planitia (Greeley and Theilig, 1978), Deuteronilus Mensae (Lucchitta, 1978), and Acidalia Planitia (Frey et al., 1979; Allen, 1980) regions of Mars (Figure 3.14), but the origin of these features is under debate. Most of these features measure approximately 600 m across the base, or double the size of the largest Myvatn crater. The Martian examples are simple, occasionally overlapping cones with craters about half their basal width. The cones are morphologically distinct from primary and secondary impact craters of the same size, as well as from the small shield volcanoes in the region. If these are really phreatic craters, they hold promise for mapping near-surface Martian ice deposits.

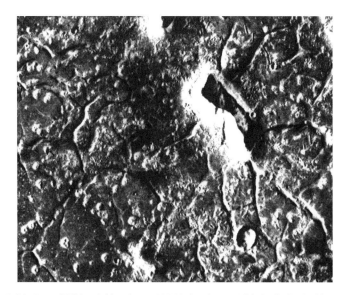

Figure 3.14. Part of Viking Orbiter image 72A02 showing possible hyaloclastic ridge and surrounding phreatic cones in Acidalia Planitia, Mars; note central ridge and alignment with nearby fissures; illumination is from the left.

3.8. HYALOCLASTIC RIDGES, TUYAS, LAHARS, AND JÖKULHLAUPS ON MARS

Possible analogues to terrestrial hyaloclastic ridges, tuyas, lahars, and jökulhlaups have been identified on Mars in the Acidalia Planitia region, north and northwest of Elysium Mons, a large volcanic construct on the east edge of the Utopia Planitia basin, and in Ares Valles at the site of the Mars Pathfinder Lander (see Chapter 4).

3.8.1. Hyaloclastic Ridges and Tuyas

Hodges and Moore (1978) noted that landforms resembling tuyas are concentrated in Acidalia Planitia. Allen (1979a) identified three tuya analogues and two possible hyaloclastite ridges in the Acidalia area. These features closely resemble Icelandic subice volcanoes in both morphology and size. Figure 3.15 shows a steep-sided, flat-topped mountain near latitude 45°N and longitude 21° in central Acidalia Planitia. A large cone with a shallow crater caps the mountain. The main plateau measures approximately 7 km by 4.8 km. By analogy, tuyas with similar structural features in Iceland are characteristically less than 10 km across. On Earth the height of the plateau approximates the local thickness of the ice during eruption, generally 200 to 1000 m. Figure 3.14 features a sharp ridge at latitude 38°N and longitude 13°. The ridge is 6 km long and 2 km at maximum width. Several smaller ridges and fissures parallel the main feature. These dimensions are well within the range of the myriad morphologically similar hyaloclastite ridges throughout central Iceland. These subice eruptive features range from <1 km to >35 km in length and are generally 200–400 m high.

Northwest of Elysium Mons, landforms interpreted to be subice volcanoes occur as ridges on the flanks of the volcano (Anderson, 1992; Chapman, 1994) and as mesas in Utopia Planitia, near the Viking 2 Lander Site (Hodges and Moore, 1978; Allen, 1979a,b). The

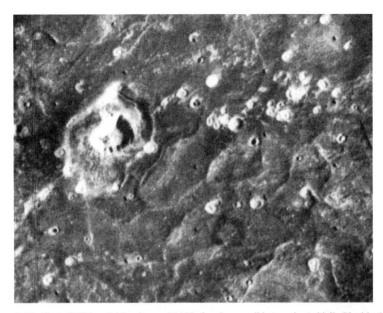

Figure 3.15. Part of Viking Orbiter image 26A28 showing possible tuya in Acidalia Planitia, Mars; illumination is from the left.

enigmatic ridges and mesas on the flank of Elysium Mons were initially included in large areas mapped as lahars (Mouginis-Mark, 1985; Christiansen, 1981; Christiansen, 1989). However, as discussed in Section 3.6, mass flows tend to form deposits of uniform thickness, not high-standing mounds and ridges. A subice volcanic origin of the ridges on the Elysium flank is supported when these features are compared with the Icelandic subglacial volcano Herdubreidartögl, a tuya whose narrow central ridge is clear evidence of its origin by fissure eruptions (Figure 3.16A; Werner et al., 1996). The rough-textured Martian mounds and ridges contain similar central, narrow, linear ridges and pits; narrow ridges also extend away from the mounds (Figure 3.16B). Möberg ridges erupted along fissures southwest of the Vatnajökull icecap (Figure 3.17A) form lineaments that appear very similar to some rough-textured ridges in the Elysium area (Figure 3.17B). Central, narrow, linear ridges and pits within the mounds and the alignment of mounds with Elysium Fossae (volcano-tectonic depressions) and linear chains of domes attest to fissure eruptions as the likely source of the mounds and ridges. Their rough texture, topography, and similarity to Icelandic features suggest that they may be formed from friable materials such as palagonite and breccia and are possible hyaloclastic ridges. The slopes of the Elysium ridges average about 2°. Curiously, the slope of Kalfstindar, Iceland (Figure 3.5), is a much steeper 18°. This discrepancy may be related to erosion of friable hyaloclasts on the much older Martian ridges (<0.8 Ma for Kalfstindar versus 200 Ma or more on Mars) and because Kalfstindar is unusually large (see Section 3.4).

Mesas in the northeastern part of Utopia Planitia may be tuyas erupted from central point sources (Figure 3.18A; Hodges and Moore, 1978; Allen, 1979a). Gaesafjöll (Figure 3.18B) is an Icelandic tuya that erupted palagonite-forming basalt from a central source until the surrounding icecap retreated and the meltwater drained; continued eruptions produced layered, subaerially extruded lavas and a central summit crater (Van Bemmelen and Rutten, 1955). One Martian mesa at about latitude 46°N, longitude 229° (Figure 3.18A) is super-

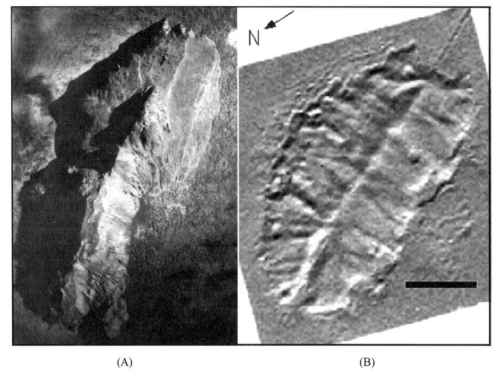

Figure 3.16. Comparison of Icelandic and Martian features; illumination is from the right. (A) Herdubreidartögl (Van Bemmelen and Rutten, 1955), a tuya 4 km wide; its narrow central ridge is evidence of emission via fissure eruptions. (B) Part of Viking Orbiter image 541A10 showing a mound interpreted as hyaloclastic ridge; note mound contains an inner ridge.

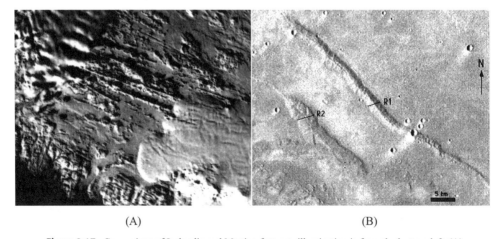

Figure 3.17. Comparison of Icelandic and Martian features; illumination is from the bottom left. (A) Hyaloclastic ridges, 2–3 km wide, erupted along fissures southwest of Vatna icecap (Van Bemmelen and Rutten, 1955). (B) Part of Viking Orbiter image 541A20 showing possible hyaloclastic ridges; note central pits and ridges extending away from mounds.

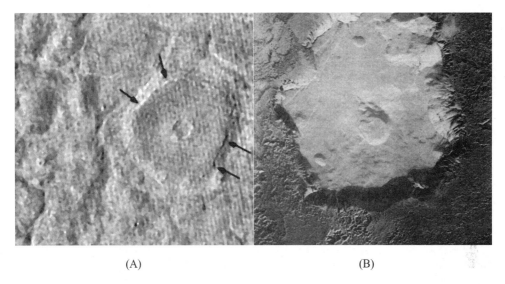

Figure 3.18. Comparison of Martian and Icelandic features. (A) Possible tuya on Mars in Utopia Planitia at about latitude 46 °N, longitude 229 °; arrows denote break in slope between cap rock and underlying material; illumination is from the upper left. (B) Gaesafjöll (Van Bemmelen and Rutten, 1955), a tuya about 5 km wide with a central summit crater; illumination is from the top.

posed on polygonal terrain; the break in slope between the cap rock (layered lavas?) and underlying slope (palagonite?) is clearly visible. The morphology of the mesas, their location in the Utopia Basin near the large volcanic center of Elysium Mons, and their alignment with other volcanic cones (Chapman, 1997) suggest that they may be tuyas.

As noted in the introduction, water may have ponded in the northern lowlands of Mars, which includes Utopia Planitia. As some additional evidence suggests, any hypothetical paleolake or ocean (and any connecting ephemeral lakes) in the basin was likely frozen to some depth (Lucchitta *et al.*, 1986, 1987; Scott and Underwood, 1991; Kargel and Strom, 1992). The thickness of the ice sheet and water level may be estimated from the height of the subglacial volcanoes (Section 3.4) in Utopia Planitia. Photoclinometric measurements of the ridges and mesas northwest of Elysium indicate that the paleo-ice sheet in Utopia may have been 200 m thick (Chapman, 1994). In addition, on images obtained by the Mars Orbiter Camera on Mars Global Surveyor, researchers at Malin Space Science Systems have interpreted small cones in the Elysium and Amazonis regions to be possible phreatic craters.

3.8.2. Lahars and Jökulhlaups

Surrounding and downslope of the hypothetical hyaloclastic ridges of northwest Elysium Mons, extending 1500 km away into central Utopia Planitia, is a lobe of coarse, rough-appearing material with inner, smooth areas. This lobe was interpreted by Christiansen (1989) to be laharic material, based on its morphologic similarity to terrestrial mass flows. Because of its proximity to the hypothetical hyaloclastic ridges, this rough-textured material was reinterpreted as jökulhlaup deposits (Chapman, 1994). The material appears to have been discharged from fissures related to the Elysium möberg ridges, as the lobe deposit can be traced back to these volcanic features. These rough deposits form topographically lower, smoother appearing deposits locally bounded by lobate scarps. The smooth areas contain nested 1-km-wide concentric craters and smooth-floored, irregular channels that may be related to collapse and dewatering of mass flow material (Christiansen, 1989).

On the north flank of Elysium Mons (west of Hecates Tholus) are local, subdued relief flows that are less lobate than other upflank lava flows. Mouginis-Mark (1985) interpreted the subdued flows to be the Martian equivalent of Icelandic jökulhlaups. Finally, deposits in Ares and Tiu-Simud Valles at the site of the Mars Pathfinder Lander were compared by Rice and Edgett (1997) to similar Icelandic outwash plains formed by jökulhlaups.

3.9. VOLCANO–GROUND WATER INTERACTIONS AND SNOW/ICE PERTURBATIONS ON MARS

Ground ice on Mars is a likely indicator of ground water at depth. In fact, volcano–ground water or ground ice interactions appear to have been pervasive throughout the geologic history of Mars. Ground water or ground ice can interact with volcanoes in several ways. Shallow intrusions can interact directly with ground water, resulting in explosive volcanism, such as maar craters. Or water might enter a magma chamber, thereby affecting the style of volcanism, perhaps producing pyroclastic activity (see Chapters 2 and 4). A magmatic intrusion at depth will alter the flow of ground water, producing convecting cells of water and a potentially long-lived hydrothermal system.

Perhaps the most convincing evidence for volcano–ground water interactions are the large channels found immediately adjacent to some volcanoes. These channels emanate from discrete collapse zones, suggesting a close relation between intrusive igneous activity and the sudden release of ground water. Examples are the channels west of Elysium Mons volcano (Mouginis-Mark, 1985; labeled 10 in Figure 4.1), Dao Vallis on the flank of the volcano Hadriaca Patera (Squyres *et al.*, 1987; Figure 4.13), and the small channels immediately to the east of Olympus Mons (Mouginis-Mark, 1990; Figure 4.2). On the largest possible scale, Baker *et al.* (1991) have suggested that the formation of the largest outflow channels was linked to a regional-scale hydrothermal system triggered by the formation of the Tharsis bulge.

On a smaller scale, fluvial valleys are found predominantly in the ancient cratered highland terrain and on the slopes of some volcanoes. The northern flank of the volcano Alba Patera is dissected by many small valleys (Gulick and Baker, 1989, 1990). Several other volcanoes, including Apollinaris Patera, Hecates Tholus, and Ceraunius Tholus (see Chapter 4), also exhibit small valleys. These valley systems conceivably could have been formed by rainfall processes, although evidence exists for ground water outflow in many cases. Furthermore, Alba Patera formed long after the putative early warm, wet period of Mars's climatic history. Fluids discharged by hydrothermal activity are thus an attractive alternative for the source of these valleys (Gulick and Baker, 1989, 1990).

3.9.1. Hydrothermal Systems

The possible influence of hydrothermal systems has long been recognized. Schultz *et al.* (1982) and Brakenridge *et al.* (1985) suggested that impact-induced hydrothermal systems could be responsible for valleys on ejecta blankets on Mars. Gulick and Baker (1989, 1990) proposed that the discharge of hydrothermal fluids to the surface was an important process for the formation of those valleys that formed on the flanks of Martian volcanoes, particularly on the younger, ash covered Hesperian- and Amazonian-aged volcanoes. Some hybrid models include both precipitation and hydrothermal systems. Gulick *et al.* (1997) proposed that a substantial amount of water could be transported during modest greenhouse periods from surfaces of frozen bodies of water to higher elevations, despite global temperatures well below freezing. This water, precipitated as snow, could ultimately form fluvial valleys if deposition sites were at or near regions of hydrothermal activity. Hydrothermal systems have also received a great deal of attention as agents for producing aqueous alteration of the SNC meteorites (Wentworth and Gooding, 1994) (see Chapter 4), for exchanging deuterium and hydrogen between the crust and atmosphere (Jakosky and Jones, 1994), for providing paleohabitats for life, and for preservation of fossils (Walter and DesMarais, 1993; Farmer and DesMarais, 1994).

An example of an idealized model of ground-water flow associated with a magmatically generated hydrothermal system as might be produced by the formation of a Martian volcano is shown in Figure 3.19. Soon after emplacement of the magma, the outer shell of the magma chamber starts to solidify, forming a low-permeability outer shell of hot rock, the thickness of which increases with time (Gasparini and Mantovani, 1984). Thermal energy is transported primarily by conduction from the magma through the shell, and then primarily by convection into the saturated, permeable country rock. This shell prevents ground water from contacting

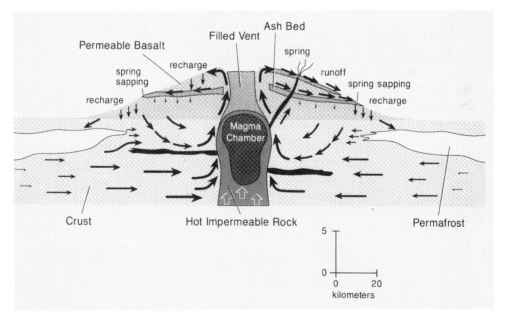

Figure 3.19. Conceptual model illustrating the ground-water flow field of a vigorous hydrothermal system associated with volcano formation on Mars. Vertical scale is exaggerated by a factor of 4 in order to illustrate details of ground-water flow in the stratigraphic layers of the volcano. Martian volcanoes are unusually large compared with their terrestrial counterparts, ranging on the order of 10^2 to 10^3 km in diameter, and those that are dissected by valleys tend to have low aspect ratios with average slopes of less than a few degrees. Our numerical model considers hydrothermal systems associated with magmatic intrusions on Mars in general. Topography, multiple intrusions, and boiling are not directly simulated in our model, although their effects on these systems are discussed elsewhere. From Gulick (1993, 1998).

the intrusion itself. Surrounding ground water that is heated by the magma forms an upwardly moving buoyant plume near the intrusion. Colder, denser ground water flows in toward the intrusion from surrounding regions and continues to replace the upwardly moving ground water as long as a thermal gradient exists. Depending on the size of the intrusion, ground water within several tens of kilometers or more could flow into the system. Ground ice above and near the intrusion would be melted, locally eliminating or thinning the permafrost zone (Gulick and Baker, 1992; Gulick, 1993). Ground water reaches the surface as liquid or vapor or both. The near-surface behavior of the hydrothermal fluids would depend on their temperature and mineral concentration, and on the atmospheric temperature and pressure.

Ground water that reaches the near-surface environment can contribute to the geomorphic modification of that surface. If local hydrologic and lithologic conditions permit, water could flow on the surface and reenter the ground water system in regions where the rate of infiltration is sufficiently high. However, if the atmospheric temperature and pressure are not favorable for fluid flow, ground water would initially start to boil and evaporate but then freeze as a result of the heat-liberating process of evaporation. This process would result in the formation of an insulating ice layer beneath which subsequent outflows of hydrothermal water may move as ice-covered streams (Wallace and Sagan, 1979; Carr, 1983; Brakenridge et al., 1985). Ground water that does not interact with the surface would help to recharge near-surface aquifers and eventually could outflow to the surface farther away from the intrusion. Whether ground water remains liquid would depend on the local lithologic conditions, the temperature and mineral concentration of the water, and the atmospheric conditions at the time of ground water outflow.

3.9.2. Permafrost

The presence of ice-rich permafrost would affect an active hydrothermal system on Mars. Upwardly moving hydrothermal fluids must melt through permafrost to reach the surface. Assuming that convective warming is much faster than conductive cooling, Gulick (1998) computed the time required to melt through a 2-km-thick permafrost zone. She found that if permafrost on Mars fills a region with a pore space of 10 to 35%, then hydrothermal fluids can melt through the permafrost in several thousand years. This time is short compared with the lifetime of moderately sized hydrothermal systems (about 100,000 years for a 50 km^3 intrusion). Therefore, the presence of an ice-rich permafrost zone should have a negligible effect on the lifetime of a hydrothermal system on Mars.

Even directly above a volcanic intrusion, the surface temperature at Mars would be primarily controlled by the balance between absorbed solar and emitted infrared radiation (Fanale et al., 1992). The surface temperature will remain below freezing and a residual ice-rich permafrost zone will remain near the surface, except for areas directly above the intrusion or where springs or seeps have formed. Figure 3.20 (from Gulick, 1998) illustrates the equilibrium thickness for a variety of heat flows and surface temperatures (applying McKay et al., 1985, Equation 2). The present-day background geothermal gradient probably provides an average of around 0.03–0.04 W m^{-2} (Fanale, 1976; Toksöz and Hsui, 1978; Davies and Arvidson, 1981). However, the average heat flow in the presence of an active terrestrial hydrothermal system can range from 2 to 5 W m^{-2}. For example, at Wairaki, New Zealand, the regional heat flow averaged over approximately 50 km^{-2} is 2.1 W m^{-2}, whereas fluxes averaged over the more intense regions are of order 500 W m^{-2} particularly in localized areas around springs (Elder, 1981). Figure 3.20 shows that the equilibrium permafrost thickness above an active hydrothermal system may be less than 100 m. Any inhomogeneities in the

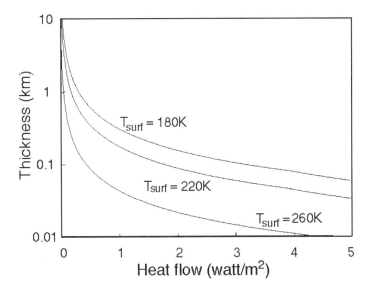

Figure 3.20. Equilibrium permafrost thickness as a function of geothermal heat flow. Permafrost thickness is shown for three mean surface temperatures. Typical estimates of current Martian heat flow as well as geomorphologic evidence indicate permafrost thicknesses of approximately 2 km. Higher surface temperatures and interior heat fluxes both produce thinner permafrost layers. Heat flow above an active hydrothermal system may be as large as 3 to 5 W m^{-2}, indicating equilibrium thicknesses of several hundred meters or less. From Gulick (1993, 1998).

subsurface, such as fractures, would permit egress of hydrothermal water to the surface. The system can adjust to the new permafrost equilibrium thickness in less than about 10,000 years, a time that is short compared with the lifetime of the hydrothermal system.

Though these theoretical ideas apply to hydrothermal systems on Mars, to date little work has been done to analyze specific potential hydrothermal sites, partly because relatively little imaging data at sufficiently high resolution are available to unequivocally identify sites of hydrothermal outflow. Also, insufficient high-resolution spectral data exist to identify hydrothermal minerals that might be associated with sites of hydrothermal activity. Once such data become available, as from Mars Global Surveyor and later orbiters, it may be possible to constrain uniquely the properties of specific hydrothermal systems on Mars.

3.9.3. The Aureole Deposits of Olympus Mons

The origin of the lobate overlapping aureole deposits around Olympus Mons remains controversial. These deposits extend from the volcano's basal scarp to distances up to 1000 km (Mouginis-Mark *et al.*, 1992) and consist of enormous lobes of closely spaced, roughly parallel arcuate ridges. Data from the laser altimeter (MOLA) aboard Mars Global Surveyor indicate that the volcano's basal scarp appears to be approximately 8 km high and the aureole deposits are roughly 3 to 4 km of relief (Smith *et al.*, 1998). Several hypotheses have been put forth to explain these landforms, including Hodges and Moore (1979), who suggested that Olympus Mons is analogous to Icelandic tuyas and its aureole deposits were formed by subglacial lava flows. They conclude that the associated ice sheet thickness was comparable to that of the aureole deposits, or 3–4 km thick. Difficulties with this hypothesis cast doubt on its validity, and mainly revolve around the preferential location of an extremely thick ice sheet late in Martian history around Olympus Mons. Subsequently, Lopes et al. (1980) and Frances and Wadge (1983) have attributed the presence of the aureole deposits as mass movement of slide material from the volcano.

3.10. CONCLUSION

Volcano/ice interactions produce meltwater. Meltwater can enter the ground water cycle and under the influence of hydrothermal systems, it can be later discharged to form channels and valleys or cycle upward to melt permafrost. Water or ice-saturated ground can erupt into phreatic craters when covered by lava. Violent mixing of meltwater and volcanic material and rapid release can generate lahars or jökulhlaups, which have the ability to transport coarse material great distances downslope from the vent. Eruption into meltwater generates unique-appearing edifices that are definitive indicators of volcano/ice interaction. These features are hyaloclastic ridges or mounds and if capped by lava, tuyas. On Earth, volcano/ice interactions are limited to alpine regions and ice-capped polar and temperate regions. On Mars, where precipitation may be an ancient phenomenon, these interactions may be limited to areas of ground ice accumulation or the northern lowlands where water may have ponded relatively late in Martian history.

3.11. REFERENCES

Allen, C. C., Volcano–ice interactions on Mars, *J. Geophys. Res.*, *84*, 8048–8059, 1979a.
Allen, C. C., Volcano–ice interactions on the Earth and Mars, *NASA TM*, *81979*, 164–326, 1979b.
Allen, C. C., Icelandic subglacial volcanism: Thermal and physical studies, *J. Geol.*, *88*, 108–117, 1980.
Allen, C. C., M. J. Jercinovic, and J. S. B. Allen, Subglacial volcanism in north central British Columbia and Iceland, *J. Geol.*, *90*, 699–715, 1982.
Anderson, D. M., Glaciation in Elysium, MSATT *Workshop Polar Regions of Mars, Lunar Planet. Sci. Inst. Tech. Rep.*, 92-08(1), 1992.
Baker, V. R., R. G. Strom, V. C. Gulick, J.S . Kargel, G. Komatsu, and V. S. Kale, Ancient oceans, ice sheets and the hydrological cycle on Mars, *Nature*, *352*, 589–594, 1991.
Björnsson, H., Explanation of jökulhlaups from Grímsvötn, Vatnajökull, Iceland, *Jökull*, *24*, 1–26, 1974.
Björnsson, H., Subglacial water reservoirs, jökulhlaups, and volcanic eruptions, *Jökull*, *25*, 1–14, 1975.
Björnsson, H., The cause of jökulhlaups in the Skaftá River, Vatnajökull, *Jökull*, *27*, 71–77, 1977.
Björnsson, H., Hydrology of ice caps in volcanic regions, *Soc. Sci. Isl.*, *45*, 137 pp., 1988.
Björnsson, H., Jökulhlaups in Iceland: Prediction, characteristics and simulation, *Ann. Glaciol.*, *16*, 95–106, 1992.
Björnsson, H., and P. Einarsson, Volcanoes beneath Vatnajökull, Iceland: Evidence from radio echo-sounding, earthquakes and jökulhlaups, *Jökull*, *40*, 147–168, 1990.
Brakenridge, G. R., H. E. Newsom, and V. R. Baker, Ancient hot springs on Mars: Origins and paleoenvironmental significance of small Martian valleys, *Geology*, *13*, 859–862, 1985.
Bull, W. B., Recognition of alluvial-fan deposits in the stratigraphic record, in Recognition of Ancient Sedimentary Environments, Soc. Econ. Paleont. and Miner. Spec. Pub., 63–83, 1972.
Carr, M. H., Stability of streams and lakes on Mars, *Icarus*, *56*, 476–495, 1983.
Carr, M. H., The Martian drainage system and the origin of networks and fretted channels, *J. Geophys. Res.*, *100*, 7479–7507, 1995.
Carrasco-Núñez, G., J. W. Vallance, and W. I. Rose, A voluminous avalanche-induced lahar from Citlaltepetl volcano, Mexico: Implications for hazard assessment, *J. Volcanol. Geotherm. Res.*, *59*, 35–46, 1993.

Chapman, M. G., Evidence, age, and thickness of a frozen paleolake in Utopia Planitia, Mars, *Icarus*, *109*, 393–406, 1994.
Chapman, M. G., Subglacial volcanoes on Mars, *GSA Abstr. Progr.*, *29*, A137, 1997.
Christiansen, E. H., Lahars in the Elysium region of Mars, *Geology*, *17*, 203–206, 1989.
Christiansen, E. H., and R. Greeley, Mega-lahars (?) in the Elysium region, Mars, *Lunar Planet. Sci. XII*, Lunar and Planetary Institute, Houston, 138–140, 1981.
Clifford, S. M., A model for the hydrologic and climatic behavior of water on Mars, *J. Geophys. Res.*, *98*, 10973–11016, 1993.
Crandell, D. R., Postglacial lahars from Mount Rainier volcano, Washington, *U.S. Geol. Surv. Prof. Pap.*, *677*, 75 pp. and 2 plates, 1971.
Crandell, D. R., and H. H. Waldron, A recent volcanic mudflow of exceptional dimensions from Mt. Rainier, Washington, *Am. J. Sci.*, *254*, 349–362, 1956.
Davies, G. F., and R. E. Arvidson, Martian thermal history, core segregation, and tectonics, *Icarus*, *45*, 339–346, 1981.
Dorava, J. M., and D. F. Meyer, Hydrologic hazards in the lower Drift River basin associated with the 1989–1990 eruptions of Redoubt Volcano, Alaska, *J. Volcanol. Geotherm. Res.*, *62*, 387–407, 1994.
Elder, J., *Geothermal Systems*, Academic Press, New York, 1981.
Fanale, F. P., Martian volatiles: Their degassing history and geochemical fate, *Icarus*, *28*, 179–202, 1976.
Fanale, F. P., S. E. Postawko, J. B. Pollack, M. H. Carr, and R. O. Pepin, Mars: Epochal climate change and volatile history, in *Mars*, edited by H. H. Kieffer, B. M. Jakosky, C. W. Snyder, and M. S. Matthews, pp. 1135–1179, University of Arizona Press, Tucson, 1992.
Farmer, J., and D. DesMarais, Exopaleontology and the search for a fossil record on Mars, *Lunar Planet. Sci.*, *XXV*, 367–368, 1994.
Fisher, R. V., and H.-U. Schmincke, *Pyroclastic Rocks*, 472 pp., Springer-Verlag, Berlin, 1984.
Frances, P. W., and G. Wadge, The Olympus Mons aureole: Formation by gravitational spreading, *J. Geophys. Res.*, *87*, 9881–9889, 1983.
Frey, H. V., B. L. Lowry, and S. A. Chase, Pseudocraters on Mars, *J. Geophys. Res.*, *84*, 8075–8086, 1979.
Furnes, H., and I. B. Fridleifsson, Pillow block breccia—Occurrences and mode of formation, *Neues Jahrb. Geol. Palaeontol. Monatsh.*, *3*, 147–154, 1979.
Furnes, H., I. B. Fridleifsson, and F. B. Atkins, Subglacial volcanics. On the formation of acid hyaloclastites, *J. Volcanol. Geotherm. Res.*, *8*, 95–110, 1980.
Gasparini, P., and M. S. M. Mantovani, in *Geophysics of Geothermal Areas: State of the Art and Future Development*, edited by A. Rapolla and G. V. Keller, pp. 9–40, Colorado School of Mines Press, Golden, 1984.
Giggenbach, W. F., N. Garcia P., A. Londoño C., L. Rodriguez, V., N. Rojas G., and M. L. Calvache V., The chemistry of fumarolic vapor and thermal-spring discharges from the Nevado del Ruiz volcanic-magmatic-hydrothermal system, Colombia, *J. Volcanol. Geotherm. Res.*, *42*, 13–39, 1990.
Greeley, R., and E. Theilig, Small volcanic constructs in the Chryse Planitia region of Mars, *NASA TM*, *79729*, 202, 1978.
Grönvold, K., *Structural and petrochemical studies in the Kerlingarfjöll region, central Iceland*, Ph.D. dissertation, unpublished, University of Oxford, 327 pp., 1972.
Gudmundsson, M. T., H. Björnsson, and F. Pálsson, Changes in jökulhlaup sizes in Grímsvötn, Vatnajökull, Iceland, 1934–91, deduced from in-situ measurements of subglacial lake volume, *J. Glaciol.*, *41*, 263–272, 1995.
Gudmundsson, M. T., F. Sigmundsson, and H. Björnsson, Ice–volcano interaction of the 1996 Gjálp subglacial eruption, Vatnajökull, Iceland, *Nature*, *389*, 954–957, 1997.
Gulick, V. C., Magmatic intrusions and a hydrothermal origin for fluvial valleys on Mars, *J. Geophys. Res.*, *103*, 19365–19388, 1998.
Gulick, V. C., *Magmatic intrusions and hydrothermal systems: Implications for the formation of Martian fluvial valleys*, Ph.D. thesis, University of Arizona, 1993.
Gulick, V. C., and V. R. Baker, Fluvial valleys and Martian paleoclimates, *Nature*, *341*, 514–516, 1989.
Gulick, V. C., and V. R. Baker, Origin and evolution of valleys on Martian volcanoes, *J. Geophys. Res.*, *95*, 14325–14344, 1990.
Gulick, V. C., and V. R. Baker, Martian hydrothermal systems: Some physical considerations, *Lunar Planet. Sci. Conf.*, *XXIII*, 463–464, 1992.
Gulick, V. C., D. Tyler, C. P. McKay, and R. M. Haberle, Episodic ocean-induced CO_2 pulses on Mars: Implications for fluvial valley formation, *Icarus*, *130*, 68–86, 1997.
Hickson, C. J., and J. G. Souther, Late Cenozoic volcanic rocks of the Clearwater–Wells Gray area, British Columbia, *Can. J. Earth Sci.*, *21*, 267–277, 1984.

Hoare, T. M., and W. L. Coonrad, A tuya in Togiak Valley, southwest Alaska, *J. Res. U.S. Geol. Surv.*, *6*(2), 193–201, 1978.

Hodges, C. A., and H. J. Moore, Tablemountains of Mars, *Lunar Planet. Sci. Conf.*, *9th*, 523–525, 1978.

Hodges, C. A., and H. J. Moore, The subglacial birth of Olympus Mons and its aureoles, *J. Geophys. Res.*, *84*, 8061–8074, 1979.

Hooke, R. L., Processes on arid-region alluvial fans, *J. Geol.*, *75*, 438–460, 1967.

Imsland, P., The geology of the volcanic island Jan Mayen, Arctic Ocean, *Nord. Volcanol. Inst. Rep.*, *7813*, 74 pp., 1978.

Jakobsson, S. P., Environmental factors controlling the palagonitization of the Surtsey tephra, Iceland, *Bull. Geol. Soc. Den.*, *27*, 91–105, 1978.

Jakobsson, S. P., and J. G. Moore, The Surtsey Research Drilling Project of 1979, *Surtsey Res. Prog. Rep.*, *9*, 76–93, 1982.

Jakobsson, S. P., and J. G. Moore, Hydrothermal minerals and alteration rates at Surtsey volcano, Iceland, *Geol. Soc. Am. Bull.*, *97*, 648–659, 1986.

Jakosky, B., and J. Jones, Evolution of water on Mars, *Nature*, *370*, 328–329, 1994.

Johannesson, H., and K. Saemundsson, *Geological Map of Iceland: Bedrock Geology*, Icelandic Institute of Natural History, Reykjavík, 2nd ed., scale 1 : 500,000, 1998.

Johnson, A. M., *Physical Processes in Geology*, 577 pp., Freeman and Cooper, San Francisco, 1970.

Jonasson, K., Rhyolite volcanism in the Krafla central volcano, north-east Iceland, *Bull. Volcanol.*, *56*, 516–528, 1994.

Jones, J. G., Intraglacial volcanoes of the Laugarvatn region, south-west Iceland, I, *Q. J. Geol. Soc. London*, *124*, 197–211, 1969.

Jones, J. G., Intraglacial volcanoes of the Laugarvatn region, south-west Iceland, II, *J. Geol.*, *78*, 127–140, 1970.

Jöns, H. P., Das relief des Mars: versuch einer zusammenfassenden ubersicht, *Geol. Rundsch.*, *79*, 131–164, 1990.

Jónsson, J., Notes on Katla volcanoglacial debris flows, *Jökull*, *32*, 61–68, 1982.

Jónsson, P., A. Snorrason, and S. Pálsson, Discharge and sediment transport in the jökulhlaup on Skeidarársandur, Iceland, in November 1996, University of Alicante (Spain), 15th International Sedimentological Congress, Abstracts, pp. 455–456, 1998a.

Jónsson, P., O. Sigurdsson, A. Snorrason, S. Víkingsson, I. Kaldal, and S. Árnason, Course of events of the jökulhlaup on Skeidarársandur, Iceland, in November 1996, University of Alicante (Spain), 15th International Sedimentological Congress, Abstracts, pp. 456–457, 1998.

Kargel, J. S., and R. G. Strom, Ancient glaciation on Mars, *Geology*, *20*, 3–7, 1992.

Kargel, J. S., V. R. Baker, J. E. Begét, J. F. Lockwood, T. L. Péwé, J. S. Shaw, and R. G. Strom, Evidence of continental glaciation in the Martian northern plains, *J. Geophys. Res.*, *100*, 5351–5368, 1995.

Kieffer, H. H., Mars south polar spring and summer temperatures: A residual CO_2 frost, *J. Geophys. Res.*, *84*, 8263–8288, 1979.

Kjartansson, G., Geology of Arnessysla [in Icelandic], *Arnesingasaga*, *I*, 1–250, 1943.

Kjartansson, G., The Moberg Formation, in *On the Geology and Geophysics of Iceland*, edited by S. Thorarinsson, Int. Geol. Congr. XXI Session, Guide to Excursion No. A2, 21–28, 1960.

Kjartansson, G., A comparison of tablemountains in Iceland and the volcanic island of Surtsey off the south coast of Iceland [in Icelandic with English summary], *Natturufraedingurinn*, *36*, 1–34, 1966.

Kokelaar, P., Magma–water interactions in subaqueous and emergent basaltic volcanism, *Bull. Volcanol.*, *48*, 275–289, 1986.

Larsen, G., M. T. Gudmundsson, and H. Björnsson, Eight centuries of periodic volcanism at the centre of the Iceland hotspot revealed by glacier tephrostratigraphy, *Geology*, *26*, 943–946, 1998.

Lonsdale, P., and R. Batiza, Hyaloclastite and lava flows on young seamounts examined with a submersible, *Geol. Soc. Am. Bull.*, *91*, 545–554, 1980.

Lopes, R. M. C., J. E. Guest, and C. J. Wilson, Origin of the Olympus Mons aureole and perimeter scarp, *Moon Planets*, *22*, 221–234, 1980.

Lucchitta, B. K., Geologic map of the Ismenius Lacus quadrangle, Mars, *U.S. Geol. Surv. Misc. Invest. Map*, I-1065, scale 1 : 5,000,000, 1978.

Lucchitta, B. K., Mars and Earth: Comparison of cold-climate features, *Icarus*, *45*, 264–303, 1981.

Lucchitta, B. K., Ice in the northern plains: Relic of a frozen ocean? MSATT *Workshop Martian Northern Plains, Lunar Planet. Sci. Inst. Tech. Rep.*, 93-04, Part 1, 9–10 [Abstr.], 1993.

Lucchitta, B. K., H. M. Ferguson, and C. Summers, Sedimentary deposits in the northern lowland plains, Mars, *J. Geophys. Res.*, *91*, 166–174, 1986.

Lucchitta, B. K., H. M. Ferguson, and C. Summers, Northern sinks on Mars? *Lunar Planet. Sci. Inst. Tech. Rep.*, 87-02, 32–33, 1987.

Major, J. J., and C. G. Newhall, Snow and ice perturbation during historical volcanic eruptions and the formation of lahars and floods, *Bull. Volcanol.*, 52, 1–27, 1989.

Mathews, W. H., "Tuyas": Flat-topped volcanoes in northern British Columbia, *Am. J. Sci.*, 245, 560–570, 1947.

McGetchin, T. R., M. Settle, and B. A. Chouet, Cinder cone growth modeled after Northeast crater, Mount Etna, Sicily, *J. Geophys. Res.*, 79, 3257–3272, 1974.

McKay, C. P., G. D. Clow, R. A. Wharton, Jr., and S. W. Squyres, Thickness of ice on perennially frozen lakes, *Nature*, 313, 561–562, 1985.

Mellon, M. T., and B. M. Jakosky, The distribution and behavior of Martian ground ice during past and present epochs, *J. Geophys. Res.*, 100, 11781–11799, 1995.

Moore, J. G., and L. C. Calk, Degassing and differentiation in subglacial volcanoes, Iceland, *J. Volcanol. Geotherm. Res.*, 46, 157–180, 1991.

Mouginis-Mark, P. J., Volcano/ground ice interactions in Elysium Planitia, Mars, *Icarus*, 64, 265–284, 1985.

Mouginis-Mark, P. J., Recent meltwater release in the Tharsis region of Mars, *Icarus*, 84, 362–373, 1990.

Mouginis-Mark, P. J., L. Wilson, and M. T. Zuber, The physical volcanology of Mars, in *Mars*, edited by H. H. Kieffer, B. M. Jakosky, C. W. Snyder, and M. S. Matthews, pp. 424–452, University of Arizona Press, Tucson, 1992.

Noe-Nygaard, A., Sub-glacial volcanic activity in ancient and recent times. Studies in the palagonite-system of Iceland No. 1, *Folia Geogr. Dan.*, 1(2), 67, 1940.

Nye, J. F., Water flow in glaciers, jökulhlaups, tunnels, and veins, *J. Glaciol.*, 76, 181–207, 1976.

Paige, D. A., The thermal stability of near-surface ground ice on Mars, *Nature*, 356, 43–45, 1992.

Parker, T. J., R. S. Saunders, and D. M. Schneeberger, Transitional morphology in the west Deuteronilus Mesae region of Mars: Implications for modification of the lowland/upland boundary, *Icarus*, 82, 111–145, 1989.

Parker, T. J., D. S. Gorsline, R. S. Saunders, D. C. Pieri, and D. M. Schneeberger, Coastal geomorphology of the Martian northern plains, *J. Geophys. Res.*, 98, 11061–11078, 1993.

Paterson, W. S. B., *The Physics of Glaciers*, 3rd ed., 4800 pp., Pergamon, Oxford, 1994.

Peacock, M. A., The vulcano-glacial palagonite formation of Iceland, *Geol. Mag.*, 63, 385–399, 1926.

Pierson, T. C., Initiation and flow behavior of the 1980 Pine Creek and Muddy River lahars, Mount St. Helens, Washington, *Geol. Soc. Am. Bull.*, 96, 1056–1069, 1985.

Pierson, T. C., Flow characteristics of large eruption-triggered debris flows at snow-clad volcanoes: Constraints for debris-flow models, *J. Volcanol. Geotherm. Res.*, 66, 283–294, 1995.

Pierson, T. C., and R. J. Janda, Volcanic mixed avalanches—A distinct eruption-triggered mass-flow process at snow-clad volcanoes, *Geol. Soc. Am. Bull.*, 106, 1351–1358, 1994.

Pierson, T. C., and R. B. Waitt, Dome-collapse rockslide and multiple sediment-water flows generated by a small explosive eruption on February 2–3, 1983, in *Hydrologic Consequences of Hot-Rock/Snowpack Interactions at Mount St. Helens Volcano, Washington 1982–84*, edited by T. C. Pierson, U.S. Geol. Surv. Open-File Rep., 96-179, 53–68, 1997.

Pierson, T. C., R. J. Janda, J.-C. Thouret, and C. A. Borrero, Perturbation and melting of snow and ice by the 13 November 1985 eruption of Nevado del Ruiz, Colombia, and consequent mobilization, flow and deposition of lahars, *J. Volcanol. Geotherm. Res.*, 41, 17–66, 1990.

Rex, D., Geochronology in relation to the stratigraphy of the Antarctic Peninsula, *Br. Antarct. Surv. Bull.*, 43, 49–58, 1976.

Rice, J. W., Jr., and K. S. Edgett, Catastrophic flood sediments in Chryse Basin, Mars, and Quincy Basin, Washington: Application of sandar facies model, *J. Geophys. Res.*, 102, 4185–4200, 1997.

Rodine, J. D., and A. M. Johnson, The ability of debris, heavily freighted with coarse clastic material, to flow on gentle slopes, *Sedimentology*, 23, 213–234, 1976.

Rórarinsson, S., The Öræfajökull eruption of 1362, *Acta Nat. Isl.*, 2(2), 102, 1958.

Rossbacher, L. A., and S. Judson, Ground ice on Mars: Inventory, distribution and resulting landforms, *Icarus*, 45, 25–38, 1981.

Saemundsson, K., Vulkanismus und tektonik das Hengill-Gebietes un südwest-Island, *Acta Isl.*, II(7), 1–103, 1967.

Saemundsson, K., Notes on the Torfajökull central volcano [in Icelandic with English summary], *Natturufraedingurinn*, 42, 81–99, 1972.

Schultz, P., R. Schultz, and J. Rogers, The structure and evolution of ancient impact basins on Mars, *J. Geophys. Res.*, 87, 9803–9820, 1982.

Scott, D. H., and J. R. Underwood, Jr., Mottled terrain: A continuing Martian enigma, *Proc. Lunar Planet. Sci. Conf.*, 21st, 627–634, 1991.

Scott, D. H., M. G. Chapman, J. W. Rice, Jr., and J. M. Dohm, New evidence of lacustrine basins on Mars: Amazonis and Utopia Planitiae, *Proc. Lunar Planet. Sci. Conf.*, *22nd*, 53–62, 1992.

Scott, K. M., Origin, behavior, and sedimentology of lahars and lahar-runout flows in the Toutle–Cowlitz River system, *U.S. Geol. Surv. Prof. Pap.*, *1477-A*, 74 pp., 1988.

Scott, K. M., J. W. Vallance, and P. T. Pringle, Sedimentology, behavior, and hazars of debris flows at Mount Rainier, Washington, *U.S. Geol. Surv. Prof. Pap.*, *1547*, 56 pp. and plate, 1995.

Sigvaldason, G. E., Structure and products of subaquatic volcanoes in Iceland, *Contrib. Mineral. Petrol.*, *18*, 1–16, 1968.

Skilling, I. P., Evolution of an englacial volcano: Brown Bluff, Antarctica, *Bull. Volcanol.*, *56*, 573–591, 1994.

Smellie, J. L., and I. P. Skilling, Products of subglacial volcanic eruptions under different ice thicknesses: Two examples from Antarctica, *Sediment. Geol.*, *91*, 115–129, 1994.

Smith, D. E., M. T. Zuber, H. V. Frey, J. B. Garvin, J. W. Head, D. O. Muhleman, G. H. Pettengill, R. J. Phillips, S. C. Solomon, H. J. Zwally, W. B. Banerdt, and T. C. Duxbury, Topography of the northern hemisphere of Mars from the Mars Orbiter Laser Altimeter, *Science*, *279*, 1686–1692, 1998.

Squyres, S. W., D. E. Wilhelms, and A. C. Moosman, Large-scale volcano–ground ice interactions on Mars, *Icarus*, *70*, 385–408, 1987.

Squyres, S. W., S. M. Clifford, R. O. Kuzmin, J. R. Zimbleman, and F. M. Costard, Ice in the Martian regolith, in *Mars*, edited by H. H. Kieffer, B. M. Jakosky, C. W. Snyder, and M. S. Matthews, pp. 523–556, University of Arizona Press, Tucson, 1992.

Staudigel, H., and H. V. Schmincke, The Pliocene seamount series of La Palma, Canary Islands, *J. Geophys. Res.*, *89*, 11195–11215, 1984.

Tanaka, K. L., D. H. Scott, and R. Greeley, Global stratigraphy, in *Mars*, edited by H. H. Kieffer, B. M. Jakosky, C. W. Snyder, and M. S. Matthews, pp. 345–382, University of Arizona Press, Tucson, 1992.

Thorarinsson, S., Hekla and Katla: The share of acid and intermediate lava and tephra in the volcanic products through the geological history of Iceland, in *Iceland and Mid-Ocean Ridges*, edited by S. Björnsson, *Soc. Sci. Isl.*, *38*, 190–197, 1967.

Thorarinsson, S., The postglacial history of the Myvatn area and the area between Myvatn and Jokulsa a Fjollum, in *On the Geology and Geophysics of Iceland*, Guide to Excursion No. A2, Int. Geol. Congr. XXI Session, Reykjavik, pp. 33–45, 1960.

Thorarinsson, S., *Vötnin Strid* [The swift flowing rivers], 254 pp., Minningarsjódur, Reykjavík, 1974.

Thorseth, I. H., H. Furnes, and M. Heldal, The importance of microbiological activity in the alteration of natural basaltic glass, *Geochim. Cosmochim. Acta*, *56*, 845–850, 1992.

Thouret, J.-C., Effects of the November 13, 1985 eruption on the snow pack and ice cap of Nevado del Ruiz, Colombia, *J. Volcanol. Geotherm. Res.*, *41*, 177–201, 1990.

Thouret, J.-C., R. Salinas, and A. Murcia, Eruption and mass-wasting-induced processes during the late Holocene destructive phase of Nevado del Ruiz volcano, Colombia, *J. Volcanol. Geotherm. Res.*, *41*, 203–224, 1990.

Toksöz, M. N., and A. T. Hsui, Thermal history and evolution of Mars, *Icarus*, *34*, 537–547, 1978.

Tómasson, H., Hamfarahlaup í Jökulsá á Fjöllum, *NáttúrufrÆdingnum*, *43*, 12–34, 1973.

Tómasson, H., The jökulhlaups from Katla in 1918, *Ann. Glaciol.*, *22*, 249–254, 1996.

Trabant, D. C., R. B. Waitt, and J. J. Major, Disruption of Drift glacier and origin of floods during the 1989–90 eruption of Redoubt volcano, Alaska, *J. Volcanol. Geotherm. Res.*, *62*, 369–386, 1994.

Van Bemmelen, M. G. and R. W. Rutten, *Tablemountains of Northern Iceland* 217 pp., Brill, Leiden, 1955.

Vilmundardottir, E. G., and S. P. Snorrason, Skaftarveita: Bedrock geology at Langisjor [in Icelandic], *Energy Authority, Reykjavik*, OS-97067 (mimeogr. rep.), 24 pp., 1997.

Voight, B., The 1985 Nevado del Ruiz volcano catastrophe: Anatomy and retrospection, *J. Volcanol. Geotherm. Res.*, *42*, 151–188, 1990.

Waitt, R. B., Swift snowmelt and floods (lahars) caused by great pyroclastic surge at Mount St. Helens volcano, Washington, 18 May 1980, *Bull. Volcanol.*, *52*, 138–157, 1989.

Waitt, R. B., Hybrid wet flows formed by hot pyroclasts interacting with snow during Crater Peak (Mt. Spurr) eruptions, summer 1992, in *The 1992 Eruptions of Crater Peak at Mount Spurr Volcano, Alaska*, edited by T. E. C. Keith, *U.S. Geol. Surv. Bull.*, *2139*, 107–118, 1995.

Waitt, R. B., Cataclysmic flood along Jökulsá á Fjöllum, north Iceland, compared to repeated colossal jökulhlaups of Washington's Channeled Scabland, University of Alicante (Spain), *Int. Sediment. Congr. Abstr.*, *15th*, 811–812, 1998.

Waitt, R. B., T. C. Pierson, N. S. MacLeod, R. J. Janda, B. Voight, and R. T. Holcomb, Eruption-triggered avalanche, flood, and lahar at Mount St. Helens—effects of winter snowpack, *Science*, *221*, 1394–1397, 1983.

Waitt, R. B., and J. E. Begét, with contrib. by J. Kienle, Provisional geologic map of Augustine volcano, Alaska, *U.S. Geol. Surv. Open-File Rep.*, *96-516*, 44 pp. and map, 1:25,000 scale, 1996.

Waitt, R. B., and N. S. MacLeod, Minor explosive eruptions dramatically interacting with winter snowpack at Mount St. Helens in March–April 1982, in *Selected Papers on the Geology of Washington, Wash. Div. Geol. Earth Resourc. Bull.*, *77*, 355–379, 1987.

Waitt, R. B., C. A. Gardner, T. C. Pierson, J. J. Major, and C. A. Neal, Unusual ice diamicts emplaced during 15 December 1989 eruption of Redoubt Volcano, Alaska, *J. Volcanol. Geotherm. Res.*, *62*, 409–428, 1994.

Walker, G. P. L., and D. H. Blake, The formation of palagonite breccia mass beneath a valley glacier in Iceland, *Q. J. Geol. Soc. London*, *122*, 45–61, 1966.

Wallace, D., and C. Sagan, Evaporation of ice in planetary atmospheres, ice-covered rivers on Mars, *Icarus*, *39*, 385–400, 1979.

Walter, M. R., and D. J. DesMarais, Preservation of biological information in thermal spring deposits: Developing a strategy for the search for fossil life on Mars, *Icarus*, *101*, 129–143, 1993.

Wentworth, S. J., and J. L. Gooding, Carbonates and sulfates in the Chassigny meteorite: Further evidence for aqueous chemistry on the SNC parent planet, *Meteoritics*, *29*, 860–863, 1994.

Werner, R., H. U. Schminke, and G. Sigvaldason, A new model for the evolution of table mountains: Volcanological and petrological evidence from Herdubreid and Herdubreidartogl volcanoes (Iceland), *Geol. Rundsch.*, *85*, 390–397, 1996.

Williams, S. N., Nevado del Ruiz volcano, Colombia: An example of the state-of-the-art of volcanology four years after the tragic November 13, 1985 eruption, *J. Volcanol. Geotherm. Res.*, *41*, 1–5, 1990.

Williams, S. N., R. E. Stoiber, N. P. Garcia, A. C. Londono, B. J. Gemmell, D. R. Lowe, and C. B. Connor, Eruption of the Nevado del Ruiz volcano, Colombia, on 13 November, 1985: Gas flux and fluid geochemistry, *Science*, *233*, 964–967, 1986.

Witbeck, N. E., and J. R. Underwood, Jr., Geologic mapping in the Cydonia region of Mars, *NASA TM*, *86246*, 327–329, 1983.

Wörner, G., and L. Viereck, Subglacial to emergent volcanism at Shield Nunatak, Mt. Melbourne Volcanic Field, Antarctica, *Polarforschung*, *57*, 27–41, 1987.

4

Volcanism on the Red Planet: Mars

Ronald Greeley, Nathan T. Bridges, David A. Crown,
Larry Crumpler, Sarah A. Fagents, Peter J. Mouginis-Mark,
and James R. Zimbelman

4.1. INTRODUCTION

Of all of the planets in the solar system, Mars is the most Earth-like in its geologic characteristics. Like Earth, it has been subjected to exogenic processes, such as impact cratering and erosion by wind and water, as well as endogenic processes, including tectonic deformation of the crust and volcanism. The effects of these processes are amply demonstrated by the great variety of surface features, including impact craters, landslides, former river channels, sand dunes, and the largest volcanoes in the solar system.

Some of these features suggest substantial changes in Mars' environment during its history. For example, as reviewed by Carr (1996), today Mars is a cold, dry desert with an average atmospheric pressure of only 5.6 mbar, which does not allow liquid water to exist on the surface. To some planetary scientists, the presence of the channels bespeaks a time when Mars was warmer and wetter. However, others have argued that these features might have formed under current conditions and that there might not have been a shift in climate.

Could the morphology of volcanoes and related features provide clues to past Martian environments? What role is played by atmospheric density in the styles of eruptions on Mars and resulting landforms? If these and related questions can be answered, then we may have a means for assessing the conditions on Mars' surface in the past and comparing the results with models of Martian evolution.

In this chapter, we outline the sources of information available for volcanism on Mars, explore the influence of the Martian environment on volcanic processes, and describe the principal volcanic features and their implications for understanding the general evolution of the Martian surface.

Environmental Effects on Volcanic Eruptions: From Deep Oceans to Deep Space.
Edited by Zimbelman and Gregg, Kluwer Academic/Plenum Publishers, New York, 2000.

4.2. BACKGROUND

Earth-based telescopic observations are unable to resolve volcanoes on Mars. However, even before volcanic features were identified in spacecraft images, astronomers noted the presence of a unique W-shaped cloud in the Tharsis region that developed around local noon and persisted through the afternoon. This cloud was "anchored" to what we now know are three enormous volcanoes. The first identification of these volcanoes came early in the Mariner 9 mission in 1971 (McCauley *et al.*, 1972). A massive global dust storm obscured the entire surface when the spacecraft arrived at Mars. As the dust slowly settled, four dark "spots" in the Tharsis region were revealed to be the summits of Olympus Mons, perhaps the largest volcano in the solar system, and the Tharsis Montes (Masursky *et al.*, 1972; Carr, 1973). By the end of the mission, additional volcanoes were identified throughout the Tharsis and Elysium regions, as well as in the southern highlands (Figure 4.1). Cameras on two Viking Orbiter spacecraft operated at Mars from 1976 to 1980, greatly improving our understanding of these and other volcanoes (Figure 4.2), as well as revealing extensive volcanic plains (Figure 4.3), small volcanoes in many areas, including the cratered highlands, and other features related to volcanic processes (Carr *et al.*, 1977).

Much of the understanding of Martian volcanoes has been derived from photogeologic studies of the surface. Through careful mapping, the global geology of Mars is now well established (Carr, 1981; Scott and Tanaka, 1986; Greeley and Guest, 1987; Tanaka and Scott, 1987). The areal density of impact craters on the mapped units allows a relative stratigraphy to be determined even where units are not in contact. The three major divisions for Mars' geologic history are the Noachian (oldest), Hesperian, and Amazonian (youngest) Periods.

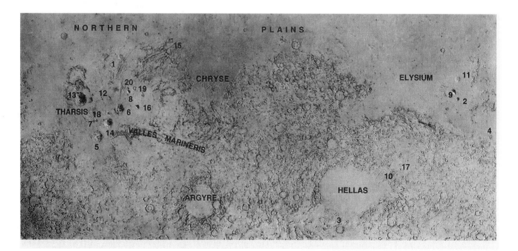

Figure 4.1. Shaded airbrush relief map of Mars (from 65 °N to 65 °S) showing the location of principal volcanic features. 1, Alba Patera; 2, Albor Tholus; 3, Amphitrites Patera; 4, Apollinaris Patera; 5, Arsia Mons; 6, Ascraeus Mons; 7, Biblis Patera; 8, Ceraunius Tholus; 9, Elysium Mons; 10, Hadriaca Patera; 11, Hecates Tholus; 12, Jovis Tholus; 13, Olympus Mons; 14, Pavonis Mons; 15, "Tempe" Patera; 16, Tharsis Tholus; 17, Tyrrhena Patera; 18, Ulysses Patera; 19, Uranius Patera; 20, Uranius Tholus. In addition, many of the plains portrayed here are thought to be of volcanic origin. From Greeley and Spudis (1981).

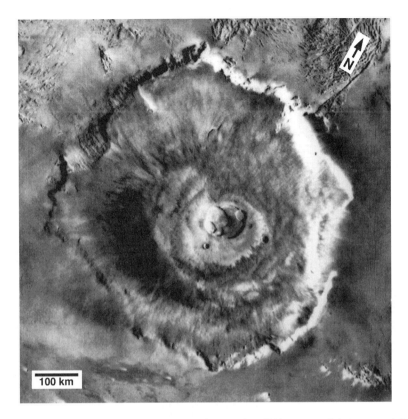

Figure 4.2. The shield volcano Olympus Mons is more than 600 km across. It is constructed from countless individual flows erupted from the summit region and the flanks of the volcano. (Viking Orbiter frame 646A28.)

Unfortunately, the absolute ages are poorly constrained; for example, different models of the impact cratering rate allow the start of the Amazonian Period to be from millions (e.g., Soderblom, 1977) to billions (e.g., Neukum and Hiller, 1981) of years in age. The ages will remain unconstrained until radiometric dates are obtained (probably from rock samples returned to Earth) for relevant stratigraphic units on Mars. Despite the uncertainty in absolute age, the stratigraphy provides a framework within which relative ages can be constrained, including those of volcanic features.

Other types of remote sensing provide information about Martian volcanoes that is complementary to photogeologic studies. Earth-based telescopic reflectance spectra indicate that bright regions are covered with dust rich in oxidized iron, whereas dark regions are less dusty and may include pyroxene-bearing mafic rocks such as basalt (Soderblom, 1992). Thermal infrared measurements by the Viking Orbiters suggest that much of the surface is coated with fine-grained material ranging from micrometer-sized dust covering bright regions to sand-sized particles mixed with larger materials in dark regions (Kieffer *et al.*, 1977). The dust coating also affects individual volcanoes, where dust might interfere with reflectance or emission from the underlying rocks, hindering remote sensing interpretations of bedrock compositions.

Figure 4.3. Volcanic plains have "resurfaced" many areas of Mars, including cratered terrain seen here northwest of the Hellas basin where lava flows have partly buried older craters. Area shown is about 170 by 150 km. (Viking Orbiter frame 95A10.)

4.3. MARS' COMPOSITION

Composition strongly affects the rheology and explosivity of magmas and lavas. As such, it must be considered to understand Martian volcanology.

4.3.1. Bulk Composition

The mass and volume of Mars are well known and yield a mean uncompressed planetary density of 3933.5 ± 0.4 kg m^{-3}. This is considerably lower than Earth's density of 5514.8 kg m^{-3}, implying that Mars' interior (core + mantle) is depleted in heavy elements, such as Fe, relative to light elements such as O, Mg, Si, and S. Further insight into Mars' composition comes from the class of meteorites known as SNCs (for the meteorites shergottites, nakhlites, and Chassigny). These basaltic to ultramafic rocks are considered to be of Martian origin based on the compositional similarity of their trapped gases to Mars' atmosphere (McSween, 1994). Although they probably do not completely represent volcanic rocks on Mars, the inferred SNC parent magmas share geochemical and isotopic characteristics that suggest melting of the same mantle source at various times (McSween, 1994). SNCs are enriched in iron relative to magnesium, which may indicate formation from evolved liquids, but can also be traced to an iron-rich mantle.

Mars' polar moment of inertia of 0.3662 ± 0.0017 suggests that it has a differentiated interior (Folkner *et al.*, 1997). Although several nonunique models that consider core size, mantle and core compositions, and temperature gradient can reproduce the observed moment of inertia, the best model involves a core ranging in composition from Fe to FeS that is a smaller fraction of the planetary mass than is Earth's core. The Martian mantle would contain the remainder of the iron and be more iron-rich than Earth's mantle (Table 4.1). The mineralogy of the upper mantle is probably lherzolitic, made up of olivine, pyroxene, and minor aluminum phases (Longhi *et al.*, 1992). This is similar to Earth's upper mantle, except that on Mars the mafic minerals would be more iron-rich. Combining the iron contents of the small Martian core and iron-rich mantle results in a total iron content that is less than that of the bulk Earth.

4.3.2. Remote Sensing

Remote sensing of Mars divides the surface into ferric-rich (oxidized) bright regions and ferrous-rich dark regions (Bell, 1996). Broad, weak absorption bands from 9.0 to 1.1 μm in the dark regions are interpreted as Fe^{2+} cations in pyroxene and olivine. Infrared absorption indicative of sulfates and possibly carbonates on the surface and in atmospheric dust is also apparent (Blaney and McCord, 1990, 1995; Pollack *et al.*, 1990; Soderblom, 1992). Additional information provided by the Mars Global Surveyor spacecraft about thermal emission features in some dark regions indicates abundant pyroxene and plagioclase, and limits the amount of olivine, quartz, carbonates, and clays to be no more than 10, 5, 10, and 20% of the surface, respectively (Christensen *et al.*, 1998). Mineralogic OH^- or H_2O has been observed in the near infrared, but the associated mineralogy is uncertain (Bell, 1996).

The ferrous–ferric heterogeneity observed from Earth and orbit is also seen on the surface. Spectra from the visible–near-infrared Imager for Mars Pathfinder (IMP) camera indicate the presence of three classes of rocks and four classes of soils (Smith *et al.*, 1997; McSween *et al.*, 1999). Many of the diagnostic spectral characteristics of the classes are explained by differences in the abundance of ferric (high red/blue ratio) and ferrous (low red/blue ratio) components.

The morphology of volcanic features provides some insight into surface chemistry. The appearance of most volcanoes and lava flows is consistent with low-viscosity basaltic volcanism (Blasius and Cutts, 1981; Hodges and Moore, 1994). Attempts to derive compositionally dependent parameters by matching flow morphology to rheology suggest compositions ranging from basalt to andesite (Hulme, 1976; Malin, 1977; Moore *et al.*, 1978; Zimbelman, 1985). Landforms suggestive of silicic volcanism are represented by a few highland structures resembling volcanic domes, a possible composite volcano (Greeley and Spudis, 1978), and festooned flows (Fink, 1980; Hodges and Moore, 1994).

4.3.3. *In Situ* Measurements

In situ measurements of the elemental composition of Martian surface materials were made by the Viking X-ray Fluorescence Spectrometer (XRF) and the Pathfinder/Sojourner Alpha Proton X-ray Spectrometer (APXS). *In situ* mineralogy was also inferred from the Pathfinder magnetic properties experiments. The soils at the Viking and Pathfinder sites are similar, although the soils at the Pathfinder site are somewhat depleted in S and enriched in Ti (Table 4.1; Banin *et al.*, 1992; Rieder *et al.*, 1997). The soils have compositions comparable to those of palagonite, a hydrated volcanic glass, and could represent the weathered products of basaltic rocks. The abundant S and Cl may be derived from volcanic gases that reacted with the surface (Banin *et al.*, 1992). The Pathfinder magnetic properties experiments indicate that the magnetized component of Martian dust is composed of claylike aggregates stained or cemented by Fe_2O_3. Some of the ferric oxide is probably maghemite (g-Fe_2O_3) formed by the

Table 4.1. Compositions of Martian Materials and Comparisons with Earth[a]

Name	SiO$_2$	TiO$_2$	Al$_2$O$_3$	MgO	FeO	CaO	Na$_2$O	K$_2$O	MnO	Cr$_2$O$_3$	P$_2$O$_5$	SO$_3$	Cl	Fe$_2$O$_3$	Reference
Pathfinder soils[b]															
A-2	51	1.2	7.4	7.9	16.6	6.9	2.3	0.2				4	0.5		Rieder et al. (1997)
A-4	48	1.4	9.1	8.3	14.4	5.6	3.8	0.2				6.5	0.6		Rieder et al. (1997)
A-5	47.9	0.9	8.7	7.5	17.3	6.5	2.8	0.3				5.6	0.6		Rieder et al. (1997)
A-8	51.6	1.1	9.1	7.1	13.4	7.3	2	0.5				5.3	0.7		Rieder et al. (1997)
A-10	48.2	1.1	8.3	7.9	17.4	6.4	1.5	0.2				6.2	0.7		Rieder et al. (1997)
A-15	50.2	1.3	8.4	7.3	17.1	6	1.3	0.5				5.2	0.6		Rieder et al. (1997)
Pathfinder rocks[b]															
A-3	58.6	0.8	10.8	3	12.9	5.3	3.2	0.7				2.2	0.5		Rieder et al. (1997)
A-7	55.5	0.9	9.1	5.9	13.1	6.6	1.7	0.5				3.9	0.6		Rieder et al. (1997)
A-16	52.2	1	10	4.9	15.4	7.4	3.1	0.7				2.8	0.5		Rieder et al. (1997)
A-17	61.2	0.7	9.9	3	11.9	7.8	2	0.5				0.7	0.3		Rieder et al. (1997)
A-18	55.3	0.9	10.6	4.9	13.9	6	2.4	0.8				2.6	0.6		Rieder et al. (1997)
Pathfinder "soil-free" rock[b]															
	62	0.7	10.6	2	12	7.3	2.6	0.7				0	0.2		Rieder et al. (1997)
Average Viking soils[c]															
	43	0.6	7.2	6		5.8						7.2	0.6	18	Banin et al. (1992)
SNC bulk															
Zagami	51.2	0.81	6.19	10.4	18.2	10.7	1.29	0.13	0.55		0.58				McCoy et al. (1992)
SNC meteorite parental magma estimates															
Nakhla (NK93)	50.2	1.0	8.6	4.0	19.1	11.9	1.2	2.8	0.4	0.1	0.7				Treiman (1993)
Nakhla (GV1)	46.7	4.2	8.1	5.1	23.3	9.7	2.1	1.2							Harvey and McSween (1992)

Nakhla (NK3)	45.8	3.1	7.2	5.7	26.2	10.4	0.8	1.4		0.1	Harvey and McSween (1992)
Nakhla (N)	48.9	1.1	2.8	5.2	26.1	13.8	1.0	0.2	0.7		Longhi and Pan (1989)
Chassigny	51.52	1.58	8.72	7.08	19.02	8.49	2.29	0.77	0.53		Johnson et al. (1991)
Mars mantle + crust models[d]
1	36.8	0.2	3.1	29.9	26.8	2.4	0.2		0.1	0.4	BVTP (1981)
2	40	0.1	3.1	27.4	24.3	2.5	0.8		0.2	0.6	Anderson (1972)
3	43.9	0.16	3.2	31.2	16.7	3	1.4				Weidenschilling (1976)
4	41.6	0.3	6.4	29.8	15.8	5.2	0.1		0.15	0.6	Morgan and Anders (1979)
5	39.4	0.6	3.1	32.7	20.8	2.7	0.5				McGetchin and Smyth (1978)
6	44.4	0.1	3	30.2	17.9	2.4	0.5		0.5	0.8	Dreibus and Wanke (1985), Longhi et al. (1992)
Earth mantle + crust model[d]
	45.1	0.2	4	38.3	7.8	3.5	0.3		0.1	0.5	Jagouz et al. (1979)

[a] Expressed in weight percent.
[b] Compositions normalized to a sum of 98%. Errors associated with each oxide can be found in Rieder et al. (1997).
[c] Compositions normalized to a sum of 95.4%.
[d] The Mars and Earth mantle + crust models come from Table 1 in Longhi et al. (1992). Model 1: 30% Orgueil C1 chondrite + 70% high-temperature component; model 2: mixture of chondrites; model 3: modified equilibrium condensation; model 4: four-component meteorite model; model 5: pyrolite + FeO; model 6: based on SNC data.

leaching of Fe^{2+} from rock by liquid water (Hviid et al., 1997), followed by oxidation and precipitation.

Data from five rocks at the Pathfinder site reveal a composition rich in Si and K and depleted in Mg relative to the soils and SNC meteorites (Figure 4.4). The rock chemistry is similar to that of icelandite, an anorogenic andesite (Figure 4.5; McSween et al., 1999). This composition closely mirrors that of the average composition of Earth's continental crust, except for a higher iron content, which may be related to the greater abundance of Fe thought to exist in the Martian mantle (Table 4.1). The S content of the rocks is unusually high for a primary igneous composition because it exceeds the S solubility in most magmas and common volcanic rocks (Rieder et al., 1997; McSween et al., 1999). There is a strong correlation between the red/blue ratio derived from IMP spectra and S content, suggesting contamination by dust derived from sulfur-rich, red soil (Bridges et al., 1997). Subtracting S from the rock composition by adjusting the other elements according to regressions of each element versus S indicates that the rock could contain as much as 62% SiO_2 (Rieder et al., 1997; McSween et al., 1999).

4.3.4. Implications

From the various data sources, we can summarize the likely compositions of Martian volcanoes and lava flows. SNC parent magmas have low alumina and high iron [Fe/(Fe + Mg)] contents relative to terrestrial basalts and in some respects more closely resemble basaltic komatiites (McSween, 1994). Basaltic shergottites probably represent

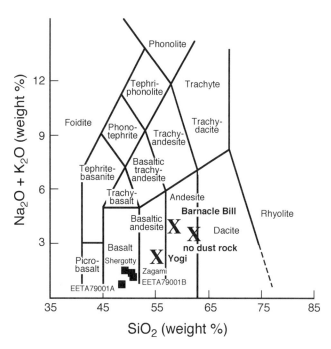

Figure 4.4. Weight percent alkalis versus silica for representative Pathfinder rocks (X, including Barnacle Bill) and SNC meteorites (squares). The "no dust" rock composition is found by plotting linear regressions of each oxide versus sulfur and then extrapolating to zero percent S.

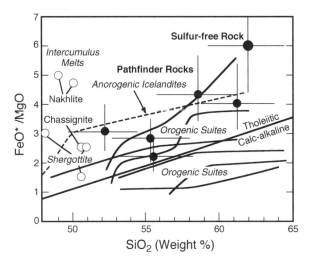

Figure 4.5. FeO*/MgO versus silica for Pathfinder rocks, SNC meteorites, and terrestrial igneous associations. Tholeiitic and calc-alkaline differentiation trends that result in andesites are shown (trends from Gill, 1981). Pathfinder rocks plot along a tholeiitic trend and have compositions close to those of anorogenic icelandites. The label "FeO*" is used to recognize that all Pathfinder iron analyses are reported as FeO even though some Fe_2O_3 is probably present. From McSween et al. (1998).

shallow intrusive rocks or thick lava flows that contained suspended crystals (McCoy et al., 1992; McSween, 1994). The morphology of most volcanoes and lava flows is consistent with low-viscosity, basaltic compositions. If the compositions measured by the Pathfinder APXS represent igneous rocks and are not strongly affected by chemical weathering, then andesites may also be present on Mars.

Although knowledge of Martian geochemistry is incomplete, it can provide insight into volcanic processes on Mars. The best estimates of physical properties of mafic Martian lavas come from analyses of the SNC meteorites. The viscosities of crystal-free, liquidus SNC magmas range from a few to tens of pascals seconds and can be even lower if dissolved water is present. These values are similar to the liquidus viscosities of terrestrial and lunar basalts. The average density of SNC parent magmas might be slightly greater (by ~5%) than those of terrestrial basalts (McSween, 1994). The lack of Martian rock samples of more evolved compositions precludes estimates of their physical properties.

An important factor controlling the viscosity and bulk density of lava is the volatile content, which affects magma ascent rate, vesicle abundance, bulk density, and viscosity. Unfortunately, the amount of water in the SNC parent magmas and the Martian mantle is unknown. The bulk water content of SNCs is 130–350 ppm, lower than most terrestrial basalts (Figure 4.6). Carr and Wanke (1992) used this value to estimate a Martian mantle water content of 35 ppm, drier than estimates for the terrestrial mantle, which range up to 10^2 ppm or more (e.g., Wood et al., 1996; Dreibus et al., 1997). However, if kaersutite amphiboles in SNC meteorites are water-bearing, SNC parent magmas may have contained up to 1.4 wt% H_2O after kaersutite crystallization and 4 wt% prior to crystallization (Johnson et al., 1991; McSween and Harvey, 1993). If the kaersutites are H-deficient (Popp et al., 1995), then the amount of water in the magma is uncertain.

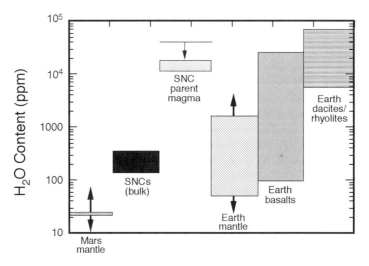

Figure 4.6. Water contents of Martian and terrestrial materials. Mantle values represent a range of estimates; the horizontal line represents the possible water content of the SNC parent mantle source prior to kaersutite crystallization. Bulk SNC and possible parent SNC water contents are from McSween and Harvey (1993) and McSween (1994). Earth basalt and dacite/rhyolite values are from compilations by Johnson *et al.* (1994) and Clemens (1984).

4.4. INFLUENCE OF MARS' ENVIRONMENT ON VOLCANISM

Many factors influence the styles of volcanic eruptions (Whitford-Stark, 1982) and the resulting volcanic morphology (Greeley, 1977; Wilson and Head, 1983). In addition to magma composition, the physical environment of Mars (Table 4.2) influences volcanic activity. For example, Mars is about half the diameter of Earth but possesses some of the largest volcanoes in the solar system. Smaller bodies cool more efficiently, so Mars probably developed a thicker lithosphere than Earth (Schubert *et al.*, 1992), perhaps accounting for the absence of plate tectonics on Mars. In turn, lack of relative motion between the lithosphere and magma source regions could have facilitated edifice construction to the large sizes observed. Furthermore, the lower mass of Mars, lower gravitational acceleration, low-density atmosphere, and cooler surface and atmosphere, work together to influence almost every aspect of ascent, storage, eruption, and emplacement of Martian magmas.

Table 4.2. Comparative Planetary Environmental Characteristics[a]

Planet	Radius (km)	Density (kg m^{-3})	Gravity (m s^{-2})	Atmospheric pressure (Pa)	Atmospheric density (kg m^{-3})	Surface temperature (K)	Atmospheric composition
Mars	3390	3930	3.73	7×10^3	0.162	225	>95% CO_2, N_2,Ar,O_2
Earth	6380	5510	9.81	$\sim 10^5$	1.23	288	75% N_2 23% O_2, Ar, CO_2
Venus	6050	5160	8.88	$\sim 10^7$	70	~ 745	>96% CO_2, N_2,SO_2,Ar,CO
Moon	1730	3340	1.62	—	—	~ 240	—

[a] Atmospheric characteristics given for mean planetary radii on Mars and Venus, and for sea level standard temperate atmosphere on Earth. Temperatures are average diurnal/seasonal values.

4.4.1. Magma Ascent

The ascent of magma is driven by buoyancy, which is a function of diapir volume, V, acceleration related to gravity, g, and the contrast between the density of the country rock, r_r, and magma, r_m, such that the buoyancy force is given by $F_b = g(r_r - r_m)V$. Assuming that density contrasts are similar for terrestrial and Martian mafic magmas, the low gravity of Mars implies a smaller buoyancy force and, hence, a smaller rise velocity for a given magma viscosity. To avoid cooling and solidification during ascent, rising magma bodies on Mars must be larger than on Earth (Wilson and Head, 1994). Rising magma can stall at rheologic barriers such as the asthenosphere–lithosphere boundary or the brittle–ductile transition at the base of the elastic lithosphere, or at a density barrier, at which buoyancy forces are reduced to zero as the country rock density decreases toward the surface (Rubin and Pollard, 1987). In both instances, magma ascent may continue through dikes as a result of brittle fracture of the country rock in response to stresses imposed by the magma body. A thick lithosphere on Mars would imply deeper rheologic barriers. In the case of neutral buoyancy zones, development of excess pressure in the magma reservoir (e.g., as a result of continued magma input or gas exsolution) could initiate dike propagation.

Consideration of gravity and compaction of Martian crustal rock suggests that neutral buoyancy zones and magma reservoirs should be deeper on Mars than on Earth (Wilson and Head, 1994). For magma to reach the surface from deep rheologic or density traps, wider dikes and greater driving pressures are required. Fracture mechanics suggests that the dimensions of dikes are inverse functions of planetary gravity (Wilson and Head, 1994). Thus, dikes on Mars should be larger and accommodate greater magma velocities. Furthermore, large magma reservoirs, inferred from both theory and the large sizes of many Martian calderas (Wood, 1984; Crumpler *et al.*, 1996b) and lava flows (Cattermole, 1987), should maintain greater driving pressures for longer durations. Together, these influences on magma motion imply systematically greater eruption rates and individual eruption volumes than on Earth.

4.4.2. Role of Volatiles

Volatiles such as H_2O and CO_2 contained in the magma are important to the style of eruption. Volatile solubility is partly a function of pressure, so as pressure decreases during magma ascent, bubbles of gas nucleate and grow by diffusion, decompression, and coalescence (Sparks, 1978). If the gas occupies a sufficiently large volume, the magma will disrupt into fragments entrained in a gas stream, resulting in explosive eruptions. The total pressure, P, at any depth, z, in the lithosphere is equal to the sum of the lithostatic and external atmospheric pressures. Thus, the low Martian gravity and atmospheric pressure combine to ensure that $P(z)$ is less than that on Earth. This has important implications for magmatic volatiles. Because of the lower pressure in the Martian lithosphere, and the lower lithostatic pressure gradient, the levels at which volatile exsolution and magma fragmentation occur are deeper and more widely separated than on Earth allowing the growth of multiple bubble populations (Wilson *et al.*, 1982). If the volatile content exceeds ~0.03 wt% H_2O (300 ppm) or a few ppm CO_2 (values significantly lower than for Earth), the expected result is a highly fragmented magma. After release by fragmentation, gas expansion is augmented by the low atmospheric pressure on Mars, which permits a greater energy release per unit mass, leading ultimately to greater eruption velocities (Wilson *et al.*, 1982).

4.4.3. Explosive Eruptions

On Earth, explosive basaltic eruptions are commonly weak because the low volatile content and viscosity of basalt produces coarsely fragmented magma erupted at low speeds, forming small lava fountains. Typically, clast ponding occurs to form lava flows; spatter and scoria deposits may also form, depending on eruption conditions. On Mars, the predicted high eruption velocities and efficient energy transfer between fine pyroclasts and the gas suggest that vigorous Plinian eruptions might have been common, even though they are rare for terrestrial basalts. In such eruptions, the emerging volcanic jet entrains and heats atmospheric gas such that the bulk density of the expanding gas–particle mixture may become less than the density of the atmosphere, promoting buoyant ascent of a tall, convecting eruption column (Wilson et al., 1978). The greater eruption velocity, together with differences in atmospheric temperature and pressure, would produce columns several times higher than on Earth. However, the limited expansion potential of low-density atmospheric gas means that buoyancy is not so readily achieved, so that unstable fountains that collapse to feed pyroclastic flows would be more common than on Earth (Wilson et al., 1982). Although pyroclastic flows have been tentatively identified on Mars, predictions of flow distances are hampered by incomplete understanding of their emplacement mechanisms, and poor knowledge of Martian topography. To first order, the higher initial velocities and weaker particle frictional interaction (caused by low gravity; Crown and Greeley, 1993) suggest systematically greater runout distances.

The nature of a pyroclastic deposit is a function of the accumulation rate and temperature of the pyroclasts on landing. These factors are related to the structure of the erupting plume or fountain, the sizes of pyroclasts, and ultimately to the volatile content and mass eruption rate of the magma (Head and Wilson, 1989). Pyroclasts ejected into the Martian atmosphere experience little drag in comparison with Earth because aerodynamic drag is a function of atmospheric density. Together with the reduced gravitational settling and initial high eruption velocities on Mars, this implies that pyroclasts will have long trajectories, allowing significant cooling during flight. Thus, Martian explosive volcanism should result in widely dispersed, poorly consolidated deposits of fine material, forming broad, low-relief edifices. If volatile-poor eruptions produce coarser clasts, they may form lava fountains, which would be taller and wider than those on Earth (Wilson and Head, 1994). The coarsest clasts would accumulate close to the vent to form flows and spatter deposits, but smaller clasts would follow long trajectories, cooling and forming broad deposits of scoria. This might explain the apparent lack of steep cinder cones on Mars (Wood, 1979; Edgett, 1990).

4.4.4. Role of Ground Water/Ice

The probable existence of ice in the regolith at higher latitudes (Clifford, 1993) and the postulated warmer, wetter climate in the past (Owen, 1992) imply ample opportunity for magma/water interactions. These could include melting and release of ground water by intrusions or lava flows, explosive generation of rootless "pseudocraters" by lavas flowing over a saturated substrate, increased eruptive vigor in sustained explosive eruptions, and formation of explosion craters (e.g., maars). Volcano/ice interactions on Mars are discussed in Chapter 3.

Some volcanic features suggestive of ground water release have been proposed for Mars (Frey et al., 1979; Mouginis-Mark et al., 1982, 1988; Mouginis-Mark, 1985; Squyres et al., 1987; Greeley and Crown, 1990; Crown and Greeley, 1993). However, the apparent scarcity of identified maars might be related to the difficulty of distinguishing explosion craters from impact craters. Models for transient explosions, in which steam is pressurized beneath a cap rock, predict ejection of clasts of country rock and juvenile magma at greater velocities than on Earth as a result of enhanced gas expansion. The higher velocities, low atmospheric drag, and low gravity suggest dispersal ranges of tens to hundreds of kilometers (Fagents and Wilson, 1996), and it is unlikely that an edifice or detectable deposit would be produced.

4.4.5. Effusive Eruptions

If magma fragmentation does not occur and the magma erupts effusively, the low gravity on Mars would lead to thicker flows, which suggests that with sufficient volumes, lavas would flow greater distances before cooling (Wilson and Head, 1994). Longer flows are also implied by the high effusion rates (Walker, 1973) and total volumes (Malin, 1980) predicted as a result of planetary environmental differences (see Section 4.1). On Earth, high effusion rates are characterized by sheetlike flows fed by long fissure vents. Low effusion rates commonly produce tube-fed lavas. These morphologies are observed on Mars, as described below.

The total heat flux from the surface of a subaerially emplaced lava is a combination of radiative, forced, and natural convective heat fluxes. The convective terms are dependent on gravity and thermal properties of the atmospheric gas, which depend on atmospheric density and viscosity. Convective heat loss should be less significant on Mars than on Earth because of differences in thermal parameters, with radiation being the dominant cooling mechanism. However, the total heat flux is somewhat less than on Earth because of the inefficiency of convective heat transfer, allowing Martian lava surface temperatures to remain tens of degrees hotter than their terrestrial counterparts for a year or more after eruption (Wilson and Head, 1994). Because of the extreme temperature dependence of lava rheology, this may enable Martian lava to remain mobile longer.

4.4.6. Summary

Unless volatile contents of Martian magmas were very low, explosive activity should have been common on Mars, forming broad, low-relief constructs. The possibility of interaction with ground water/ice enhances this probability. Because of Mars' unique environment, there is no need to invoke magmas of more evolved compositions to explain explosive features, although large Martian magma chambers could promote the generation of small volumes of more evolved magma via fractional crystallization. There are some indications that explosive activity may commonly have accompanied the extensive effusions. For example, the "stealth" terrain (Muhleman *et al.*, 1991; Edgett *et al.*, 1997), the dune fields west of the Tharsis volcanoes (Edgett, 1997), and the fine "dust" that apparently mantles many effusive features could all have explosive volcanic origins. In any case, Mars' large edifices, voluminous flows, and inferred pyroclastic deposits are consistent with predictions of larger magma chambers and dikes, greater eruption rates and volumes, and vigorous, explosive activity.

4.5. LARGE SHIELD VOLCANOES

4.5.1. Tharsis Region

The best-known volcanoes and greatest concentration of central-vent constructs occur in the Tharsis region (Figure 4.1). The Tharsis volcanoes have morphologies typical of basaltic shield volcanoes on Earth, but are much larger. The largest, Olympus Mons, is >500 km across and 25 km high. As measured from the seafloor, Mauna Loa volcano is the largest mountain on Earth, yet it is dwarfed by the Martian volcanoes. The flanks of the Martian volcanoes have maximum slopes of 5° with shallower summit and basal slopes, similar to the subaerial part of Mauna Loa volcano. The flanks of Olympus Mons are composed of interwoven lava flows that spill onto the surrounding plains. Many of the flows have medial lava channels and partly collapsed lava tubes similar to those on terrestrial volcanoes (Greeley, 1973; Carr and Greeley, 1980). The forms and dimensions of individual flows suggest basaltic compositions, although basaltic andesite is a viable alternative (e.g., Moore *et al.*, 1978). Some of the flows and lower flanks of the Tharsis volcanoes are buried by plains-forming flows erupted from the shields or other vents (Scott and Tanaka, 1986).

Olympus Mons is surrounded by a scarp 3 to 6 km high (Figure 4.2), the origin of which is controversial. Enormous fan-shaped deposits northwest of the volcano are interpreted as landslides, perhaps lubricated by volatiles along the detachment surface (e.g. Tanaka, 1985).

Other volcanoes in the Tharsis area include *tholi* (Figure 4.7) and the unique feature Alba Patera, discussed below. Tholi are commonly smaller than 200 km in diameter and have

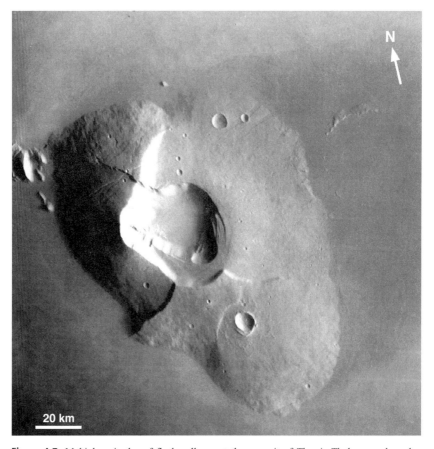

Figure 4.7. Multiple episodes of flank collapse at the summit of Tharsis Tholus may have been caused by failure of an unconsolidated basal layer equivalent to the lower units of Olympus Mons and Apollinaris Patera. The isolated massifs northwest of the cone may be large slump blocks that are embayed by lava flows from Ascraeus Mons to the west. (Viking Orbiter frame 858A23.)

a variety of morphologies. Most tend to have slopes steeper than those on the shields and have calderas more than half as wide as the overall construct, suggesting that they may be the summits of partly buried volcanoes. Ceraunius Tholus displays sinuous channels carved into its flanks, interpreted to be caused by erosive pyroclastic flows (Reimers and Komar, 1979).

4.5.2. Elysium Region

The Elysium region is also a central-vent volcanic province on Mars. Two Elysium volcanoes were compared with constructs in the Tibesti highlands in Africa (Malin, 1977) and illustrate contrasting morphologies. Hecates Tholus is ~160 by 175 km wide and ~6 km high and its flanks are cut by narrow, shallow valleys. Elysium Mons is 500 by 700 km wide and ~13 km high and displays numerous flank flows (Figure 4.8). Differences in the numbers of superposed impact craters around the summit of Hecates Tholus, including a near absence of craters within a semiannulus 2 km west of the summit (Figure 4.9), led Mouginis-Mark *et al.* (1982) to suggest that mantle-forming Plinian eruptions occurred in the recent geologic past.

Some of the Elysium volcanoes, such as Hecates Tholus, have highly dissected flanks (Mouginis-Mark *et al.*, 1982, 1984; Gulick and Baker, 1990). It is also evident that the valley-forming events predate the eruption of the surrounding lava plains because the valleys are truncated by the plains (Figure 4.10). The dissection could reflect either erosion or volcanic flank deposits. Reimers and Komar (1979) proposed that lava or fast-moving pyroclastic flows may have carved the valleys. However, many of the valleys form midway on the flanks (Mouginis-Mark *et al.*, 1982) and have branches characteristic of surface runoff (Gulick and Baker, 1990). Local fluvial erosion could have resulted from hydrothermal release of water

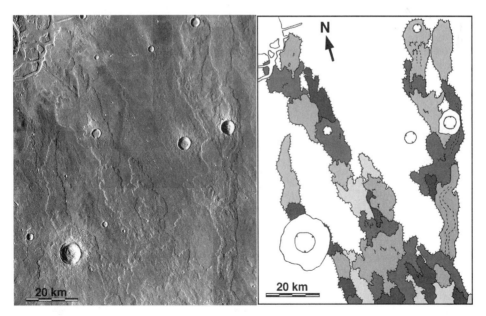

Figure 4.8. Left: View of segmented lava flows northwest of Elysium Mons. The distal ends of the flows are more than 300 km from the summit of Elysium Mons, and are estimated to be 40- to 60-m-thick flows (Mouginis-Mark and Tatsumura-Yoshioka, 1998). (Viking Orbiter frames 651A08–12.) Right: Geomorphic interpretation of the flows showing individual lobe segments; copyright, American Geophysical Union.

Figure 4.9. The summit area of Hecates Tholus has one of the best candidate ash fall deposits identified on Mars. The absence of small (<2 km) impact craters around the caldera suggests that a Plinian airfall deposit mantles the summit to a depth of ~100 m (Mouginis-Mark *et al.*, 1982). (Viking Orbiter frame 651A19.)

(Brackenridge *et al.*, 1985). The valleys show no evidence of lava flow fronts or other volcanic features (as found on the Tharsis volcanoes), and comparable valleys are found on the volcano Alba Patera. Therefore, Mouginis-Mark *et al.* (1988) proposed that the flanks are composed of easily eroded material, such as ash, and that the valleys formed by water-lubricated mass movements. In turn, this would imply that ash-producing eruptions occurred on Hecates Tholus, but not Elysium Mons, which lacks valleys.

4.5.3. Calderas

Calderas occur at the summits of all of the larger volcanoes on Mars. The calderas on the summits of Olympus Mons and Arsia Mons are representative of the morphologic range in Martian calderas (Crumpler *et al.*, 1996a). Variable distributions of strain at the surface are predicted in association with differences in shape, lateral dimensions, aspect ratio (width/height), and depth of individual magma chambers (Gudmundsson, 1988). An

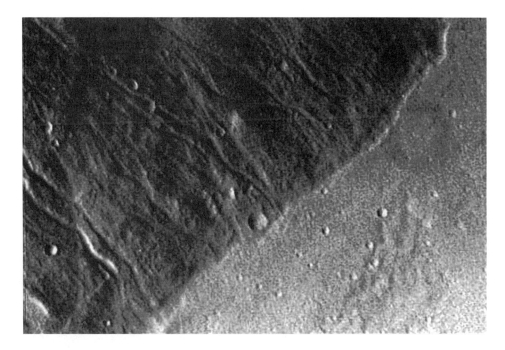

Figure 4.10. Narrow valleys in networks are found on the flanks of Hecates Tholus, Ceraunius Tholus, and Alba Patera. These valleys suggest that the flank materials are easy to erode, and could be ash deposits. The valleys shown here on the southeast flank of Hecates Tholus predate the emplacement of lava flows (lower right of figure) from Elysium Mons. (Viking Orbiter frame 651A22.)

additional factor in their morphologic diversity and larger dimensions may be the greater predicted depth and larger dimension of magma chambers on Mars relative to Earth (Wilson and Head, 1990).

The Olympus Mons caldera is irregular in planform, consisting of overlapping, nested, and scalloped margins with steep walls (Figure 4.11), and provides a record of the evolution of magma chamber size and location. The sequence in the complex is a trend from large calderas initially, followed by smaller local collapses (Mouginis-Mark and Robinson, 1992). However, the reverse trend is seen at Ascraeus Mons where the last caldera collapse formed a more circular and larger feature than previous calderas within the complex.

The Arsia Mons caldera is larger and more circular than typical terrestrial forms (Figure 4.11D). It has concentric fractures, ring grabens, and pit craters exterior to the caldera, and the walls are terraced. Topographic profiles (Smith *et al.*, 1998) show that the Arsia Mons caldera is relatively shallow and flat-floored. In contrast, the entire Pavonis Mons caldera complex appears to have sagged over a region 100 km across. Nested within the area is a single circular, steep-walled caldera 45 km across and as deep as 5 km.

The greater predicted depth and larger dimensions of magma chambers on Mars relative to Earth (Wilson and Head, 1990) are consistent with the larger dimensions of Martian calderas. Magma chambers form only if a zone of magma accumulation can develop (Wilson and Head, 1994) and if magma replenishment and thermal environment are sufficient for chamber growth (Crumpler *et al.*, 1996a). The necessary magma replenishment rate is controlled by the lithospheric thermal gradient and the shape and size of the magma chamber.

Figure 4.11. Martian calderas. (A) The summit caldera complex of Olympus Mons. Several smaller calderas are nested inside the larger, more circular caldera. (Viking Orbiter frame 890A68.). (B) The Ascraeus Mons caldera complex is similar to the Olympus Mons caldera, but has smaller calderas that overlap the margins of the largest caldera. (Viking Orbiter frames 892A 11,32).

Volcanism on Mars 93

(C)

Figure 4.11. Martian calderas. (C) The Uranius Patera caldera is the result of the coalescence of several smaller calderas. (Viking Orbiter frames 857A43–46).

Crumpler *et al.* (1996b) suggest that the larger Martian calderas require replenishment rates that exceed Hawaiian rates.

4.5.4 Alba Patera

Alba Patera, north of the three Tharsis shield volcanoes, is unique on Mars (Carr *et al.*, 1977). It is larger in areal extent than Olympus Mons but lacks the relief of the shield volcanoes. It has flank slopes of $\sim 1°$ (Smith *et al.*, 1998) and is surrounded by a band of fractures suggesting that most flank flows were emplaced before the regional stress pattern was imposed. Impact crater densities indicate that Alba Patera is older than either the Tharsis or Elysium shield volcanoes (Neukum and Hiller, 1981). However, some young flows may represent eruptions that postdate the fractures (Schneeberger and Pieri, 1991). Materials

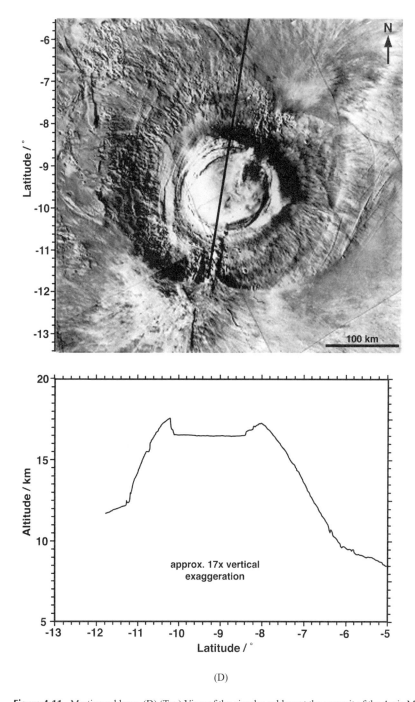

Figure 4.11. Martian calderas. (D) (Top) View of the circular caldera at the summit of the Arsia Mons shield volcano. The line represents the track of the MGS laser altimeter (MOLA; Smith *et al.*, 1998). (Part of Viking Orbiter mosaic MC-17NW.) (Bottom) MOLA topographic profile of Arsia caldera. The caldera is 123 km from rim to rim.

(E)

Figure 4.11. Martian calderas. (E) The summit caldera complex of Pavonis Mons (shown here) is similar to Arsia Mons in that there is a large circular subsidence, but is also similar to Olympus Mons in that there is a nested smaller caldera with terraced walls. (Viking Orbiter frame 643A54.)

surrounding the central zone could indicate both effusive and explosive volcanism (Mouginis-Mark et al., 1988).

The western flanks of Alba Patera are composed of multiple tube- and channel-fed lava flows, many of which display partly collapsed lava tube segments. These flows were compared with similar features on Mt. Etna (Greeley and Spudis, 1981) and probably represent long-duration eruptions. Other flows are more sheetlike and lack channels and lava tubes. More detailed analyses of the Alba Patera flows are provided by Cattermole (1987). He also provides evidence for the existence of mild spatter and pyroclastic activity (Cattermole, 1986). Some flank areas have valley networks, suggesting fluvial erosion into pyroclastic deposits (Mouginis-Mark et al., 1988). Alba Patera possesses two discrete calderas (Wood, 1984). The northernmost caldera is incomplete, and is filled on the eastern side by lava flows erupted from the southern caldera. This may indicate local topography or large-scale modifications to the magma plumbing system of the volcano during late-stage activity.

4.5.5. Discussion

Shield volcanoes on Mars display morphologies that are common on basaltic shields on Earth. Features typical of Martian volcanoes are large (>50 km) caldera complexes, which suggest that comparable-sized magma chambers were within 20 km of the volcano summits (Zuber and Mouginis-Mark, 1992).

Most of the youngest valley networks on Mars are found on volcanoes. This has been suggested as evidence against the idea that the early Martian climate was warm and wet and that it changed to the present cold and dry conditions (Carr and Chuang, 1997). For example, Hecates and Ceraunius Tholi lack well-preserved lava flows on their flanks but have valley networks that predate the emplacement of the surrounding plains lavas. The valleys could be fluvial features formed in unconsolidated pyroclastic deposits produced during explosive eruptions of the volcanoes. Alba Patera also has valley networks on its flanks, and may be transitional in eruptive style between early explosive volcanism and more recent effusive activity on Mars (Mouginis-Mark et al., 1988).

The environment may have played a role in the development of the basal units of several Martian volcanoes. Geomorphic evidence suggests that large bodies of water may have existed on Mars in the Amazonian Period (Baker et al., 1991; Parker et al., 1993). This is a controversial idea compared with the more generally accepted view of a "dry Mars" but it enables some of the features found with some volcanic landforms to be reinterpreted in a self-consistent manner. For example, the Olympus Mons escarpment is enigmatic. Early interpretations suggested erosion of cratered material on which the volcano was constructed (Head et al., 1976) by the wind (King and Riehle, 1974). More recently, a tectonic origin was proposed for the scarp (Borgia et al., 1990). Apollinaris Patera and Hecates Tholus also have a scarp, although lower in height (0.5–1.5 km) and smaller (Robinson et al., 1993) than the Olympus Mons scarp. All three volcanoes have the lowest base elevations (<2 km above mean Mars datum) of Martian volcanoes and are adjacent to the putative shoreline of the hypothesized Martian seas (Baker et al., 1991), so the basal scarps could be wave-cut features.

Considerable debate has also focused on the formation of the Olympus Mons aureole deposits. A central issue is the long size of the lobes. Ideas for the formation of the aureole include gravity spreading (Francis and Wadge, 1983; Tanaka, 1985), thrust faults and landslides (Harris, 1977; Lopes et al., 1982), emplacement as ash flows (Morris, 1982), and subglacial lava flows (Hodges and Moore, 1979). Each mechanism requires low shear strength of the materials to facilitate sliding, and no equivalent runout slides have been identified on Earth. The closest analogies (Tanaka, 1985) are Hawaiian submarine landslides (Moore et al., 1989). The Olympus Mons aureole deposits occur at low elevations (<1 km above Mars datum). In the model for shallow seas on Mars (Baker et al., 1991), the entire northwest side of Olympus Mons would have been submerged, facilitating the formation of submarine slides, suggesting that they could be generated by the collapse of the submerged basal materials of the volcano.

Collapse of other Martian volcanoes might also be explained by the presence of unconsolidated basal materials. Tharsis Tholus (Figure 4.7) is relatively small and its summit is dissected by multiple sets of arcuate scarps (Robinson and Rowland, 1994). The scarps appear to represent a complex form of sector collapse comparable to that seen on volcanoes in Hawaii and Réunion Island (McGuire, 1996; Crumpler et al., 1996b). Although the lower flanks of Tharsis Tholus are buried by younger flows, remnants of the edifice can be found to the northeast of the volcano, suggesting that the segments of the flanks have moved horizontally. By analogy with the terrestrial examples, it is possible that the basal unit of Tharsis Tholus is structurally weak, perhaps similar to the base of Olympus Mons (Mouginis-Mark, 1993).

4.6. MARTIAN HIGHLAND PATERAE

Highland paterae are low-relief, areally extensive central-vent volcanoes found in ancient cratered terrains on Mars, and are thought to be the oldest central-vent volcanoes on the planet (Plescia and Saunders, 1979). They are characterized by low shield morphology, central caldera complexes, and radial channels and ridges. The largest, best-imaged, and most extensively studied highland paterae are found around the Hellas basin.

Previous analyses of highland paterae include initial studies of Martian volcanoes based on Mariner 9 images, geologic mapping, surveys of volcanic features using Viking Orbiter images, and process-oriented studies. In addition, particular focus was placed on channels characteristic of patera flanks which may provide information on the environment in which the eruptions occurred.

Interpretations of the style(s) of eruption associated with highland paterae have changed with time. Mariner 9 data suggested that Amphitrites, Hadriaca, and Tyrrhena Paterae were basaltic shield volcanoes, formed from low-viscosity lavas (Potter, 1976; Peterson, 1977, 1978; King, 1978). Their formation by pyroclastic eruptions was thought improbable because the extensive erosion expected in ancient pyroclastic deposits was not observed in Mariner 9 images (Peterson, 1978).

Viking Orbiter images resulted in reevaluation of patera origins (Carr, 1981). The geomorphology of Tyrrhena Patera and its erosional characteristics led Greeley and Spudis (1981) to propose that phreatomagmatic activity, caused by magma rising through water-rich megaregolith, was the dominant eruptive mechanism. Francis and Wood (1982) suggested that the highland paterae could be mafic pyroclastic structures with eruptions driven by volatiles from sources at great depths. They argued that large-scale, evolved magmatic activity was improbable and that phreatomagmatic activity at a scale large enough to produce the paterae was unlikely, based on terrestrial analogues. Later examinations of Tyrrhena and Hadriaca Paterae, including geomorphic analyses and corresponding assessments of eruption mechanisms, suggest that they consist of pyroclastic flow deposits. An evaluation of the energy required to emplace the patera flank materials indicates that eruptions driven by either magmatic volatiles or ground water are viable (Greeley and Crown, 1990; Crown and Greeley, 1993).

4.6.1. Tyrrhena Patera

The flanks of Tyrrhena Patera consist of layered, friable deposits dissected by extensive erosional channels radial to the summit region (Figure 4.12). The $\sim$50-km caldera complex includes ring fractures and two connected depressions. A large channel extends from the depressions to the southwest where it joins a flank flow unit $\sim$1000 km long (Greeley and Crown, 1990). The flank flow may have been supplied by this channel and contains numerous $\sim$100-km-long lava flows and parts of a leveed channel system that may have spanned $\sim$600 km. These lava flows are the only clear evidence of effusive volcanic activity. Based on crater statistics, the flows were emplaced in the Late Hesperian/Early Amazonian Epochs, long after the earlier shield-building materials (Crown *et al.*, 1992; Gregg *et al.*, 1998). Viking Orbiter images with resolutions as high as $\sim$10 m pixel^{-1} show no evidence of primary flow features within the flank deposits, which exhibit erosional surfaces at the limits of resolution (Crown and Greeley, 1993).

4.6.2 Hadriaca Patera

The summit of Hadriaca Patera (Figure 4.13) is marked by a well-defined, nearly circular caldera $\sim$77 km across. The caldera appears to be filled with late-stage lavas. A large, curvilinear wrinkle ridge and small domelike features in the eastern part of the caldera, together with the scalloped rim suggest a complex history. The flanks of the volcano are asymmetric in plan view ($\sim$300 $\times$ 560 km) and exhibit layering and remnant mesas. The flank channels are trough-shaped, lack tributaries, and are continuous over long distances.

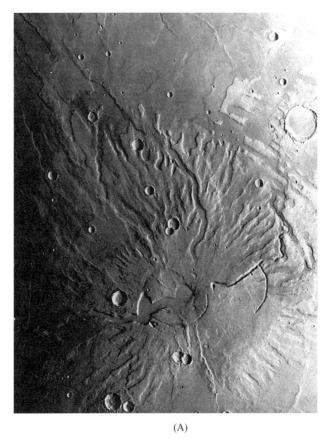

(A)

Figure 4.12. Volcano flanks. (A) View of Tyrrhena Patera showing the complex summit region surrounded by its eroded flanks. A large rille extends to the southwest and connects the caldera complex to the flank flow unit consisting of numerous lava flow lobes. The summit caldera complex is ∼50 km across. North is to the upper right corner of the image. (Viking Orbiter frame 087A14.)

Amphitheater-headed channels are common. The erosional morphology of the channeled flanks of Hadriaca and Tyrrhena Paterae is attributed to a combination of ground water sapping and surface runoff, with more extensive surface dissection evident at Hadriaca Patera in the form of V-shaped valley interiors (Gulick and Baker, 1990; Crown and Greeley, 1993). Crater statistics suggest that Hadriaca Patera formed in the Early Hesperian Epoch and has a slightly younger surface than Tyrrhena Patera (Crown *et al.*, 1992).

4.6.3. Amphitrites/Peneus Patera Complex

The Amphitrites/Peneus Patera complex is defined by two large (∼120 km in diameter) circular depressions and surrounding dissected and ridged plains. Amphitrites Patera exhibits nested, shallow depressions that are surrounded by a distinctive radial channel system similar to those at Tyrrhena and Hadriaca Paterae. Peneus Patera consists of concentric fractures surrounded and flooded by ridged plains. The degradation of small impact craters in the

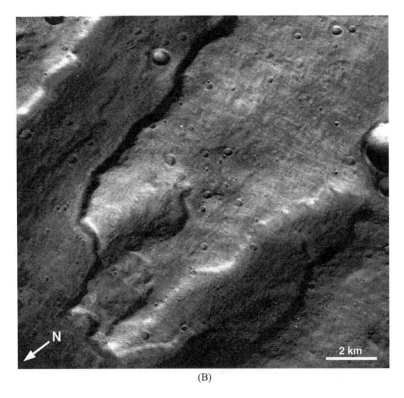

Figure 4.12. Volcano flanks. (B) The eroded flanks of Tyrrhena Patera show nested channels indicative of headward erosion. Lineations potentially related to wind erosion are observed orthogonal to the channel orientation. The frame is ∼15 km across. North is to the bottom left corner. (Viking Orbiter image 794A01.)

region suggests a complex and extensive history of modification. The Amphitrites/Peneus Patera complex was initially interpreted as an overlapping complex of low shield volcanoes (Potter, 1976; Peterson, 1977, 1978) with up to six source calderas. This complex could be the source of the ridged plains of Malea Planum (Greeley and Guest, 1987; Tanaka and Scott, 1987).

4.6.4 Tempe Patera

Tempe Patera is found in the northern hemisphere. It consists of a shallow, 16-km-diameter depression surrounded by smooth deposits that are extensively degraded by radial, poorly defined shallow troughs (Plescia and Saunders, 1979; Wise, 1979; Hodges and Moore, 1994). Its morphology is generally similar to that of Tyrrhena Patera, although it is much smaller. Plescia and Saunders (1979) indicate that Tempe Patera is as old as, or slightly older than, Tyrrhena Patera. A potentially similar feature ∼100 km across found in the Thaumasia region was interpreted to be a volcano by Scott and Tanaka (1981) and Scott (1982) but does not have the distinct central vent characteristics of other highland paterae.

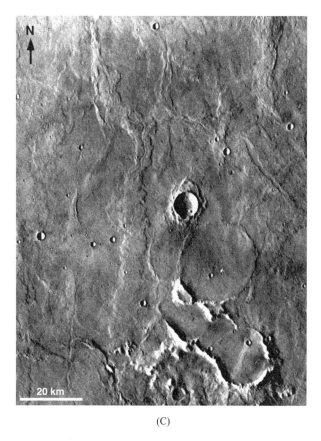

(C)

Figure 4.12. Volcano flanks. (C) View of a lava flow lobe within the flank flow southwest of the Tyrrhena Patera summit. The upper, more irregular part of the flow lobe has a channel in its interior that appears to have fed a wider, ~60-km-long unchannelized zone. North is to the top of the image. (Part of Viking Orbiter image 413S13.)

4.6.5. Summary

The geologic evolution and styles of eruptive activity for the Martian highland paterae have significant implications for interpreting the volcanic and climatic history of Mars. Highland paterae are the oldest preserved central-vent volcanoes on Mars and could represent a transition from earlier plains-forming volcanic activity to large discrete eruptive centers (Greeley and Spudis, 1981). In addition, most paterae represent basin-related, highland volcanic centers that provide a basis for comparison with Martian shields and tholi, and they may provide evidence for differences in magma source regions between the highlands and the northern lowlands and/or temporal changes in Martian magmas or their emplacement conditions (Crown and Greeley, 1993).

Use of the paterae in comparisons with other volcanoes or for inferences about the environment depends on interpretations of their eruptive style. The interpretation of phreatomagmatic activity for Tyrrhena and Hadriaca Paterae is consistent with the timing and extent of aqueous erosion documented for the region. The lack of volcanoes with morphologic characteristics similar to highland paterae in younger volcanic provinces and the

(A)

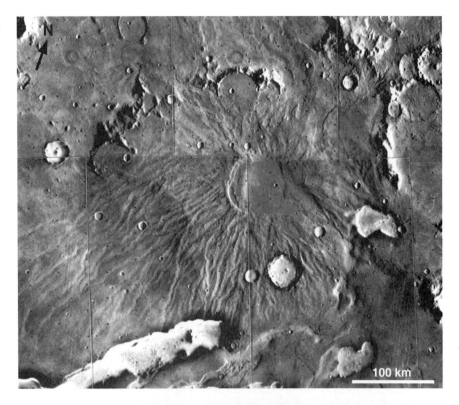

(B)

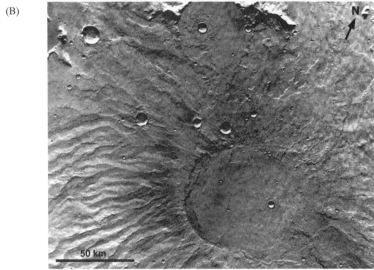

Figure 4.13. Hadriaca Patera. (A) View of the summit caldera (∼77 in diameter) and surrounding channeled flanks of Hadriaca Patera. Collapse features associated with Dao Vallis are observed to the east (right) and south (bottom) of Hadriaca Patera and truncate its flanks. North is to the upper right corner of the image. (Part of Viking Orbiter photo-mosaic 211-5456.) (B) Detailed view of the summit of Hadriaca Patera. The western rim (left) of the caldera suggests multiple episodes of collapse, whereas the rim to the north and east may be covered by eruptive products. A large scarp in the southeast may be the margin of lava ponded in the caldera; small domelike features are observed on the surface of caldera-filling materials to the east. North is to the upper right corner of the image. (Viking Orbiter image 410S02.)

inferred transition from explosive to effusive volcanism at Tyrrhena and Hadriaca Paterae may reflect a global change in Martian near-surface and atmospheric conditions that limited or depleted the volatiles necessary to generate hydrovolcanic eruptions.

If magmatic volatiles drove explosive eruptions at the highland paterae, the generation of less volatile-rich magmas over time could explain the transitions thought to occur in the eastern Hellas region, as well as the observed difference in volcano morphology between the highlands and the Tharsis and Elysium provinces (Francis and Wood, 1982; Crown and Greeley, 1993). Heterogeneities in magma source regions may also be a factor. If the early suggestions that the paterae consist of fluid lavas prove to be true, lava flows associated with the paterae could indicate significantly different thermal environments, magma bodies, emplacement conditions, or erosional regimes than are typical of the younger volcanoes on the Martian surface.

4.7. SMALL VOLCANIC CONSTRUCTS

Clusters of kilometer-sized conical hills, many with summit craters, are found in the northern plains and southern highlands. Fields of knobs, buttes and eroded hills in the Aeolis (Greeley and Spudis, 1981), Thaumasia (Scott, 1982), and Tempe (Plescia and Saunders, 1979) regions are suggested to be volcanoes. In each case, image resolution and erosion of the features make their origin unclear. For example, the thousands of small hills in Acidalia and Utopia Planitiae (Frey and Jarosewich, 1982) could be cinder cones, but they could equally well be pingos or volcanic pseudocraters formed from lava flows.

Martian surface conditions might also complicate recognition of small volcanoes in low-resolution images. For example, the lower Martian gravity could cause pyroclastic eruptions to spread cinders over a wider area than for similar eruptions on Earth. At least some of the "lava shields" in Tempe Terra (Plescia, 1981) could have formed in this manner. The small hills aligned across the floor of Arsia Mons caldera (Carr et al., 1977) may also be cinder cones. Small ridges on Alba Patera were interpreted as spatter ridges produced by low-level fire-fountaining (Cattermole, 1986), but diagnostic features are too small to be seen on available data.

With evidence for volcanism and ground ice in close proximity (Allen, 1979; Hodges and Moore, 1979; Mouginis-Mark, 1985; Squyres et al., 1987), it is unclear why more maars have not been identified. One possibility is that the resultant craters would be morphologically similar to small impact craters and have been misinterpreted. As with cinder cones, the low Martian gravity would cause greater ejecta dispersal, leading to low, wide crater rims. High-resolution images could shed light on volcano/ground ice interactions, as well as other aspects of Martian volcanic features.

4.8. VOLCANIC PLAINS

Volcanic plains of several types are found on Mars. The most widespread are ridged plains that resemble the lunar maria. They are characterized by wrinkle ridges and commonly fill older impact craters, but rarely display vents or vent structures. As with the lunar maria, the ridged plains on Mars are thought to represent flood lavas erupted at high rates of effusion from fissures that were buried by their own products. Lack of distinctive flow features such as lobes and flow fronts suggests that the lavas were extremely fluid, perhaps comparable to komatiites erupted on Earth in the Precambrian. The lack of flow features, however, also means that their interpretation as volcanic is uncertain.

Some plains areas, as shown in Figure 4.3, are clearly volcanic, as evidenced by distinct flow lobes, and constitute a second type of volcanic plains. They form widespread sheets and typically lack lava channels and apparent lava tubes at the resolution of Viking images. As with the ridged plains, these flows could have been emplaced from massive eruptions; however, the lavas might not have been as fluid, but rather possessed greater viscosities, resulting in preservation of flow margins and fronts.

A third type of volcanic plains is associated with some of the large constructs. For example, lava flows can be traced from the flanks of Olympus Mons and onto the surrounding plains. These units also have distinctive flow fronts and side lobes. Together with ridged plains and the sheet flows described above, these constitute plains of *probable* volcanic origin.

Martian plains of *possible* volcanic origin are found throughout much of the northern lowlands. Flow lobes and ventlike structures are seen in high-resolution images in some areas. However, imaging resolution and quality are insufficient to map most of the lowland plains consistently; these plains probably are composed of materials of many diverse origins, including volcanic.

Plains of probable volcanic origin constitute more of the Martian surface than any other terrain (Greeley and Spudis, 1981; Greeley and Schneid, 1991). Although they appear to have attained their greatest areal extent in the early Hesperian (Tanaka *et al.*, 1988; Greeley and Schneid, 1919), this is likely to be an artifact of preservation. For example, all of the northern lowlands are of probable volcanic origin but subsequent burial and reworking with sediments may have obscured their volcanic origin. Moreover, the likely thermal history of Mars suggests extensive volcanism in the Noachian Period but intense impact cratering has destroyed this record. The morphologies of small areas of Noachian age in the highlands (chiefly in the Arabia to Noachis Terra regions; Figure 4.14) are consistent with lava plains emplacement.

Ridged plains that could be comprised of lava flows are abundant around many volcanic centers, such as Hesperia Planum in the vicinity of Tyrrhena Patera (Figure 4.15). In addition, lobate flows associated with volcanic centers, and channels similar to lunar sinuous rilles are evidence for volcanic plains surrounding volcanic centers in some highland settings.

Extensive Hesperian volcanic plains were emplaced as precursors to the Tharsis and Elysium volcanic regions. The intravolcano plains in the Tharsis area were further surfaced by lobate and digitate lava flows, many of which represent late-stage eruptions (Figures 4.11, 4.16 and 4.17).

Morphologies of some lava plains flows might have been affected by local variations in the Martian environment. Mouginis-Mark and Tatsumura-Yoshioka (1998) mapped the area north and west of Elysium Mons, where 59 large lava flows were identified. Some of these flows can be divided into discrete segments that could represent emplacement as a series of pulses (Figure 4.8). Each breakout of a new flow segment appears to have occurred at the distal end of the earlier lobe. Lengths range from 9 to 41 km and maximum thicknesses range from 40 to 125 m. The surface of each flow is flat, lacks festoon ridges, and is interpreted to be a'a (Bruno *et al.*, 1992). There is no evidence for lava tubes, roofed channels, or lobes emerging from the sides of earlier lobes, so that some (unknown) limit imposed by properties of the flow prevented their lateral growth. The relatively flat surfaces suggest that flow inflation was not part of the emplacement. Similar flow segmentation was described by Gregg and Fink (1997) in laboratory simulations under conditions that they call the rifting regime; the simulated lava temporarily had a sufficient hydraulic pressure to rupture the crust.

Figure 4.14. Possible volcanic plains of Noachian age, showing the "flooded" appearance of impact craters and mare-type wrinkle ridges typical of volcanic plains on the Moon. Image is 22 km wide, centered at 33°N latitude, 306.5°W longitude. (Viking Orbiter frames 641A14, 16–18.)

Another type of segmented flow occurs northwest of the Alba Patera caldera. Called "M-type flows" by Lopes and Kilburn (1990), they appear to be topographically constrained, so that individual flows might have been forced by adjacent high points to override earlier flow lobes. However, image resolution for these flows is ~80 m pixel^{-1} and direct comparison with those in the Elysium Planitia field is not possible. Although Viking Orbiter image resolution varies over the planet, it appears that segmented lava flows are rare on Mars and they could indicate unusual eruptions and/or emplacement conditions. For example, the flat preflow terrain, abundant permafrost, or pulsing of the flow could have produced the segmented flows (Mouginis-Mark and Tatsumura-Yoshioka, 1998). Other morphologic evidence also suggests that northwest Elysium Planitia has been affected by subsurface volatiles. Many of the graben have outflow deposits suggestive of surface water flow (Mouginis-Mark, 1985) and an area of chaos just north of the lava flows appears to have formed by collapse following the removal of ground ice (Carr and Schaber, 1977), perhaps leading to lahar deposition (Christiansen and Greeley, 1981; Christiansen, 1989). Alternately, this area has been interpreted as glacial outflow deposits associated with subice eruptions (see Chapter 3).

4.9. CONCLUSION

Despite more than three decades of exploration by spacecraft, many fundamental questions regarding Martian volcanism remain unanswered. The following is not a complete

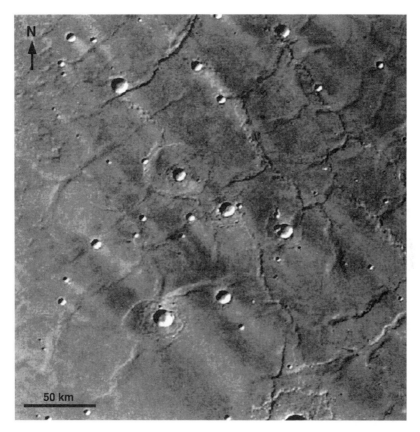

Figure 4.15. Hesperian-age ridged plains east of Tyrrhena Patera. Plains of similar morphology are among the most extensive probable volcanic areas on Mars. Image is 270 km wide, centered at 21.5°S latitude, 249°W longitude. (Viking Orbiter frames 391S51, 87A14–17.)

compilation, but includes some of the issues regarding the role of the Martian environment in controlling the styles of volcanism and the resulting volcanic features:

1. Do the changes in volcanic morphology and inferred styles of volcanism for the highland paterae reflect a change in Martian environment in which surface or near-surface volatiles were depleted (Greeley and Spudis, 1981), or do the changes represent magma evolution or other factors? How has the atmospheric pressure/density changed over time, and how would this affect eruptive styles and volcanic morphology?
2. Do the channels on some of the volcanoes, such as Hecates Tholus (Mouginis-Mark *et al.*, 1982), represent local climatic regimes that allowed precipitation; alternatively, could the channels represent a form of volcanism or flow emplacement not previously recognized?
3. Do the wrinkle ridges represent deformation of volcanic materials, or are they structural features that can form in rock units of any type? As discussed by Greeley and Spudis (1981), mare-type ridges are commonly used to infer marelike basalts on planetary surfaces by lunar analogy, but the criterion of the presence of such ridges is far from definitive.

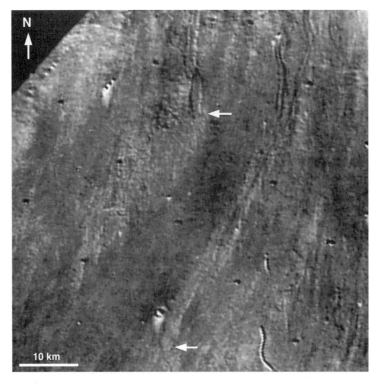

Figure 4.16. Amazonian-age volcanic plains in the Tharsis region of Mars south of Arsia Mons. Lobate fronts of individual digitate lava flows are indicated by arrows. Image is 60 km wide, centered at 17°S latitude, 115°W longitude. (Viking Orbiter frame 56A28.)

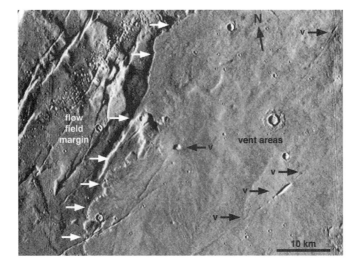

Figure 4.17. Volcanic centers and surrounding lava plains in the Tempe–Mareotis region of Mars. Similar plains occur in several areas surrounding the Tharsis region and appear to have formed by eruption of sheets of lava from numerous small vents (black arrows). (Viking Orbiter frame 627A27.)

4. When did the last eruptions occur on Mars? Although the SNC meteorites indicate that some eruptions may have occurred 120 Ma (McSween, 1994), the exact source of the SNCs on Mars is conjectural (Plescia, 1993) and they probably do not represent the latest eruptions. Mars Global Surveyor images will enable assignments of relative ages to flows and other volcanic features, but the absolute chronology will only be determined when samples are returned to Earth from specific volcanic areas.

5. What is the chemical diversity of lavas on Mars, and is there a correlation between volcano morphology and chemistry? Most investigators attribute differences among volcanic features to interaction of the magma with near-surface volatiles or other eruption parameters (Mouginis-Mark *et al.*, 1982; Crown and Greeley, 1993; Robinson *et al.*, 1993; Wilson and Head, 1994). Thermal Emission Spectrometer (Christensen *et al.*, 1992) data from Mars Global Surveyor, other remote sensing data, and measurements made by landers might resolve the issue of chemical diversity of Martian volcanic rocks.

6. What is the topography of Martian volcanoes and lava flows? The slope and thickness of individual flows are very poorly constrained and estimates of lava rheology (e.g., Zimbelman, 1985; Cattermole, 1987; Lopes and Kilburn, 1990; Glaze and Baloga, 1998) at the time of emplacement may be significantly in error. Similarly, the summit caldera volumes and edifice heights might also be in error, which has important implications for the volume of the magma chamber, the size of a caldera-forming event (Zuber and Mouginis-Mark, 1992), and the dynamics that initiated the eruption (Wilson and Head, 1994).

Some of these issues will be resolved with additional measurements and observations from current and ongoing missions, as outlined above. The international space exploration program has a focus on Mars in the late 1990s and well into the next century. Approved and planned missions have the potential to return a great wealth of data on the history of Mars and its evolution through time. Combined with missions that will return a series of samples from Mars, laboratory experiments, theoretical considerations, and study of relevant terrestrial analogues should ensure that the next decade sees a significant improvement in our understanding of Mars, its volcanic history, and the role of the environment in controlling volcanic processes.

ACKNOWLEDGMENTS

R. G. and S. A. F. were supported partly by the NASA Planetary Geology and Geophysics Program; J. R. Z. was supported partly by NASA grants NAGW-1390 and NAG5-4586. L. S. C. was supported by NASA grant NAG5-4309. D. A. C. was supported partly by NASA grants NAG5-4037 and NAG5-3642. We appreciate reviews provided by M. Chapman, J. A. Crisp, J. W. Head, and L. Wilson.

4.10. REFERENCES

Allen, C. C., Volcano–ice interactions on Mars, *J. Geophys. Res.*, *84*, 8048–8059, 1979.
Anderson, D. L., The internal composition of Mars, *J. Geophys. Res.*, *77*, 789–795, 1972.
Baker, V. R., G. Komatsu, R. G. Strom, V. C. Gulick, J. S. Kargel, and V. S. Kale, Ancient oceans, ice sheets, and the hydrologic cycle of Mars, *Nature*, *352*, 589–594, 1991.
Banin, A., B. C. Clark, and H. Wänke, Surface chemistry and mineralogy, in *Mars*, edited by H. H. Kieffer, B. M. Jakosky, C. W. Snyder, and M. S. Matthews, pp. 594–625, University of Arizona Press, Tucson, 1992.

Bell, J. F., Iron, sulfate, carbonate, and hydrated minerals on Mars, in *Mineral Spectroscopy: A Tribute to Roger G. Burns, Geochemical Society Special Publication 5*, edited by M. D. Dyar, C. McCammon, and M. W. Schaefer, pp. 359–380, 1996.

Blaney, D. L., and T. B. McCord, An observational search for carbonates on Mars, *J. Geophys. Res.*, *95*, 10159–10166, 1990.

Blaney, D. L., and T. B. McCord, Indications of sulfate minerals in the Martian soil from Earth-based spectroscopy, *J. Geophys. Res.*, *100*, 14433–14441, 1995.

Blasius, K. R., and J. A. Cutts, Topography of Martian central volcanoes, *Icarus*, *45*, 87–112, 1981.

Borgia, A., J. Burr, W. Montero, L. D. Morales, and G. E. Alvarado, Fault propagation folds induced by gravitational failure and slumping of the central Costa Rica volcanic range: Implications for large terrestrial and Martian volcanic edifices, *J. Geophys. Res.*, *95*, 14357–14382, 1990.

Brackenridge, G. R., H. E. Newsom, and V. R. Baker, Ancient hot springs on Mars: Origins and paleoenvironmental significance of small Martian valleys, *Geology*, *13*, 859–862, 1985.

Bridges, N. T., R. C. Anderson, J. A. Crisp, T. Economou, and R. Reid, Separating dust and rock APXS measurements based on multispectral data at the Pathfinder landing site, *EOS (Trans. Am. Geophys. Union)*, *78*, F402–F403, 1997.

Bruno, B. C., G. J. Taylor, S. K. Rowland, P. G. Lucey, and S. Self, Lava flows are fractals, *Geophys. Res. Lett.*, *19*, 305–308, 1992.

BVTP, *Basaltic Volcanism on the Terrestrial Planets*, Pergamon Press, New York, 1981.

Carr, M. H., Volcanism on Mars, *J. Geophys. Res.*, *78*, 4049–4062, 1973.

Carr, M. H., *The Surface of Mars*, 232 pp, Yale University Press, New Haven, 1981.

Carr, M. H., *Water on Mars*, 229 pp., Oxford University Press, London, 1996.

Carr, M. H., and F. C. Chuang, Martian drainage densities, *J. Geophys. Res.*, *102*, 9145–9152, 1997.

Carr, M. H., and R. Greeley, *Volcanic Features of Hawaii: A Basis for Comparison with Mars*, NASA SP-403, 211 pp., 1980.

Carr, M. H., and G. G. Schaber, Martian permafrost features, *J. Geophys. Res.*, *82*, 4039–4065, 1977.

Carr, M. H., and H. Wanke, Earth and Mars: Water inventories as clues to accretional histories, *Icarus*, *98*, 61–71, 1992.

Carr, M. H., R. Greeley, K. R. Blasius, J. E. Guest, and J. B. Murray, Some Martian volcanic features as viewed from the Viking orbiters, *J. Geophys. Res.*, *82*, 3985–4015, 1977.

Cattermole, P., Linear volcanic features at Alba Patera–Probable spatter ridges, *J. Geophys. Res.*, *91*, E159–E165, 1986.

Cattermole, P., Sequence, rheological properties, and effusion rates of volcanic flows at Alba Patera, Mars, *J. Geophys. Res.*, *92*, E553–E560, 1987.

Christensen, P. R., D. L. Anderson, S. C. Chase, R. N. Clark, H. H. Kieffer, M. C. Malin, J. C. Pearl, J. Carpenter, N. Bandiera, F. G. Brown, and S. Silverman, Thermal Emission Spectrometer Experiment: Mars Observer mission, *J. Geophys. Res.*, *97*, 7719–7734, 1992.

Christensen, P. R., D. L. Anderson, S. C. Chase, R. T. Blancy, R. N. Clark, B. J. Conrath, H. H. Kieffer, R. O. Kuzmin, M. C. Malin, J. C. Pearl, T. L. Roush, and M. D. Smith, Results from the Mars Global Surveyor Thermal Emission Spectrometer, *Science*, *279*, 1692–1698, 1998.

Christiansen, E. H., Lahars in the Elysium region of Mars, *Geology*, *17*, 203–206, 1989.

Christiansen, E. H., and R. Greeley, Megalahars(?) in the Elysium region, Mars, *Lunar Planet. Sci.*, *12*, 138–140, 1981.

Clemens, J. D., Water contents of silicic to intermediate magmas, *Lithos*, *17*, 273–287, 1984.

Clifford, S. M., A model for the hydrologic and climatic behavior of water on Mars, *J. Geophys. Res.*, *98*, 10973–11016, 1993.

Crown, D. A., and R. Greeley, Volcanic geology of Hadriaca Patera and the eastern Hellas region of Mars, *J. Geophys. Res.*, *98*, 3431–3451, 1993.

Crown, D. A., K. H. Price, and R. Greeley, Geologic evolution of the east rim of the Hellas basin, Mars, *Icarus*, *100*, 1–25, 1992.

Crumpler, L. S., J. W. Head, and J. C. Aubele, Magma chambers associated with calderas on Mars: Significance of long-term magma replenishment rates, *Lunar Planet. Sci.*, *26*, 305–306, 1996a.

Crumpler, L. S., J. W. Head, and J. C. Aubele, Calderas on Mars: Characteristics, structural evolution, and associated flank structures, in *Volcanic Instability on Earth and Other Planets*, Geological Society of London Special Publication, *110*, 307–347, 1996b.

Dreibus, G., and H. Wanke, Mars, a volatile-rich planet, *Meteoritics*, *20*, 367–381, 1985.

Dreibus, G., E. Jagoutz, and H. Wanke, Water in the Earth's mantle, *Geol. Geofiz.*, *38*, 269–275, 1997.

Edgett, K. S., Possible cinder cones near the summit of Pavonis Mons, Mars, *Lunar Planet. Sci. Conf.*, *21*, 311–312, 1990.

Edgett, K. S., Aeolian dunes as evidence for explosive volcanism in the Tharsis region of Mars, *Icarus*, *130*, 96–114, 1997.

Edgett, K. S., B. J. Butler, and J. R. Zimbelman, Geologic context of the Mars radar "Stealth" region in southwestern Tharsis, *J. Geophys. Res.*, *102*, 21545–21567, 1997.

Fagents, S. A., and L. Wilson, Numerical modeling of ejecta dispersal around the sites of volcanic explosions on multiple regression, *Icarus*, *123*, 284–295, 1996.

Fink, J. H., Surface folding and viscosity of rhyolite flows, *Geology*, *8*, 250–254, 1980.

Folkner, W. M., C. F. Yoder, D. N. Yuan, E. M. Standish, and R. A. Preston, Interior structure and seasonal mass redistribution of Mars from radio tracking of Mars Pathfinder, *Science*, *278*, 1749–1752, 1997.

Francis, P. W., and G. Wadge, The Olympus Mons aureole: Formation by gravitational spreading, *J. Geophys. Res.*, *88*, 8333–8344, 1983.

Francis, P. W., and C. A. Wood, Absence of silicic volcanism on Mars: Implications for crustal composition and volatile abundance, *J. Geophys. Res.*, *87*, 9881–9889, 1982.

Frey, H., and M. Jarosewich, Subkilometer Martian volcanoes: Properties and possible terrestrial analogs, *J. Geophys. Res.*, *87*, 9867–9879, 1982.

Frey, H., B. L. Lowry, and S. A. Chase, Pseudocraters on Mars, *J. Geophys. Res.*, *84*, 8075–8086, 1979.

Gill, J., *Orogenic Andesites and Plate Tectonics*, 390 pp., Springer-Verlag, Berlin, 1981.

Glaze, L. S., and S. M. Baloga, Dimensions of Puu Oo lava flows on Mars, *J. Geophys. Res.*, *103*, 13659–13666, 1998.

Greeley, R., Mariner 9 photographs of small volcanic structures on Mars, *Geol. Soc. Am.*, *1*, 175–180, 1973.

Greeley, R., Volcanic morphology, in *Volcanism of the Eastern Snake River Plain, Idaho: A Comparative Planetary Geology Guidebook*, NASA CR-154621, 1977.

Greeley, R., and D. A. Crown, Volcanic geology of Tyrrhena Patera, Mars, *J. Geophys. Res.*, *95*, 7133–7149, 1990.

Greeley, R., and J. E. Guest, Geologic map of the eastern equatorial region of Mars, *U.S. Geol. Surv. Misc. Invest. Ser. Map, I-1802B*, 1987.

Greeley, R., and P. D. Spudis, Volcanism in the cratered terrain hemisphere of Mars, *J. Geophys. Res.*, *5*, 453–455, 1978.

Greeley, R., and P. D. Spudis, Volcanism on Mars, *Rev. Geophys.*, *19*, 13–41, 1981.

Greeley, R., and B. D. Schneid, Magma generation on Mars: Amounts, rates, and comparisons with Earth, Moon, and Venus, *Science*, *254*, 996–998, 1991.

Gregg, T. K. P., and J. H. Fink, Variations in lava flow width related to effusion rates through laboratory simulations, *Lunar Planet. Sci. Conf.*, *28*, 461–462, 1997.

Gregg, T. K. P., D. A. Crown, and R. Greeley, Geologic map of MTM quadrangle -20252, Tyrrhena Patera region of Mars, *U.S. Geol. Surv. Misc. Invest. Ser. Map, I-2556*, 1998.

Gudmundsson, A., Formation of collapse calderas, *Geology*, *16*, 808–810, 1988.

Gulick, V. C., and V. R. Baker, Origin and evolution of valleys on Martian volcanoes, *J. Geophys. Res.*, *95*, 14325–14344, 1990.

Harris, S. A., The aureole of Olympus Mons, *J. Geophys. Res.*, *82*, 3099–3107, 1977.

Harvey, R. P., and H. Y. McSween, The parent magma of the Nakhlite meteorites: Clues from melt inclusions, *Earth Planet. Sci. Lett.*, *82*, 3099–3107, 1977.

Head, J. W., and L. Wilson, Basaltic pyroclastic eruptions: Influence of gas-release patterns and volume fluxes on fountain structure, and the formation of cinder cones, spatter cones, rootless flows, lava ponds and lava flows, *J. Volcanol. Geotherm. Res.*, *37*, 261–271, 1989.

Head, J. W., M. Settle, and C. A. Wood, Origin of the Olympus Mons escarpment by erosion of pre-volcano substrate, *Nature*, *263*, 667–668, 1976.

Hodges, C. A., and H. J. Moore, The subglacial birth of Olympus Mons and its aureoles, *J. Geophys. Res.*, *84*, 8061–8074, 1979.

Hodges, C. A., and H. J. Moore, Atlas of volcanic landforms on Mars, *U.S. Geol. Surv. Prof. Pap.*, *1534*, 61–70, 1994.

Hulme, G., The dependence of the rheological properties and effusion rate of an Olympus Mons lava, *Icarus*, *27*, 207–213, 1976.

Hviid, S. F., M. B. Madsen, H. P. Gunnlaugsson, W. Goetz, J. M. Knudsen, R. B. Hargraves, P. Smith, D. Britt, A. R. Dinesen, C. T. Mogensen, M. Olsen, C. T. Pederson, and L. Vistisen, Magnetic properties experiments on the Mars Pathfinder lander: Preliminary results, *Science*, *278*, 1768–1770, 1997.

Jagoutz, E., H. Palme, H. Baddenhausen, K. Blum, M. Cendales, G. Dreibus, B. Spettel, V. Lorenz, and H. Wanke, The abundance of major, minor, and trace elements in the Earth's mantle as derived from primitive ultramafic nodules, *Proc. Lunar Planet. Sci.*, *10*, 2031–2050, 1979.

Johnson, M. C., M. J. Rutherford, and P. C. Hess, Chassigny petrogenesis: Melt compositions, intensive parameters, and water contents of Martian (?) magmas, *Geochem. Cosmochim. Acta*, *55*, 349–366, 1991.

Johnson, M. C., A. T. Anderson, Jr., and M. J. Rutherford, Pre-eruptive volatile contents of magmas, in *Volatiles in Magmas, Reviews in Mineralogy*, edited by M. R. Carroll and J. R. Holloway, Mineralogical Society of America, Washington, *30*, 281–330, 1994.

Kieffer, H. H., T. Z. Martin, A. R. Peterfreund, B. M. Jakosky, E. D. Miner, and F. D. Palluconi, Thermal and albedo mapping of Mars during the Viking primary mission, *J. Geophys. Res.*, *82*, 4249–4291, 1977.

King, E. A., Geologic map of the Mare Tyrrhenum quadrangle of Mars, *U.S. Geol. Surv. Misc. Invest. Ser. Map*, I-1073, 1978.

King, J. S., and J. R. Riehle, A proposed origin for the Olympus Mons escarpment, *Icarus*, *23*, 300–317, 1974.

Longhi, J., and V. Pan, The parent magmas of the SNC meteorites, *Proc. Lunar Planet. Sci.*, *19*, 451–464, 1989.

Longhi, J., E. Knittle, J. R. Holloway, and H. Wanke, The bulk composition, mineralogy and internal structure of Mars, in *Mars*, edited by H. H. Kieffer, B. M. Jakosky, C. W. Snyder, and M. S. Matthews, pp. 184–208, University of Arizona Press, Tucson, 1992.

Lopes, R. M. C., and C. R. J. Kilburn, Emplacement of lava flow fields:Application of terrestrial studies to Alba Patera, Mars, *J. Geophys. Res.*, *95*, 14383–14397, 1990.

Lopes, R. M. C., J. E. Guest, K. H. Hiller, and G. P. O. Neukum, Further evidence for mass movement origin of the Olympus Mons aureole, *J. Geophys. Res.*, *87*, 9917–9928, 1982.

Malin, M. C., Comparison of volcanic features of Elysium (Mars) and Tibesti (Earth), *Geol. Soc. Am. Bull.*, *88*, 908–919, 1977.

Malin, M. C., Lengths of Hawaiian lava flows, *Geology*, *8*, 306–308, 1980.

Masursky, H., R. M. Batson, J. F. McCauley, L. A. Soderblom, R. L. Wildey, M. H. Carr, D. J. Milton, D. E. Wilhelms, B. A. Smith, T. B. Kirby, J. C. Robinson, C. B. Levoy, G. A. Briggs, T. C. Duxbury, C. H. Acton, B. C. Murray, J. A. Cutts, R. P. Sharp, S. Smith, R. B. Leighton, C. Sagan, J. Veverka, M. Noland, G. DeVaucoulerus, M. Davies, and A. T. Young, Mariner 9 television reconnaissance of Mars and its satellites: Preliminary results, *Science*, *175*, 294–304, 1972.

McCauley, J. F., M. H. Carr, J. A. Cutts, W. K. Hartmann, H. Masursky, D. J. Milton, R. P. Sharp, and D. E. Wilhelms, Preliminary Mariner 9 report on the geology of Mars, *Icarus*, *17*, 289–327, 1972.

McCoy, T. J., G. J. Taylor, and K. Keil, Zagami: Product of a two-stage magmatic history, *Geochim. Cosmochim. Acta*, *56*, 3571–3582, 1992.

McGetchin, T. R., and J. R. Smyth, The mantle of Mars: Some possible geological implications of its high density, *Icarus*, *34*, 512–536, 1978.

McGuire, W. J., Volcano instability: A review of contemporary themes, in *Volcano Instability on Earth and Other Planets*, pp. 1–23, Geol. Soc. London Spec. Publ., 1996.

McSween, H. Y., What we have learned about Mars from SNC meteorites, *Meteoritics*, *29*, 757–779, 1994.

McSween, H. Y., and R. P. Harvey, Outgassed water on Mars: Constraints from melt inclusions in SNC meteorites, *Science*, *259*, 1890–1892, 1993.

McSween, H. Y., Jr., S. L. Murchie, J. A. Crisp, N. T. Bridges, R. C. Anderson, J. F. Bell III, D. T. Britt, J. Bruckner, G. Dreibus, T. Economou, A. Ghosh, M. P. Golombek, J. P. Greenwood, J. R. Johnson, H. J. Moore, R. V. Morris, T. J. Parker, R. Rieder, R. Singer, and H. Wanke, Chemical, multispectral, and textural constraints on the composition and origin of rocks at the Mars Pathfinder landing site, *J. Geophys. Res.*, *104*, 8679–8715, 1999.

Moore, H. J., D. W. G. Arthur, and G. G. Schaber, Yield strengths of flows on the Earth, Mars, and Moon, *Proc. Lunar Planet. Sci.*, *9*, 3351–3378, 1978.

Moore, J. G., D. A. Clauge, R. T. Holcomb, P. W. Lipman, W. R. Normark, and M. E. Torresan, Prodigious submarine landslides on the Hawaiian ridge, *J. Geophys. Res.*, *94*, 17465–17484, 1989.

Morgan, J. W., and E. Anders, Chemical composition of Mars, *Geochim. Cosmochim. Acta*, *43*, 1601–1610, 1979.

Morris, E. C., Aureole deposits of the Martian volcano Olympus Mons, *J. Geophys. Res.*, *87*, 1164–1178, 1982.

Mouginis-Mark, P. J., Volcano/ground ice interactions in Elysium Planitia, Mars, *Icarus*, *64*, 265–284, 1985.

Mouginis-Mark, P. J., The influence of oceans on Martian volcanism, *Lunar Planet. Sci.*, *24*, 1021–1022, 1993.

Mouginis-Mark, P. J., and M. S. Robinson, Evolution of the Olympus Mons caldera, Mars, *Bull. Volcanol.*, *54*, 347–360, 1992.

Mouginis-Mark, P. J., and M. Tatsumura-Yoshioka, The long lava flows of Elysium Planitia, Mars, *J. Geophys. Res.*, *103*, 19389–19400, 1998.

Mouginis-Mark, P. J., L. Wilson, and J. W. Head, Explosive volcanism on Hecates Tholus, Mars; Investigation of eruption conditions, *J. Geophys. Res.*, *87*, 9890–9904, 1982.

Mouginis-Mark, P. J., L. Wilson, J. W. Head, H. Brown-Steven, J. L. Hall, and K. D. Sullivan, Elysium Planitia, Mars: Regional geology, volcanology, and evidence for volcano–ground ice interactions, *Earth Moon Planets*, *30*, 149–173, 1984.

Mouginis-Mark, P. J., L. Wilson, and J. R. Zimbelman, Polygenic eruptions on Alba Patera, Mars, *Bull. Volcanol.*, *50*, 361–379, 1988.

Muhleman, D. O., A. W. Grossman, B. J. Butler, and M. A. Slade, Radar images of Mars, *Science*, *253*, 1508 1513, 1991.

Neukum, G., and K. Hiller, Martian ages, *J. Geophys. Res.*, *86*, 3097–3121, 1981.

Owen, T., The composition and early history of the atmosphere of Mars, in *Mars*, edited by H. H. Kieffer, B. M. Jakosky, C. W. Snyder, and M. S. Matthews, pp. 818–834, University of Arizona Press, Tucson, 1992.

Parker, T. J., D. S. Gorsline, R. S. Saunders, D. C. Pieri, and D. M. Schneeberger, Coastal geomorphology of the Martian northern plains, *J. Geophys. Res.*, *98*, 11061–11078, 1993.

Peterson, J. E., Geologic map of the Noachis quadrangle of Mars, *U.S. Geol. Surv. Misc. Invest. Ser. Map*, *I-910*, 1977.

Peterson, J. E., Volcanism in the Noachis–Hellas region of Mars, 2, *Proc. Lunar Planet. Sci.*, *9*, 3411–3432, 1978.

Plescia, J. B., The Tempe volcanic province of Mars and comparisons with the Snake River Plains of Idaho, *Icarus*, *45*, 586–601, 1981.

Plescia, J. B., An assessment of volatile release from recent volcanism in Elysium, Mars, *Icarus*, *104*, 20–32, 1993.

Plescia, J. B., and R. S. Saunders, The chronology of the Martian volcanoes, *Proc. Lunar Planet. Sci.*, *10*, 2841–2859, 1979.

Pollack, J. B., T. Roush, F. Witteborn, J. Bregman, D. Wooden, C. Stoker, O. B. Toon, D. Rank, B. Dalton, and R. Freedman, Thermal emission spectra of Mars (5.4–10.5 µm): Evidence for sulfates, carbonates, and hydrates, *J. Geophys. Res.*, *95*, 14595–14628, 1990.

Popp, R. K., D. Virgo, and W. Phillips-Michael, H deficiency in kaersutitic amphiboles: Experimental verification, *Am. Mineral.*, *80*, 1347–1350, 1995.

Potter, D. B., Geologic map of the Hellas quadrangle of Mars, *U.S. Geol. Surv. Misc. Invest. Ser. Map*, *I-941*, 1976.

Reimers, P. E., and P. D. Komar, Evidence for explosive volcanic density currents on certain Martian volcanoes, *Icarus*, *39*, 88–110, 1979.

Rieder, R., T. Economou, H. Wänke, A. Turkevich, J. Crisp, J. Brückner, G. Dreibus, and H. Y. McSween, Jr., The chemical composition of Martian soil and rocks returned by the mobile alpha proton X-ray spectrometer: Preliminary results from the X-ray mode, *Science*, *278*, 1771–1774, 1997.

Robinson, M. S., and S. K. Rowland, Evidence for large scale sector collapse at Tharsis Tholus, in *Conf. Volcano Instability Earth and Other Planets*, p. 44, Geological Society of London, 1994.

Robinson, M. S., P. J. Mouginis-Mark, J. R. Zimbelman, S. S. C. Wu, K. K. Ablin, and A. E. Howtington-Kraus, Chronology, eruption duration, and atmospheric contribution of the Martian volcano Apollinaris Patera, *Icarus*, *104*, 301–323, 1993.

Rubin, A. M., and D. D. Pollard, Origins of blade-like dikes in volcanic rift zones, in *Volcanism in Hawaii*, edited by R. W. Decker, T. L. Wright, and P. H. Stauffer, pp. 1449–1470, *U.S. Geol. Surv. Prof. Pap.*, *1350*, 1987.

Schneeberger, D. M., and D. C. Pieri, Geomorphology and stratigraphy of Alba Patera, Mars, *J. Geophys. Res.*, *96*, 1907–1930, 1991.

Schubert, G., S. C. Solomon, D. L. Turcotte, M. J. Drake, and N. H. Sleep, Origin and thermal evolution of Mars, in *Mars*, edited by H. H. Kieffer, B. M. Jakosky, C. W. Snyder, and M. S. Matthews, pp. 147–183, University of Arizona Press, Tucson, 1992.

Scott, D. H., Volcanoes and volcanic provinces: Martian western hemisphere, *J. Geophys. Res.*, *87*, 9839–9851, 1982.

Scott, D. H., and K. L. Tanaka, Mars: A large highland volcanic province revealed by Viking images, *Proc. Lunar Planet. Sci.*, *12*, 1449–1458, 1981.

Scott, D. H., and K. L. Tanaka, Geologic map of the western equatorial region of Mars, *U.S. Geol. Surv. Misc. Invest. Map*, *I-1802-A*, 1986.

Smith, D. E., M. T. Zuber, H. V. Frey, J. B. Garvin, J. W. Head, D. O. Muhleman, H. J. Zwally, W. B. Banerdt, T. C. Duxbury, G. H. Pettengill, R. O. Phillips, and S. C. Solomon, Topography of the northern hemisphere of Mars from the Mars Orbiter Laser Altimeter, *Science*, *279*, 1686–1692, 1998.

Smith, P. H., J. F. Bell III, N. T. Bridges, D. T. Britt, L. Gaddis, R. Greeley, H. Kueller, K. E. Herkenhoff, R. Jaumann, J. R. Johnson, R. L. Kirk, M. Lemmon, J. N. Maki, M. C. Malin, S. L. Murchie, J. Oberst, T. J. Parker, R. J. Reid, R. Sablotny, L. A. Soderblom, C. Stoker, R. Sullivan, N. Thomas, M. G. Tomasko, W. Ward, and E. Wegryn, Results from the Mars Pathfinder camera, *Science*, *278*, 1758–1765, 1997.

Soderblom, L. A., Historical variations in the density and distribution of impacting debris in the inner solar system: Evidence from planetary imaging, in *Impact and Explosion Cratering*, edited by D. J. Roddy, R. O. Pepin, and R. B. Merrill, pp. 629–633, Pergamon Press, New York, 1977.

Soderblom, L. A., The composition and mineralogy of the Martian surface from spectroscopic observations: 0.3 mm to 50 mm, in *Mars*, edited by H. H. Kieffer, B. J. Jakosky, C. W. Snyder, and M. S. Matthews, pp. 557–593, University of Arizona Press, Tucson, 1992.

Sparks, R. S. J., The dynamics of bubble generation and growth in magmas: A review and analysis, *J. Volcanol. Geotherm. Res.*, 3, 1–37, 1978.

Squyres, S. W., D. E. Wilhelms, and A. C. Moosman, Large-scale volcano–ground ice interactions on Mars, *Icarus*, 70, 385–408, 1987.

Tanaka, K. L., Ice-lubricated gravity spreading of the Olympus Mons aureole deposit, *Icarus*, 62, 191–206, 1985.

Tanaka, K. L., and D. H. Scott, Geologic map of the polar regions of Mars, *U.S. Geol. Surv. Misc. Invest. Ser. Map, I-1802-C*, 1987.

Tanaka, K. L., N. K. Isbell, D. H. Scott, R. Greeley, and J. E. Guest, The resurfacing history of Mars: A synthesis of digitized, Viking-based geology, *Proc. Lunar Planet. Sci.*, 18, 665–678, 1988.

Treiman, A. H., The parent magma of the Nakhla (SNC) meteorite, inferred from magmatic inclusions, *Geochim. Cosmochim. Acta*, 57, 4753–4767, 1993.

Walker, G. P. L., Lengths of lava flows, *Philos. Trans. R. Soc. London Ser. A*, 274, 107–118, 1973.

Weidenschilling, S. J., Accretion of the terrestrial planets. II, *Icarus*, 27, 161–170, 1976.

Whitford-Stark, J. L., Factors influencing the morphology of volcanic landforms: An Earth–Moon comparison, *Earth Sci. Rev.*, 18, 109–168, 1982.

Wilson, L., and J. W. Head, A comparison of volcanic eruption processes on Earth, Moon, Mars, Io, and Venus, *Nature*, 302, 663–669, 1983.

Wilson, L., and J. W. Head, Factors controlling the structures of magma chambers in basaltic volcanoes, *Lunar Planet. Sci.*, 21, 1343–1344, 1990.

Wilson, L., and J. W. Head, Mars: Review and analysis of volcanic eruption theory and relationships to observed landforms, *Rev. Geophys.*, 32, 221–263, 1994.

Wilson, L., S. J. Sparks, T. C. Huang, and N. D. Watkins, The control of volcanic column heights by eruption energetics and dynamics, *J. Geophys. Res.*, 83, 1829–1836, 1978.

Wilson, L., J. W. Head, and P. J. Mouginis-Mark, Theoretical analysis of Martian volcanic eruption mechanisms, in *The Planet Mars, ESA SP-185*, pp. 107–113, Leeds, UK, 1982.

Wise, D. U., Geologic map of the Arcadia quadrangle of Mars, *U.S. Geol. Surv. Misc. Invest. Ser. Map, I-1154*, 1979.

Wood, B. J., A. Pawley, and D. R. Frost, Water and carbon in the Earth's mantle, *Philos. Trans. R. Soc. London Ser. A*, 354, 1495–1511, 1996.

Wood, C. A., Monogenetic volcanoes of the terrestrial planets, *Proc. Lunar Planet. Sci. Conf.*, 10, 2815–2840, 1979.

Wood, C. A., Calderas: A planetary perspective, *J. Geophys. Res.*, 89, 8391–8406, 1984.

Zimbelman, J. R., Estimates of rheologic properties for flows on the Martian volcano Ascraeus Mons, Proc. Lunar Planet. Sci. Conf. 16, in *J. Geophys. Res.*, 90 (suppl.), D157–D162, 1985.

Zuber, M. T., and P. J. Mouginis-Mark, Caldera subsidence and magma chamber depth of the Olympus Mons volcano, Mars, *J. Geophys. Res.*, 97, 18295–18307, 1992.

5

Volcanism on Earth's Seafloor and Venus

Eric B. Grosfils, Jayne Aubele, Larry Crumpler,
Tracy K. P. Gregg, and Susan Sakimoto

5.1. INTRODUCTION

The surface of Venus, obscured by dense cloud cover, is similar in many ways to the seafloor that lies hidden beneath the deep waters of Earth's oceans. Although both are difficult to observe, decades of research indicate that each surface is dominated primarily by basaltic volcanism. This is not surprising as Earth and Venus are similar in size, bulk density, and position in the solar system, and the probability of similar elemental abundances and internal heat sources implies corresponding similarity between their interior melting, magma production, and surface volcanism. Even though Earth's seafloor and Venus are dissimilar in many ways, both environments are characterized by significantly elevated pressure at the surface resulting, respectively, from the burden imposed by the overlying ocean water and the weight of the dense atmosphere. This provides volcanologists with an excellent opportunity to examine how elevated surface pressure affects the development and behavior of volcanic systems.

We begin this chapter with an overview of the types of data collected to date that reveal the surfaces of the seafloor and Venus, then summarize what is known about the environmental conditions at or near each surface. In addition, since comprehensive reviews are presented elsewhere (e.g., Venus: Head *et al.*, 1992, and Crumpler *et al.*, 1997; seafloor: Fornari and Embley, 1995, and Chadwick and Perfit, 1998), we provide only a brief overview of the volcanic styles observed in each setting so as to establish a context within which to consider the behavior of several specific volcanic processes and features. The bulk of the chapter focuses on four "case study" examinations of how the high-pressure environments of Venus and the seafloor specifically affect intrusive processes (1: magma stalling related to neutral buoyancy), effusive eruptions (2: formation of small to intermediate volcanoes; 3: emplacement of large larva flows), and explosive events (4: pyroclastic eruptions). These detailed presentations serve to illustrate the similarity between key volcanic styles observed on Venus and Earth's seafloor, and help demonstrate the role played by the near-surface, high-pressure environment present in both settings.

Environmental Effects on Volcanic Eruptions: From Deep Oceans to Deep Space.
Edited by Zimbelman and Gregg, Kluwer Academic/Plenum Publishers, New York, 2000.

5.1.1. Venus

Available Data. Current knowledge about Venus has resulted from Earth-based radar studies, planetary flyby missions, atmospheric entry probes (including balloons), landers, and global radar reconnaissance of the surface by orbiter spacecraft. In the 1960s modern Earth-based radar observatories, chiefly Goldstone (Goldstein *et al.*, 1978) and Arecibo (Campbell *et al.*, 1976), provided radar images of the surface at regional scales and resolutions of tens to several kilometers showing radar-bright and -dark areas up to several thousand kilometers across. In addition to mapping the large-scale radar patterns on the surface, studies using the radar data provided the first reliable estimates of the rotation rate (243 days, retrograde). Earth-based observations continued to be a source of the highest-resolution image data until the late 1980s. Radar signals of different wavelength and incidence angle show differences in scattering properties that relate to differences in the radar reflectivity and roughness of the surface at each wavelength and incidence angle. As a result these early data, acquired with radar wavelengths and polarizations that differ from later spacecraft data, still provide unique contributions to our understanding of Venus. Useful high-resolution Earth-based data are limited to one hemisphere, however, because of the peculiarities of the orbital period of Venus in which the same side of Venus points toward Earth at each inferior conjunction (closest approach to Earth). As a result, Earth-based radar image data are confined to the hemisphere centered on the prime meridian of Venus; the prime meridian was defined in part from this phenomenon. Earth-based data were ultimately acquired at image resolutions as high as 2 km pixel^{-1} (Campbell *et al.*, 1991) over about one-quarter of the surface.

The U.S. Pioneer-Venus mission in 1978 built on the early Earth-based radar results and initially obtained a global map of the surface altimetry at a resolution of 100 km (Pettengill *et al.*, 1980). In addition to showing that the topography on Venus is relatively flat compared with that of Earth, the Pioneer-Venus altimetry established the principal large-scale surface characteristics, which include: (1) a unimodal global surface hypsometry (in contrast to the bimodal hypometry of Earth), (2) the presence of distinct radar-bright highland areas from 1 to 2 km above the mean surface altitude, and (3) the presence of a network of deep linear valleys along upland regions comparable to rift valleys on Earth, with linear mountain belts bordering at least one of the highlands. Although a few continentlike highland regions rise from 1 to several kilometers above the mean planetary radius, they are spatially scattered and cover 15% of the surface of Venus, or roughly half the surface area covered by continents on Earth. The Pioneer-Venus data also include reflectivity and emissivity characteristics, as well as a band of radar imaging with a resolution of approximately 10 km along the equatorial region.

The Russian Venera 15/16 missions in 1983 first used synthetic aperture radar to acquire information on the distribution of landforms over the northern 25% of the surface at a resolution of 2 km (Barsukov *et al.*, 1986). These data were obtained at a lower radar incidence angle than later Magellan images and provide better insight into small-scale relief, but are relatively insensitive to the variations in radar reflectivity and roughness compared with Magellan data. Venera 15/16 data were used to established the presence of numerous volcanoes ranging from a few kilometers across to volcanoes as large as several hundred kilometers, radiating patterns of fractures, and circular features [coronae, features defined primarily by an annulus of concentric fractures and ridges up to several hundred kilometers across (Barsukov *et al.*, 1986; Stofan *et al.*, 1997)]. The Russian Venera lander missions also provided detailed optical images (Figure 5.1) and chemistry for a few sites on Venus.

Figure 5.1. Optical images of the surface of Venus as seen by the Venera 13 lander. Chemical analysis of the surrounding materials yielded basaltic chemical compositions.

Radar technology and spacecraft technology paced each other in resolution with alternately Earth-based and then spacecraft-based observations providing the highest-resolution information during the 1970s through the late 1980s. This process culminated with the Magellan mission, which obtained global (98%) synthetic aperture image, emissivity, reflectivity, and altimetry data at resolutions as high as 120 m wide (image mode). Imaging was done on successive orbits in swaths several thousand kilometers long and 20 km wide. The swaths or strips were mosaicked to produce images of larger areas much like mosaics of sonar image strips used in seafloor studies. Strangely, the analogy to Earth's seafloor extends to the overall similarity in necessary image characteristics between side-scanning sonar imaging and side-looking synthetic aperture radar imaging.

Environmental Conditions at and Near the Surface. The surface environment of Venus is dominated by three fundamental characteristics: high surface pressure, a dense atmosphere of carbon dioxide, and a high surface temperature. These facts, initially deduced from Earth-based telescopic data, were confirmed by spacecraft observations (Table 5.1) made by U.S. and Soviet flybys, atmospheric entry probes, and eight Soviet Venera landers (Zahn *et al.*, 1983). Altogether there have been 18 short-lived entry and landing probes that have made direct measurements of the surface environment. The Russian Venera landers 4 through 14, the landers and balloons of Vega 1 and 2, and the four U.S. Pioneer-Venus entry probes all sampled and measured the atmosphere at several entry points and to variable depths. Pressures and temperatures like those found at the Earth's surface occur at 60 km altitude on Venus, far above the highest peaks. The high density and thermal opacity of the atmosphere result in a rapid increase in temperature and pressure with depth: The temperature increases rapidly from 120 K at 100 km altitude, rises to 273 K at 60 km altitude, and increases nearly linearly thereafter to 733 K at the mean surface elevation (see below). The ambient atmospheric pressure correspondingly increases from an Earth surface-like 0.1 MPa at 60 km to 4.7 MPa at 10 km and rises thereafter to 9.2 MPa at the mean surface altitude. The surface pressure on Venus is therefore similar to that encountered on Earth at a depth of 1 km in the oceans (Figure 5.2), and the surface temperature is only 400 to 500°C less than the liquidus for rhyolite.

Measurements of the surface rock and soil chemical composition were made by Venera and Vega lander instruments (X-ray fluorescence and gamma-ray spectrometers) at several points (Zahn *et al.*, 1983). In general, the measured samples had basaltic affinities. Analyses

Table 5.1. Chronologic Summary of Spacecraft Missions to Venus

Spacecraft	Country	Arrival	Type	Observations made
Mariner 2	USA	14-Dec-62	flyby	surf. temp., magnetic field
Venera 4	USSR	18-Oct-67	entry probe	atm. to 25 km altitude
Mariner 5	USA	19-Oct-67	flyby	surf. temp., magnetic field
Venera 5	USSR	16-May-69	entry probe	atm. to 12 km altitude
Venera 6	USSR	17-May-69	entry probe	atm. to 16 km altitude
Venera 7	USSR	15-Dec-70	lander	atm. to surface, temp./pressure
Venera 8	USSR	22-Jul-72	lander	atm., light, composition
Mariner 10	USA	5-Feb-74	flyby	cloud tops, magnetic field
Venera 9	USSR	25-Oct-75	lander	atm., images, surf. comp.
Venera 10	USSR	25-Oct-75	lander	atm., surf. images
Pioneer-Venus	USA	4-Dec-78	orbiter	global topo., 100 km res.
Pioneer-Venus	USA	9-Dec-78	entry probe	atm. to surface
Venera 11	USSR	21-Dec-78	lander	atm. comp.
Venera 12	USSR	25-Dec-78	lander	atm. comp.
Venera 13	USSR	1-Mar-82	lander	surf. color images/comp.
Venera 14	USSR	5-Mar-82	lander	surf. color images/comp.
Venera 15	USSR	Oct-83	orbiter	radar, 2 km res.
Venera 16	USSR	Oct-83	orbiter	radar, 2 km res.
Vega 1	USSR	15-Dec-84	lander/balloon	surf. comp., atm.
Vega 2	USSR	21-Dec-94	lander/balloon	surf. comp., atm.
Galileo	USA	10-Feb-90	flyby	cloud top images
Magellan	USA	10-Aug-90	orbiter	radar, 120 m res.

at Venera 9, 10, 14, and Vega 1 and 2 sites are comparable to the tholeiitic rocks common on Earth's seafloor, while analyses at Venera 8 and 13 suggest somewhat more alkalic compositions at these two sites.

The mean surface elevation of Venus is defined as the mean planetary radius (MPR), or 6051.84 km (Ford and Pettengill, 1992). Over 80% of the surface lies within 1 km of this radius. The maximum deviation of the surface from this elevation is 3 km below MPR in the deepest rift valley and nearly 12 km above MPR on the highest summit of Maxwell Montes. Depending on where an eruption occurs, atmospheric pressure and temperature can differ substantially, thus potentially influencing the exsolution of gases and the rate of heat transfer from cooling lavas and pyroclasts. Volcanic vents on Venus are, in general, slightly elevated with respect to MPR: The mean summit elevation of large volcanoes is 1.5 km above MPR. Similarly, many volcanic edifices occur within upland elevations (2 to 4 km) (Keddie and Head, 1994), particularly in association with the Beta–Atla–Themis region of higher-than-average volcanic center abundance (Crumpler *et al.*, 1993) and "volcanic rises" such as Atla Regio. The ambient conditions from 1.5 to 4 km above MPR range from 8.1 to 7.1 MPa and 715 to 700 K.

Because of the high surface temperature, extremely dry surface, inferred low silica compositions, and relatively flat topography on Venus, the question of the long-term strength and rock creep rates has arisen repeatedly (Weertman, 1979; McGill *et al.*, 1983), particularly as regards the lifetime of the few steep and high mountain belts (Smrekar and Solomon, 1992). The primary limit on these studies thus far is that laboratory measurements of presumably "dry" basalt and diabase (Mackwell *et al.*, 1998) with which creep rate estimates are made may still not be as dry as the rocks on Venus. Assumptions about the general mineralogy of the mantle and upper crust of Venus have been used to estimate the relatively

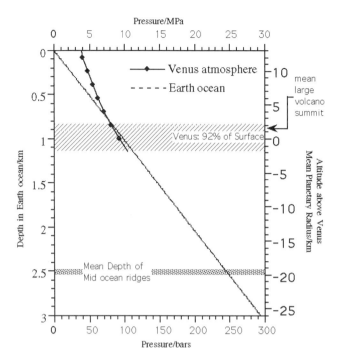

Figure 5.2. Comparison of the pressure with depth in Earth's oceans and Venus's atmosphere. Atmospheric pressure at the surface of Venus is equivalent to approximately 1 km depth in the sea.

strength of the crust and mantle under strain over geologic time scales (Zuber, 1987; Bindschadler and Parmentier, 1990). Like the Earth, Venus is predicted to have a brittle upper crust and ductile lower crust. The estimated ductility of the lower crust may be low because of the lack of internal water (Banerdt *et al.*, 1997) such that the overall crustal strength may be comparable to the oceanic crust on Earth. In addition to the uncertainties about the influence of water, regional and global geothermal gradients are not known and can only be approximated.

Finally, within the altitude range between 70 and 40 km the main cloud decks of sulfuric acid droplets occur. This is far above the highest points on the surface. As a result, the presence of swirling mist of sulfur should not be expected at any volcanic summits. Nonetheless, Vega surface composition measurements were rich in sulfur, illustrating the likely importance of sulfur in the surface chemistry. An additional observation relevant to the chemistry of the surface is the fact that mountainous regions more than 5 km above MPR have high radar reflectivity. The dependence of the reflectivity on altitude, regardless of the underlying terrain characteristics, suggests some form of vapor transport of a volatile and radar reflective element with the hot plains to the relatively "cool" mountaintops (Wood, 1997).

Overview of Volcanic Styles. Venus has perhaps more volcanoes and volcanic centers than any known planet other than Earth. Volcanoes are widely distributed over the surface and are morphologically diverse. The differences in morphology are significant in that they may be related to variations in eruptive style, vent morphology, and magma body emplacement depths, as well as to possible variations in composition and gas content of the magmas (e.g., Head *et al.*, 1992; Crumpler *et al.*, 1997).

Large, voluminous flow fields (akin to flood basalt deposits on Earth in size; see Section 2.6) are common, and massive eruptions of lava have flooded the vast lowland areas resulting in extensive plains with very low relief. Sinuous channels occur within the lowlands (Figure 5.3) and are interpreted, with reservation, to be lava channels that fed the extensive sheets of lava that cover the lowlands (Baker *et al.*, 1992). The reservation arises from their great length (up to thousands of kilometers), specifically whether silicate melts can remain fluid for the amount of time necessary to form the observed channels (Gregg and Greeley, 1993).

Volcanic edifices on Venus have low relief (Schaber, 1991; Keddie and Head, 1994) compared with large volcanoes on other terrestrial planets. The largest volcanoes are several hundred kilometers in diameter, yet rarely exceed a few kilometers in relief (Figure 5.4). The simplest of the large volcanoes are little more than radial arrays of digitate lava flows that together define an edifice with several hundred meters of relief. When large volcanoes are sometimes associated with stellate fracture patterns or coronae, in some instances they clearly postdate these features while in others the volcano is the older feature. A small percentage of large volcanoes are the site of subsequent calderalike collapse and others are surmounted by steep domes and cones. Both of these attributes imply that large volcanoes are frequently associated with relatively shallow magma bodies. These bodies either may undergo deflation through eruption or lateral intrusion, producing collapses, or may undergo chemical or volumetric evolution leading to secular changes in the details of late eruption styles.

Somewhat less understood are the domical to rounded volcanic edifices on Venus (a.k.a. "pancake" or "steep-sided" domes). Domical and flat-topped, steep-sided domes range up to 60 or more kilometers in diameter. The principal and most anomalous characteristics of many of these are their relative circularity, frequent presence of polygonal patterns of fractures on their summit suggestive of deformation through wholesale inflation and deflation of the edifice, and absence of evidence for individual lava flows. Together the latter two characteristics may imply that the edifices were erupted as a single endogenous mass (McKenzie *et al.*, 1992a) like that associated with viscous domes on Earth (see Chapter 2). Some steep-sided domes appear to have slumped at one or more points along their margins (Bulmer and Guest, 1996) (Figure 5.5). In the simplest context, the pancake domes have been interpreted as indicators of both relatively primitive silica-poor lavas, in the case of large volcanoes with low relief, and more evolved intermediate to silica-rich compositions in the case of the more domical intermediate-size volcanoes. In practice, morphology is at best a poor constraint on chemical composition for volcanoes on Venus (Grimm and Hess, 1997).

The single most abundant type of volcanic center on Venus is the small shield volcano (Aubele and Slyuta, 1990) (Figure 5.6). These are centers of eruption less than 20 km in diameter and may occur singly or in clustered volcanic fields called *shield fields* (Crumpler *et al.*, 1997). In addition to true shield morphologies, small volcanoes commonly exhibit the morphologies noted in intermediate and large volcanoes, including steep-sided, modified domical, flat-topped, and even conical.

In addition to surficial features, there are many centers where both radiating and circular fractures suggest magma movement at shallow depths. Coronae often appear to be associated with nearby surface volcanism or occur within volcanic rises. They are unique to Venus, although it has been suggested that some large volcanic features on other planets may be related to coronae (Watters and Janes, 1995), and are interpreted to be the surface expression of rising plumelike magma bodies. Stellate fracture patterns, frequently accompanied by volcanism, are also common in association with rifts and linear arrays of fractures. These are analogous to patterns of radiating dike swarms mapped in continental shields on Earth (McKenzie *et al.*, 1992b; Grosfils and Head, 1994; Ernst *et al.*, 1995). Unlike the deeply

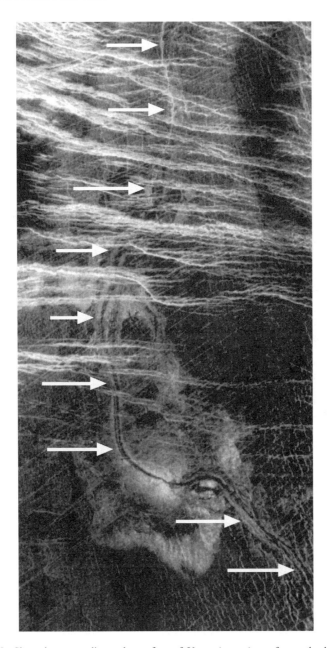

Figure 5.3. Channels, or canali, on the surface of Venus (arrows) are frequently deformed by subsequent fractures, ridges, and swells, as seen in this example. Image width is 150 km, centered at 35°S latitude, 251°W longitude. From FMIDR 35S250.

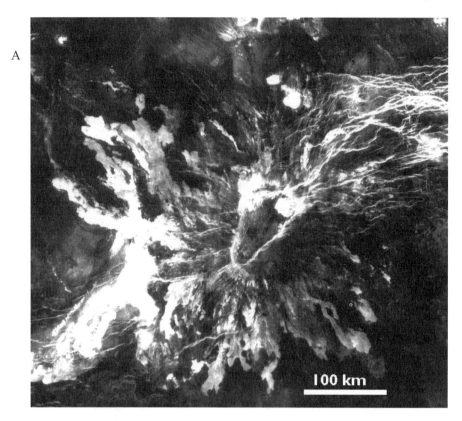

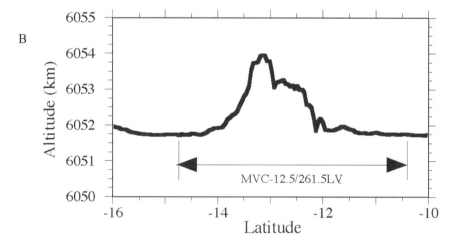

Figure 5.4. (A) Magellan SAR image of a typical large volcanic edifice on Venus. This example is centered at 12°S latitude, 261°W longitude. From FMIDR 10S267. (B) Altimetric profile along a Magellan ground track across the large volcano in A.

Volcanism on Earth's Seafloor and Venus

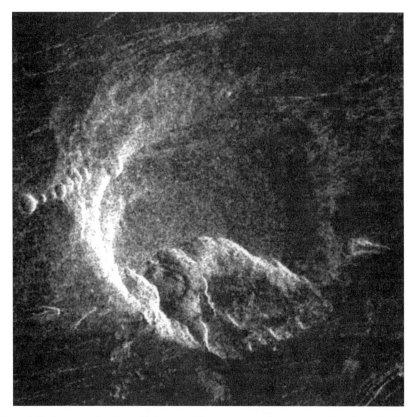

Figure 5.5. Example of a steep-sided dome with an apparent slump on one flank. Magellan SAR image, centered at 16.1°S latitude, 211.8°W longitude. Image width is approximately 40 km. From FMIDR 15S214.

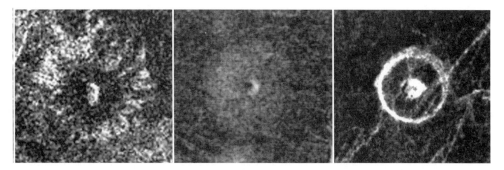

Figure 5.6. Three examples of common small volcanoes on Venus. Small shield volcanoes with distinct flows (left) are relatively less common than small volcanoes with featureless flanks (center). Flat-topped small volcanoes are common on both Venus (right) and on the seafloor.

eroded patterns of radiating dike swarms on Earth, however, the stellate patterns on Venus are characterized by individual graben and small scarps that are interpreted to be the surface expression of shallow dikes (Grosfils and Head, 1994).

Volcanic and magmatic centers are not distributed along linear belts such as in the plate boundary settings of volcanism on Earth (see Chapter 2). Tectonism on Venus is best described as being more like the intraplate tectonism on Earth. There is no direct evidence for large-scale motion of the surface over the past 500 million years of the type that typifies plate tectonics on Earth (Solomon et al., 1992; Hansen et al., 1997), but neither are volcanoes randomly distributed. Instead, some large volcanoes are clustered in areas of slightly elevated relief and at the intersection of several extensive regional fracturing and rifting trends. Volcanoes are notably concentrated (two to three times higher than the global mean) within the broad area lying between Beta, Atla, and Themis Regions (Crumpler et al., 1993). In these "volcanic rises" and similar areas the associated positive gravity anomalies, relief, rifting, and concentrated volcanism have been interpreted (e.g., Smrekar et al., 1997) as the surface manifestation of "hot spots" or localized mantle upwelling. In general, volcanic centers are correspondingly lower in abundance in the low-lying plains, particularly those plains characterized by numerous small ridges, and in the relatively high-standing areas of tesserae [regions characterized by two or more sets of intersecting structural lineaments and enhanced surface roughness at centimeter- to meter-scale e.g. (Barsukov et al., 1986)].

5.1.2. Earth's Seafloor

Available Data. Very little of the seafloor has been imaged at a resolution sufficient to distinguish unequivocally among various volcanic deposits; for a complete description of deep-sea imaging systems, see Kleinrock (1992). Because the seafloor is exposed to essentially no visible light from above, acoustic techniques provide the greatest coverage but at the expense of spatial resolution (Figure 5.7). Hull-mounted multibeam sonars are used to obtain water depths. These systems map a swath as the ship steams along a track. The horizontal spacing of the depth points is proportional to the water depth so that in water depths of 2000 m or so the spacing is ~100 m. The vertical resolution of these systems is ~10 m. There are also side-scan sonar instruments that are towed near the sea surface (e.g., HMR1 in Figure 5.7). These systems provide side-scan sonar imagery and phase bathymetry, but the phase bathymetry has a lower resolution (~50 m vertical resolution) than the multibeam sonar systems.

In addition to the near-surface instrument systems, deeply towed, high-resolution, side-scan systems have also been developed (e.g., 120 kHz in Figure 5.7). The 120-kHz system is towed 100 m or so above the seafloor. Similar to radar, sonar reveals textural information about the seafloor: Flat regions with high sonar return (similar to radar backscatter) are considered rough, and those with low sonar return are believed to be smooth or covered by sediment. Because the sonar systems used in the oceans are side-looking, strong echo amplitudes can also be the result of slopes that face the instrument. The deeply towed side-scan systems also collect phase bathymetry, although this is just coming on line. As an example, the resolution of the 120-kHz phase bathymetry is a few meters. Ship-based sonar systems, such as SeaMARC and SeaBeam, are used primarily for bathymetric surveys, and have absolute vertical resolution as high as 20 m. Sonar packages, such as GLORIA and AMS-120, towed behind the ship a few tens of meters above the seafloor, have correspondingly higher resolution at the cost of less areal coverage. These tow packages can resolve features as little as 1 m across or less, depending on the precise altitude at which the package is towed.

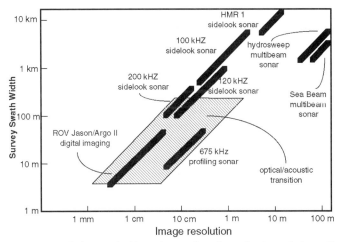

Figure 5.7. Image resolution acquired by various seafloor data collection techniques. From Bowen *et al.* (1993).

Whilst it is possible to identify volcanic features such as eruptive fissures and in some places lava flow margins using the high-resolution side-scan sonar systems, there is an inherent uncertainty with sonar imagery: Without visual information, for example, it is often not possible to distinguish unambiguously between such things as flow margins and talus. Visible wavelength imagery, in the form of video and still-camera photographs from towed packages and remotely operated vehicles, as well as direct observation from manned submersibles, are the best ways to identify volcanic features and lava flow morphologies and boundaries on the seafloor clearly.

Most of the studies in the oceans have been focused on plate boundaries: midocean ridges (MORs; see Table 5.2), transform faults, back-arc basins, and trenches. MORs have been given a lot of attention in the last 10 years since all of the oceanic crust is generated at them (see Chapter 2), yet less than 1% of Earth's MOR system has been mapped in detail (Fornari and Embley, 1995). Nonetheless, through comparison of existing data, we can begin to characterize the basic volcanic styles observed on the seafloor. In this chapter we concentrate primarily on volcanic products formed at the fast-spreading East Pacific Rise (EPR; see Table 5.2) and the slow-spreading Mid-Atlantic Ridge (MAR; see Table 5.2), although in Section 5.2.2 we include some discussion about the distribution of off-axis seamounts.

Environmental Conditions at and Near the Seafloor. As on the surface of Venus, volcanism on the deep seafloor takes place under high pressures. A good "rule of thumb" for the seafloor is that ambient pressure increases ~ 1 MPa for every 100 m of depth. Earth's

Table 5.2. Acronyms and Full Spreading Rates for Representative Midocean Ridges

Midocean ridge	Acronym	Relative spreading rate	Absolute spreading rate
N. East Pacific Rise	NEPR	fast	$9\text{–}13\,\text{cm}\,\text{a}^{-1}$
S. East Pacific Rise	SEPR	superfast	$16\text{–}19\,\text{cm}\,\text{a}^{-1}$
Juan de Fuca Ridge	JFR	intermediate	$6\text{–}9\,\text{cm}\,\text{a}^{-1}$
Mid-Atlantic Ridge	MAR	slow	$1\text{–}3\,\text{cm}\,\text{a}^{-1}$

MORs globally average about 2500 m deep (e.g., Fornari and Embley, 1995) with a corresponding ambient pressure of ~25 MPa The ambient fluid in both environments (primarily carbon dioxide on Venus and water on Earth) has a density at least one order of magnitude greater than air at sea level (Seiff, 1983; Lide, 1990; see Chapter 9). At Earth's MORs, like on the surface of Venus, the ambient fluid acts to cool the surface of lava flows convectively (Figure 5.8), and convection is much more efficient than radiation or conduction at cooling a lava flow surface (e.g., Head and Wilson, 1896; Aubele and Slyuta, 1990; Gregg and Greeley, 1993). In stark contrast to Venus, the ambient temperature at MORs (away from localized hydrothermal vents) is only ~2°C. However, results of numerical modeling (e.g., Griffiths and Fink, 1992; Gregg and Sakimoto, 1996) indicate that for a basaltic lava erupting at ~1200°C, the temperature difference between 2 and 450°C is insignificant because the glass transition temperature of basalt is ~730°C (Ryan and Sammis, 1981). Thus, both on the

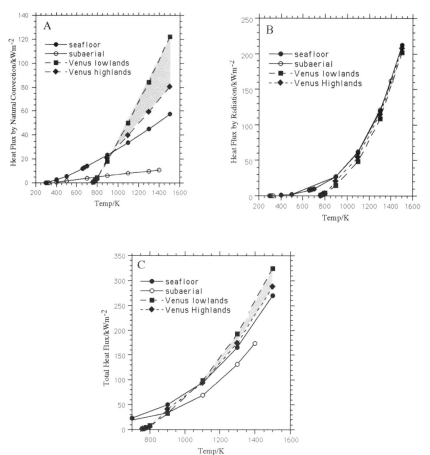

Figure 5.8. Comparison of the heat flow rates for lava flows by natural convection and radiation on Venus (highlands and lowlands) and on Earth (subaerial and seafloor). Heat flux as a function of temperature for (A) natural convection, (B) radiative heat loss, and (C) combined natural convection and radiant heat loss. Heat flux from a lava flow surface on Venus and on the seafloor are comparable for lava surface temperatures above 700 K. The similarity is primarily the result of the efficient thermal transfer of water and dense carbon dioxide. After Aubele and Slyuta (1990), Figure 15); Venus and subaerial Earth data from Head and Wilson (1986); seafloor plot derived from numerical methods in Head and Wilson (1986).

surface of Venus and at Earth's midocean ridges, volcanoes encounter high pressures and dense, convecting ambient fluids.

Overview of Volcanic Styles. Given the high ambient pressures at MORs, explosive eruptions are rare, and lava flow forms (including seamounts) are the most common volcanic deposits, although there are rare exceptions. Volcanic "ash" has been retrieved from the EPR near 9°N (Haymon *et al.*, 1993). Based on analyses of the geologic setting, Haymon *et al.* (1993) concluded that a lava flow completely engulfed an actively venting hydrothermal chimney. The vent fluids accumulated within the lava flow, ultimately generating a small explosion as they forced their way through the overriding lava. Hyaloclastites are more common on the summits of seamounts, where water depth (<1700 m) and corresponding pressures are less, allowing water to flash to steam on contact with molten basalt (Smith and Batiza, 1989; Clague *et al.*, 1990). It also appears that some hyaloclastites may be the result of volatile-rich (probably CO_2) lavas rising rapidly through the oceanic crust (e.g., Clague *et al.*, 1990).

Variations in volcanic morphology at MORs are controlled by the relative dominance of volcanism and tectonism. At the slow-spreading ($\sim$2.5 cm yr^{-1} MAR, the axis is dominated by a broad rift valley 30–40 km wide and 1–2 km deep. The inner floor of this median valley is the primary site of crustal construction and in most places contains elongate axial volcanic ridges up to several hundred meters high, several kilometers across, and many tens of kilometers in length (Smith and Cann, 1993). These axial volcanic ridges are composed of smaller individual volcanic features that pile on one another to form the large ridges. The fast-spreading ($\sim$10 cm yr^{-1} EPR, by contrast, is distinguished by an axial high a few kilometers wide with only a few hundred meters of relief. The top of the rise may contain an axial trough, <250 m wide, which has been interpreted as a volcanic collapse feature, formed over the zone of primary eruptive fissuring and repeated intrusion (Fornari *et al.*, 1998). Although on a small scale (a few tens of meters) the volcanic morphology of the two ridges is similar, with small domes and hummocks occurring at both, there are no large constructional edifices on the axis of the EPR. Large edifices (>50 m in relief) only start to form a few kilometers from the rise axis. The intermediate-spreading Juan de Fuca Ridge (JFR) shows characteristics of both fast- and slow-spreading centers (Chadwick *et al.*, 1998).

Finally, results from numerical modeling (Gregg *et al.*, 1996) indicate that eruptions on the EPR are frequent and rapid, occuring every 5–10 years and lasting only a few hours. In contrast, eruptions on the MAR are most likely less frequent. Based on spreading rate and assuming that dikes are $\sim$1 m in width, a dike would be injected every 40–50 years at the MAR. Whether such a dike is associated with a volcanic eruption or not is not known. Based on the observed flow morphologies at the FAMOUS region of the MAR near 37°N, Ballard and Moore (1978) suggest that eruptions occur every 1000 years or so there, and it appears that the eruptions are longer-lived (Gregg and Fink, 1995).

5.2. INSIGHTS INTO VOLCANIC STYLE: INTRUSIONS, EFFUSIVE ERUPTIONS, AND EXPLOSIVE EVENTS

The high pressures characteristic of both Earth's seafloor and the surface of Venus affect the behavior of volcanic systems relative to what would be expected subaerially on Earth or on other bodies that lack a thick, dense atmosphere. Here we examine four distinct styles of volcanic behavior, comparing observations of the seafloor with Venus and exploring the role played by the near-surface environmental conditions. We begin by examining whether neutral buoyancy contributes significantly to shallow magma stalling, a process that appears to occur frequently based on the abundant evidence that reservoir formation is common on both Venus and the seafloor. Next, we compare the formation of small volcanic edifices, presumably reservoir-derived features, focusing specifically on evidence of geomorphologic modifications induced by the near-surface environment. Third, we consider textural and other observational evidence that near-surface conditions play a critical role in the emplacement of large lava flows. Finally, we conclude the section by discussing retardation of explosive volcanic eruptions resulting from limited exsolution of magma volatiles and inhibition of particulate dispersion. Taken together, examination of these four styles of volcanism underscores the similarity between key volcanic styles observed on Venus and Earth's seafloor, emphasizing the physical significance of their high-pressure environments.

5.2.1. Neutral Buoyancy Stalling and Magma Reservoir Formation

The presence of shallow magma reservoirs at the EPR on Earth has been inferred from numerous seismic studies (e.g., Sinton and Detrick, 1992) and more generally from study of ophiolite exposures. At the EPR large sections of the axis are underlain by a more or less continuous seismic reflector that is interpreted to be the top of a melt lens a few kilometers wide and a few tens of meters thick. No equivalent seismic reflector has been imaged at the MAR primarily because of the difficulties of conducting a seismic reflection study in such rough terrain. At the EPR, therefore, melt generated at depth that ascends toward the seafloor regularly stalls at shallow levels with the oceanic crust. Similarly, the presence of abundant volcanic edifices, radiating dike swarms, calderas, and other Venusian surface features that are likely to be reservoir-derived (Head *et al.*, 1992; Crumpler *et al.*, 1997) implies that magma stalling and reservoir formation at shallow depth is also a common occurrence on Venus. Since shallow reservoir formation appears to occur frequently within the basaltic crusts of both Earth's seafloor and Venus, it is important to investigate the physical mechanism(s) governing magma stalling at shallow depth in these environments.

Seafloor. Termination of magma ascent through crustal material prior to eruption can be induced by several factors, including: (1) gravitational forces resulting from density differences between the rising magma and surrounding host rock (Ryan, 1987; Walker, 1989), (2) stress and/or rheologic variations as a function of depth within the crust (Rubin, 1990; Holliger and Levander, 1994), and (3) the presence of mechanically weak planes within layered deposits (Wojcik and Knapp, 1990). While the relative importance of these factors is not yet fully understood and may vary from site to site, the first two possibilities currently seem to be the most plausible for the EPR. For instance, some seismic data appear to indicate that the depth to the magma lens beneath the EPR varies as a function of spreading rate (Purdy *et al.*, 1992), suggesting that rheologic variations induced by hydrothermal circulation may control the stalling depth (e.g., Phipps Morgan and Chen, 1993). Alternatively, it has been argued that mafic magmas typically stall beneath the EPR within or near the depth range where the density of the ascending magma is thought to match that of the surrounding host rock (Ryan, 1993), i.e., where the magma becomes neutrally buoyant. Similar stalling behavior is observed for magmas beneath many volcanoes (Rubin and Pollard, 1987; Wilson *et al.*, 1992) and within continental crust (Glazner and Ussler, 1988; Glazner and Miller, 1997) elsewhere on Earth. If there is indeed a repeated coincidence between the depth at which ascending magma stalls beneath the EPR and that at which it becomes neutrally buoyant (thus achieving gravitational stability), this suggests that density-dependent forces play a role in facilitating the development of shallow mafic magma reservoirs beneath many MOR spreading centers. Because of this, understanding the neutral buoyancy mechanism for magma stalling and assessing its validity relative to other proposed mechanisms has served as the focus of a great deal of research for more than a decade.

To assess the significance of the role played by neutral buoyancy in the process of magma stalling, it is necessary to know both the density structure of the crust as a function of depth and the density of the ascending magma (Ryan, 1994). The density structure of the *in situ* oceanic crust is determined primarily through interpretation of seismic data (Christensen, 1982). The density of the crust as a function of depth is thus fairly well constrained, although increasingly sophisticated seismic measurements will undoubtedly continue to refine our understanding.

At present, our understanding of magma density is not as well constrained. The density of an ascending basaltic magma varies as a function of time because of the evolving interaction between fractional crystallization processes, which provide insight into the crystal-free liquid phase density for varying water contents, and convective entrainment of olivine crystals back into the liquid. Although the crystal-free liquid phase density evolution related to fractional crystallization is well known from experimental data and theoretical models (Ryan, 1994, and references therein), the amount of convective olivine crystal entrainment that occurs is not as well understood. In the absence of convective entrainment, a typical density for the ascending crystal-free magma as it nears the surface is expected to be approximately $2.7\,g\,cm^{-3}$ (Hooft and Detrick, 1993). Alternatively, if 20% of the magma is occupied by entrained olivine crystals, the typical density will increase to slightly more than $2.8\,g\,cm^{-3}$ (Ryan, 1994). This difference in densities is critically important to ongoing evaluation of the neutral buoyancy mechanism. If a magma with very few olivine crystals entrained is the norm, comparison with oceanic crustal density profiles indicates that the depth at which the magma is expected to become neutrally buoyant is much shallower than the depth at which MOR magma reservoirs actually form, suggesting that density-dependent mechanisms may not control magma stalling beneath the ridges. On the other hand, if a few tens of percent olivine crystal entrainment is the norm, there is a strong coincidence between magma reservoir locations and the depth at which the magma will become neutrally buoyant, suggesting that density-dependent mechanisms may govern the process of magma stalling beneath MOR spreading centers. At present, because it seems reasonable to expect that some degree of convective crystal entrainment occurs within ascending magma (Turner and Campbell, 1986), there is reason to conclude that neutral buoyancy mechanisms could contribute significantly to magma stalling and reservoir formation within oceanic crust, as has been inferred for a diverse array of volcanic environments elsewhere on Earth. Nevertheless, since our understanding of crystallization in magma bodies is continuing to evolve (e.g., Marsh, 1998), any conclusions about whether or not magma stalling occurs as a result of neutral buoyancy forces beneath the EPR must be regarded as tentative until we improve our insight into the amount of crystal entrainment that actually takes place.

Venus. Global examination of Magellan radar data reveals abundant geomorphologic evidence from which the existence of shallow magma reservoirs on Venus can be inferred (Head *et al.*, 1992; Crumpler *et al.*, 1997). Since neutral buoyancy may contribute significantly to magma stalling in most major volcanic environments on Earth, research efforts to date have focused on assessing whether this same mechanism could also govern magma stalling at shallow depth within the Venusian crust.

The density structure of the shallow Venusian crust, inferred to be basaltic from surface measurements, is not known. Since a mathematical description of pore space (vesicle) compaction as a function of depth can be used to explain the crustal density structure observed in Hawaii and Iceland, a similar model for Venus (modified to take into account the substantial variation in atmospheric pressure that occurs as a function of elevation) has been proposed (Head and Wilson, 1992). This model, which considers a range of volatile contents ($CO_2 \leq 0.7$ wt%; $H_2O \leq 1.0$ wt%), systematically compares variations in magma density during ascent with variations in host rock density related to pore compaction. This allows one to estimate the depth at which (or if) the magma reaches a state of neutral buoyancy. Based on their calculations, Head and Wilson (1992) propose that, for a given volatile content, (1) formation of magma reservoirs at low elevations (planetary radii < 6051 km) is strongly inhibited by the high atmospheric pressure, (2) at intermediate elevations (6051–6053 km) and above, decreasing atmospheric pressure promotes reservoir formation as a result of magma neutral buoyancy, and (3) as elevation increases, the depth to the level of neutral buoyancy increases, which gradually promotes formation of larger reservoirs, correspondingly greater eruption volumes (Blake, 1981), and lateral intrusion over surface eruption (Parfitt et al., 1993). If these predictions are correct, a few reservoir-derived volcanic features should exist at low elevations, there should be a general increase in the size of volcanic edifices with elevation (primarily at intermediate altitudes), and a gradual transition from extrusion- to intrusion-dominated, reservoir-derived volcanism should be observed from intermediate to upper (>6053 km) elevations.

To test the Head and Wilson (1992) model, studies to date have focused on evaluating how specific sets of volcanic features (e.g., large volcanoes) are distributed as a function of altitude, and if these distributions individually and collectively are consistent with the neutral buoyancy model predictions. If there are no altitude-dependent effects caused by variations in atmospheric pressure (or other factors), and any given set of features is simply distributed randomly as a function of available surface area at a particular altitude (smooth curve in Figure 5.9), then 12% of the set should lie below a planetary radius of 6051 km, 79% should fall at intermediate elevations of 6051–6053 km, and the remaining 9% should be located at elevations greater than 6053 km.

The observed distributions of intermediate volcanoes (Ristau et al., 1998), large volcanoes (Keddie and Head, 1994), and giant radiating dike swarms (Grosfils and Head, 1995) as a function of altitude are shown in Figure 5.9. Several important aspects of these data agree with the predictions of the Head and Wilson (1992) model. There is a clear gradation from intermediate volcanoes to large volcanoes to radiating dike swarms with increasing altitude, consistent with predictions that volcanic edifices should increase in size with altitude and that there should be a transition from extrusion- to intrusion-dominated volcanism at higher elevations. In addition, chi-square analyses of each of the three sets of features reveal that in each case there is more than a 99% chance that the observed distribution does not coincide with one that is random as a function of the area present at each elevation. Finally, there is an almost total absence of large volcanoes and radiating dike swarms at low altitudes (<6051 km). While approximately 30 intermediate volcanoes form below altitudes of 6051 km, the bulk of these do so at the very upper end of the range (i.e., only 7 of 274 intermediate volcanoes form below 6050.75 km), consistent with the idea that higher than usual magma volatile contents may permit formation of a limited number of small, very shallow reservoirs at the planet's lowest elevations. Taken together, these observations provide compelling evidence that neutral buoyancy may have played an important role in magma stalling and reservoir formation on Venus.

While the evidence collected to date is consistent with available model predictions, further testing of the neutral buoyancy hypothesis is highly desirable. For example, future

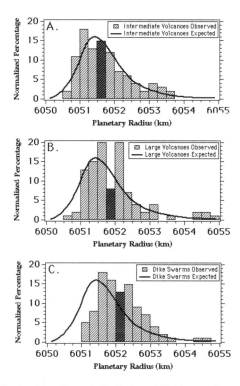

Figure 5.9. (A) Graph of the observed distribution of 274 intermediate volcanoes as a function of altitude (adapted in part from Ristau *et al.*, 1998), with the bin containing the median altitude for the population indicated by darker shading. Expected intermediate volcano population if distributed uniformly by surface area is shown by the solid line. (B) Graph of the observed distribution of 123 large volcanoes as a function of altitude (adapted from Keddie and Head, 1994) (C) Graph of the observed distribution of giant radiating dike swarms as a function of altitude (adapted from Grosfils and Head, 1995).

studies should attempt to assess the possibility of bias in the elevation measurements related to the enhanced concentration of volcanic features in the anomalous Beta–Atla–Themis zone (Crumpler *et al.*, 1993), and whether the distribution of additional reservoir-derived volcanic features (e.g., large calderas) are also consistent with the neutral buoyancy model predictions.

5.2.2. Small Volcanic Constructs

Venus. Small hills (less than 20 km in diameter) on the surface of Venus were originally identified in Earth-based radar images (e.g., Campbell *et al.*, 1989). They were first interpreted to be "volcanic" in studies based on Venera 15/16 images (Aubele and Slyuta, 1990; Garvin and Williams, 1990). These features were originally called *domes* in order to follow the established lunar nomenclature (Barsukov *et al.*, 1986), a term that is not to be confused with the volcanologic connotation of a steep mass of generally viscous lava (see Chapter 2). These small volcanoes were first studied in detail using Venera 15/16 synthetic aperature data (Aubele and Slyuta, 1990) and interpreted to be dominantly effusive shield volcanoes. Subsequent studies based on Magellan data (Guest *et al.*, 1992; Head *et al.*, 1992;

Crumpler *et al.*, 1997) confirm the initial interpretation. Estimates of the total number of small volcanoes on Venus range from the hundreds of thousands to millions, and these edifices preferentially occur in large numbers on the lowland to intermediate elevation plains.

Small volcanoes on Venus are commonly shield-shaped, with a modal diameter of 4 km, variable flank slopes generally much less than 10° and a summit pit (volcanic crater) ranging in diameter from a few hundred meters to 1 or 2 km. However, a range of morphologic diversity does occur in the small volcanoes. In addition to shield-shaped, observed morphologies include conical and domical edifices, edifices produced by radially patterned lava flows, steep-sided domical edifices, flat-topped edifices, edifices with large summit craters, and scalloped or fluted edifices.

Although the small volcanoes of Venus are extremely numerous and widely distributed throughout the planet's plains, they are not randomly distributed. Enhance or clustered concentrations of small volcanoes are one of the most common volcanic features on Venus. Some clusters show a predominance of a specific morphologic type, but in most cases clusters exhibit a range of sizes and morphologies. The edifices within clusters generally show a lack of obvious alignment. Planetwide, there are two types of small volcano concentrations. The first type consists of equant clusters of hundreds of volcanoes in areas approximately 150 km in diameter, often with associated lava flows or larger volcanic edifices. These clusters of small volcanoes have been called *dome fields* or *shield fields* (Senske *et al.*, 1991; Aubele and Crumpler, 1992; Head *et al.*, 1992) following the terrestrial volcanologic usage of the term *volcanic field* (any prominent cluster of volcanic vents). These clusters may represent a distinct type of volcanic center resulting from a specific style and rate of magma reservoir volcanism (Crumpler *et al.*, 1997). The second type of concentration of small volcanoes consists of widely distributed regions of small volcanoes in association with a specific plains unit or units. This type has been informally called *shield plains* and may represent one or more anomalous periods of enhanced formation of small volcanoes as a result of thinner lithosphere, higher heat flow, or widespread mantle melting (Aubele, 1996).

Seamounts. The high-pressure ambient environments on Venus and on the deep seafloor have encouraged many workers to search for analogues to Venusian volcanic constructs on Earth's seafloor (Aubele and Slyuta, 1990; Sakimoto, 1994; Bridges, 1995; Smith, 1996). Like the surface of Venus, however, investigating the seafloor is technologically challenging. The problems of identification and interpretation of volcanic edifices on Venus and on the seafloor are comparable. On Venus, information on morphology is more easily extracted from synthetic aperture radar, but altimetry data are not available for small features. On the seafloor, however, most seamounts are identified and classified using bathymetry data that are readily available. There are minimal high-resolution side-scan sonar data sets (such as TOBI) that reveal some textural information, and visual data (collected via manned submersible, remotely operated vehicles, or deep tow cameras) are limited to a handful of seamounts.

From studies of seamounts several basic observations can be made. Recently, Smith (1996) compiled shape statistics for 2014 seamounts from the Pacific and Atlantic oceans. Analyses of these and similar data (Searle, 1983; Smith, 1988; Smith and Cann, 1992; Bemis and Smith, 1993; Magde and Smith, 1995; Sheirer and Macdonald, 1995) reveal that most seamounts can be adequately approximated by a flat-topped cone. The height-to-basal diameter is defined as x_d. Small volcanoes on Venus exhibit sizes and aspect ratios that are similar to those of seamounts (Aubele and Slyuta, 1990; Smith, 1996). For example, the heights and diameters of three selected small volcanoes on Venus, for which radarclinometry is available (Guest *et al.*, 1992), yield x_d values of 0.10, 0.11, and 0.13—the same as those obtained for seamounts.

The morphologies recognized for the small volcanoes are also very similar to those described from sonar images of seamounts (Smith and Jordan, 1988; Aubele and Slyuta, 1990; Smith and Cann, 1992; Bemis and Smith, 1993; Bridges and Fink, 1994; Sakimoto, 1994; Bridges, 1995; Smith et al., 1995; Smith, 1996). Venusian volcanoes described as "flat-topped" (Guest et al., 1992; Aubele, 1993; Crumpler et al., 1997) or "fluted, scalloped margin and modified domes" (Guest et al., 1991; Bulmer et al., 1992, 1993; Head et al., 1992; Bulmer and Guest, 1996; Crumpler et al., 1997) are distinctively common morphologies in seamounts (Simkin and Batiza, 1984). In particular, flat-topped seamounts imaged near Hawaii (U.S. Exclusive Economic Zone mapping project) and near the northern EPR (Scheirer and Macdonald, 1995) have shapes that are identical to the broad, circular, flat-topped "pancake" domes of Venus (Sakimoto, 1994; Bridges, 1995).

Seamount morphology and planform shape appear to vary with size of the construct and are dependent on (1) the size and shape of the conduit, (2) the presence or absence of a summit caldera, and (3) the presence of faults, fractures, or rifts (Fornari et al., 1987). Smith and others (Smith and Jordan, 1988; Smith and Cann, 1992; Bemis and Smith, 1993; Smith et al., 1995; Smith, 1996) have investigated the morphologic characteristics of seamounts in a range of plate-tectonic settings, using primarily SeaBeam bathymetric data, to determine how the effects of spreading rate, crustal thickness, and magma supply combine to generate observed seamount size-frequency distributions. By comparing these results with observations of Venusian shield fields, future studies may be able to constrain the mode of formation for the Venusian constructs.

The size distribution of both seamounts and small Venusian volcanoes is exponential. Aubele and Slyuta (1990) found that the diameter distribution of small volcanoes identified from Venera data is well described by an exponential size-frequency model. Their abundances and characteristic diameters (estimated from a maximum likelihood fit of a model curve to the data) fall within those estimated for the Pacific seamounts (Smith, 1996).

Seamounts are not randomly distributed (Bemis and Smith, 1939). In addition to varying geographically within and between oceans, they cluster in groups and form linear chains. Furthermore, seamounts originate both on and near MORs as well as in the middle of the tectonic plates, and small-sized seamounts greatly outnumber large seamounts (e.g., Smith and Jordan, 1988). Near the fast-spreading northern EPR on crust less than 2 Ma, Scheirer and Macdonald (1995) concluded that most seamounts larger than 200 m in relief are associated with chains that are oriented between the relative and absolute plate motions. No seamounts form directly at the ridge. Instead they appear to form primarily within a region 5–15 km (0.1–0.3 Ma) from the axis, and seamount abundances seem to be correlated with the shape of the rise axis. Larger numbers of seamounts are found near the areas of the axis that are shallowest and broadest, presumably indicating a robust magma supply. By contrast, at the MAR seamounts are formed within the median valley floor and transported off-axis. A study of seamounts on the northern MAR on crust 0–28 Ma (Jaroslow et al., 2000) indicates little or no off-axis volcanism in their study area (27–29°N).

Although most seamounts observed in the oceans were built near or on the MORs, it is clear that, in the Pacific Ocean at least, seamounts are added to the crust as a natural part of its evolution (Batiza, 1982; Smith and Jordan, 1988). Seamount building is also associated with "anomalous" regions of the ocean basins affected by large hot spots such as Hawaii, the Canaries, and the Superswell region of the South Pacific (McNutt and Fischer, 1987). Bemis and Smith (1993) investigated seamount distribution in three arbitrarily defined regions of the southern Pacific. Region I is north of the Marquesas fracture zone on crust ~24–80 Ma old. Region II is part of the Superswell region (McNutt and Fischer, 1987), an anomalously shallow area of seafloor, which may be indicative of underlying mantle that is hotter than normal—perhaps similar to regions of coronae or shield fields on Venus (e.g., Head *et al.*, 1992; Crumpler *et al.*, 1997). Region III is south of the Marquesas fracture zone, adjacent to the EPR on crust 0–18 Ma old. They found that the density of seamounts in both the Superswell area (Region II) and the region adjacent to the EPR (Region III) is higher than expected in other areas of the Pacific.

Seamounts are extremely numerous on the seafloor and their large numbers suggest that, like the equally abundant small volcanoes of Venus, they play a significant role in the construction and evolution of oceanic crust. For both seamounts and the small volcanoes of Venus, however, the relationship between the volcanic edifices and the plains on which they occur is not yet completely understood.

5.2.3. Lava Flow Emplacement

Lava Plains and Large Flow Fields. While there are many volcanic features in common for Venus and the terrestrial seafloor, perhaps the most similarities are found between the seafloor and the Venusian plains. Immediately apparent is the extent: Both the Venusian plains and the terrestrial seafloor are the result of the predominate form of volcanism on their respective planets, and encompass some of the youngest surfaces. Additionally, in both regimes the individual flow margins are difficult or impossible to pick out, source vents are not apparent, and individual flow surfaces are relatively smooth. Figure 5.10 shows examples of the five types of surface textures on the seafloor flows (see Fink and Griffiths, 1992, and Griffiths and Fink, 1992), which Gregg and Fink (1996) relate to effusion rates. All of these

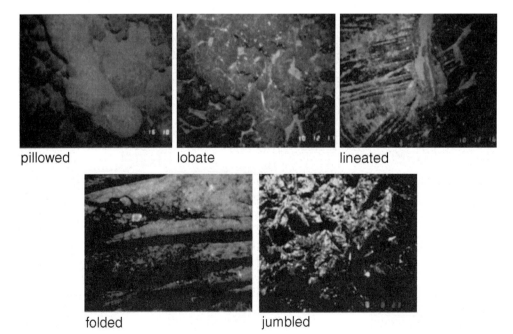

Figure 5.10. Examples of submarine lava flow surface textures from the Juan de Fuca Ridge axial volcano. Field of view is about 4 m across for all but the folded flow, which is ~2 m across. (Courtesy of R. W. Embley.)

flow types are relatively smooth compared with subaerial flows. Griffiths and Fink (1992) and others have shown that this may be related to early and rapid crust growth enhanced by the convective thermal losses to the seawater. This rapid crust growth (relative to that of subaerial flows) may then be followed by relatively quiet lava emplacement and slower cooling under an insulating crust (Gregg and Fornari, 1998). The rapid chilling in the submarine environment is thought to allow less small-scale surface roughness, and might also suppress some flow-to-flow differences in small-scale surface texture. The thick atmosphere of Venus may have a similar effect on the cooling of Venusian flows (Figure 5.8) (e.g., Griffiths and Fink, 1992; Gregg and Greeley, 1993). On Venus, the immediate convective thermal losses and crust formation should be about midway between the terrestrial subaerial and terrestrial submarine rates. The cooling then is likely to be rate-limited by the conductive losses through the crusts, and therefore by the temperature gradient between the flow and the ambient temperature. Since the ambient temperatures on Venus are significantly higher (see Section 5.1.1) than those of the seafloor, the resulting conductive losses through the crust will be slower. Well-insulated, slow-moving Venusian flows should thus cool less per unit distance traveled than terrestrial subaerial or seafloor flows with similar amounts of crust.

In addition to the volcanic plains, Venus has numerous very large flow fields where the flow margins are more apparent, and source and distribution (channel) regions can be inferred or identified (e.g.. Crumpler *et al*., 1997). Mapping of Venusian flow fields such as those shown in Figure 5.11 (e.g., Roberts *et al*., 1992; Zimbelman, 1998) has led to recognition of flow unit margins that are tens to hundreds of kilometers long—much longer than the terrestrial average for individual basalt flows, but similar to the lengths seen in terrestrial flood basalt fields (e.g., Cashman *et al*., 1998; Reidel, 1998). Although the margins of individual

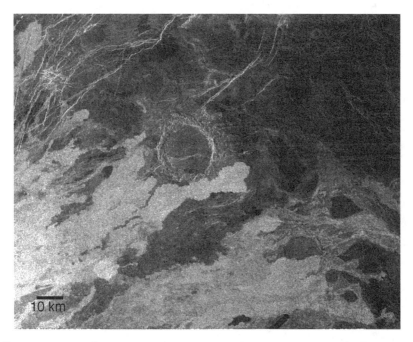

Figure 5.11. Image from Magellan FMIDR 10N188. This image, near the eastern flank of Sapas Mons volcano, shows radar-bright and radar-dark lava flows tens to hundreds of kilometers long, a lava-filled impact crater 20 km in diameter and a portion of the underlying lava plains.

Venusian plains flows may be difficult to identify, the flow fields have more apparent margins and, if this is not simply an artifact of missing flow margins and lumping individual flows into one unit, the Venusian eruptions must be volumetrically much larger than most terrestrial (nonflood) basaltic eruptions. Either they were emplaced very quickly to travel the longer distances before cooling (e.g., Walker, 1973) or, more likely, if crust formation was enhanced as thought, they were emplaced in processes that involved more thermal insulation from the environment than typical terrestrial subaerial flows. For example, they might be emplaced in partially roofed channels, tube flows, or by the slow emplacement and inflation of roofed sheets.

Additional arguments for slower flow emplacement can be made from the roughness of the Venusian and seafloor flows. The radar backscatter values for Venusian flows suggest that, at approximately the wavelength of the Magellan radar (12.5 cm), the volcanic plains flows are considerably smoother on average than subaerial flows on Earth. Most Venusian flows have radar backscatter properties that suggest they are nearly all as smooth or smoother than pahoehoe flows (see Section 2.4.1) in Hawaii (e.g., Campbell and Campbell, 1992) at this wavelength. In Hawaii, as on the seafloor, the smoother, often inflated sheet lavas are usually emplaced more slowly and at lower effusion rates than the rougher channelized and/or a'a lavas (Rowland and Walker, 1990; Self *et al.*, 1996). Extending this rough correlation suggests that some Venusian flow may have been subject to inflation and slow emplacement.

Channels and Valleys. The plains and flow fields of Venus are widely distributed with more than 200 volcanic channels and valleys. Some can be related to specific flows and source regions, like those in Mylitta Fluctus (Roberts *et al.*, 1992), but many have indistinct sources and ends. They range from a few to thousands of kilometers long (Baker *et al.*, 1992, 1997; Komatsu *et al.*, 1993), and many (e.g., the canali type) are exceptional for their remarkably constant width over hundreds of kilometers. Their morphology is well summarized in Baker *et al.* (1997), and ranges from simple meandering channels to complex channel networks with cutoff bends, distributary deltas, and tributary networks (see Figure 5.12).

Several authors have suggested that the channels may be the result of highly fluid lavas at sustained high discharges (e.g., Baker *et al.*, 1992, 1997; Bussey *et al.*, 1995), with some suggesting thermal erosion and exotic or evolved lava types (Baker *et al.*, 1992, 1997; Komatsu *et al.*, 1993; see Chapter 8). Alternatively, the calculations of Gregg and Greeley (1993) and Gregg and Sakimoto (1996) suggest that the channels ought to crust over rapidly from the atmospheric convective cooling, and then proceed in a fashion analogous to terrestrial tube flow, where they could continue for several thousand kilometers at a leisurely pace without cooling enough to halt (Gregg and Sakimoto, 1996). While the thermal arguments for this are similar to those made for tube flows (Keszthelyi, 1995a,b; Sakimoto and Baloga, 1995; Sakimoto and Zuber, 1998), and allow compositions and flow rates similar to those in long basaltic terrestrial flows, the problem of either supporting the width of the canali roofs (for a single tube), or maintaining the relatively constant channel width (for tube networks) remains to be explained. The generally low regional slopes of the Venusian plains where some of the longer canali are found are conducive to slow, gravity-driven flow, as is common in many terrestrial long lava flows (Atkinson *et al.*, 1975; Stephenson and Griffin, 1976). The high discharge rates suggested may be difficult to confine in a channel, and it would be difficult to maintain sufficient driving force to overcome resistive losses over long flow lengths if the flow rate is driven by the effusion rate and hydrostatic head at the flow source (Sakimoto *et al.*, 1997).

Thus far, there is little evidence for channelized flow on the terrestrial seafloor, although there is growing evidence for large flow extents (e.g., Davis, 1982), and Gregg and Fornari

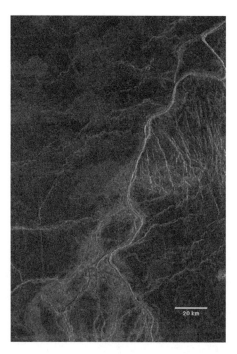

Figure 5.12. Image from Magellan FMIDR F45N019. This image from south of Ishtar Terra illustrates both the lava plains (upper left) and a portion of a complex lava channel (upper right to lower left).

(1998) argue that the rapid formation of flow crust should favor the development of insulated flow like that in tubes or crusted sheets. In addition, Smith and Cann (1999) argue that tubes are important for transporting lava at the MAR from the summit of the axial volcanic ridges down the flanks. These recent studies may indicate that undiscovered canali-like features lie unseen on the seafloor.

Transitional or Enigmatic Flow Types. Figure 5.13 shows a flow field that, though unusual, is one of a handful of radar-bright flows that are thicker than most Venusian surface flows and appear to have been emplaced on top of the plains. Moore *et al.* (1992) suggest that this flow field (among other features) might be evidence for more viscous and thus more silicic volcanism than that of the plains. Gregg and Fink (1996), on the other hand, suggest that little or no extra silica is required to form the feature if it is considered to be a folded-type flow similar to those found on the seafloor. We have little synoptic data for seafloor volcanism thus far, so this Venusian–seafloor comparison is not yet well constrained. These and other enigmatic Venusian flows may thus lack readily identifiable counterparts in the solar system for a while longer.

5.2.4. Explosive Volcanism

As on the deep seafloor, the high ambient pressures at the surface of Venus hinder the generation of pyroclastic deposits. Even so, a handful of possible pyroclastic deposits have been identified on Venus (Head *et al.*, 1991; Guest *et al.*, 1992; Moore *et al.*, 1992; Wenrich and Greeley, 1992; Campbell *et al.*, 1998). It is important to note, however, the inherent

Figure 5.13. Magellan image of radar-bright flows from FMIDR 37S164. Image is approximately 400 km by 400 km.

difficulty in unequivocally identifying pyroclastic deposits using radar in an anhydrous environment (Campbell and Shepard, 1997). Perhaps not surprisingly, therefore, most work to date has not focused on locating pyroclastic deposits on Venus, but on the use of analytical and numerical models to determine the conditions under which an explosive eruption might occur (Head and Wilson, 1986; Fagents and Wilson, 1995; Campbell et al., 1998).

In these models, it is generally assumed that CO_2 is the primary volatile component in Venusian magmatic systems; interestingly, CO_2 is also the dominant volatile in MOR basalts (e.g., Tormey et al., 1987; Perfit et al., 1994). While the details of each model differ, all conclude that ~ 1.5–5.0 wt% CO_2 is required to cause sufficient magmatic disruption to allow an explosive eruption on Venus because of the high ambient pressures. These volatile contents are quite high for terrestrial basalts, as even "volatile-rich MOR basalts contain $\ll 1.5$ wt% CO_2 (Tormey et al., 1987). Additionally, gas-rich MOR basalts still contain <10 vol% vesicles (T. K. P. Gregg, unpublished data) because the high ambient pressure inhibits volatile exsolution. At the summits of seamounts (<1700 m depth), vesicle contents as high as 20 vol% have been reported (Batiza and Vanko, 1984). Even if sufficient amounts of CO_2 were available to generate an explosive eruption on Venus, the high surface temperature coupled with the relatively high density of CO_2 (as compared with H_2O strongly suggests that a Plinian eruption column could not be sustained (Fagents and Wilson, 1995; Campbell et al., 1998). These model results led Campbell et al. (1998) to propose that pyroclastic flows, generated via collapse of an eruption column, may be the preferred method for depositing pyroclastic materials on Venus (as opposed to Plinian airfall deposits). On Earth, large pyroclastic flow deposits (ignimbrites) can be difficult to identify because they have subtle topographic expression and commonly bury their source vent (see Chapter 2), and there is no *a priori* reason to suspect that pyroclastic flows would be any easier to identify on Venus. Thus, while there is not overwhelming evidence for the generation and existence of explosive volcanism on Venus because of its high surface pressures and temperatures, the presence of Venusian pyroclasts should not be summarily dismissed.

5.3. CONCLUSION

Examination of intrusive, effusive, and explosive volcanic processes occurring near the surface of both Earth's seafloor and Venus indicates the importance of the unusual environmental condition they have in common: a high surface pressure related to the presence of a dense, overlying fluid capable of rapid convection. While differences in size can occur, there are otherwise strong similarities between the morphologies, textures, spatial distributions, and emplacement mechanisms that characterize analogous sets of volcanic features occurring on each planet. Excitingly, our understanding of the volcanic processes that dictate these similarities has taken tremendous strides forward in the past decade. Of equal importance, however, is the knowledge that similar types of problems currently define the extent to which we understand the formation of volcanic features found in both locations. For instance, while it is reasonable to infer from existing data that neutral buoyancy may have played an important role in magma stalling and the development of shallow reservoirs, limitations in our understanding of magma density evolution during ascent as well as the shallow crustal density structure (particularly on Venus) currently restrict our ability to evaluate the importance of this process relative to other potential mechanisms. Similarly, while emplacement of pyroclastic materials may be possible in both environments, in order to search for evidence of these types of explosive events it is first necessary to improve our ability to identify the associated deposits remotely. During the decades ahead, which clearly present many challenges, it seems equally clear that sustaining our comparative study of Venus and the seafloor is likely to enhance our understanding of how high surface pressures affect volcanic processes, which by extension will continue to advance significantly our understanding of volcanism within the solar system.

5.4. REFERENCES

Atkinson, A., T. J. Griffin, and P. J. Stephenson, A major lava tube system from Undara Volcano, North Queensland, *Bull. Volcanol.*, *39*, 266–293, 1975.

Aubele, J. C., Venus small volcano classification and description. *Lunar Planet. Sci.*, *XXIV*, 47–48, 1993.

Aubele, J. C., Akkruva small shield plains: Definition of a significant regional plains unit on Venus, *Lunar Planet, Sci.*, *XXVII*, 49–50, 1996.

Aubele, J. C., and L. S. Crumpler, Shield fields: Concentrations of small volcanoes on Venus, *LPI Contrib.*, *789*, 7–8, 1992.

Aubele, J. C., and E. N. Slyuta, Small domes on Venus: Characteristics and origin, *Earth Moon Planets, 50/51*, 493–532, 1990.

Baker, V. R., G. Komatsu, T. J. Parker, V. C. Gulick, J. S. Kargel, and J. S. Lewis, Channels and valleys on Venus: Preliminary analysis of Magellan data, *J. Geophys. Res. 97* (E8), 13421–13444, 1992.

Baker, V. R., G. Komatsu, V. Gulick, and T. J. Parker, Channels and valleys, in *Venus II*, edited by Bougher *et al.*, pp. 757–793, University of Arizona Press, Tucson, 1997.

Ballard, R. D., and J. G. Moore, *Photographic Atlas of the Mid-Atlantic Ridge Rift Valley*, Springer-Verlag, Berlin, 1978.

Banerdt, W. B., G. E. McGill, and M. T. Zuber, Plains tectonics on Venus, in *Venus II*, edited by Bougher *et al.*, pp. 901–968, University of Arizona Press, Tucson, 1997.

Barsukov, V. L., A. T. Basilevsky, G. A. Burba, N. N. Bobinna, V. P. Kryuchkov, R. O. Kuzmin, O. V. Nikolaeva, A. A. Pronin, L. B. Ronca, I. M. Chernaya, V. P. Shashkina, A. V. Garanin, E. R. Kushky, M. S. Markov, A. L. Sukhanov, V. A. Kotelniokov, O. N. Rziga, G. M. Petrov, Y. N. Alexandrov, A. I. Sidorenko, A. F. Bogomolov, G. I. Skrypnik, M. Y. Bergman, L. V. Kudrin, I. M. Bokshtein, M. A. Kronrod, P. A. Chochia, Y. S. Tyuflin, S. A. Kadnichansky, and E. Akim, L., The geology and geomorphology of the Venus surface as revealed by the radar images obtained by Veneras 15 and 16, *Proc. Lunar Planet. Sci. Conf., XVI, J. Geophys. Res., 91*, D378–D398, 1986.

Batiza, R., Abundance, distribution and sizes of volcanoes in the Pacific Ocean and implications for the origin of non-hotspot volcanoes, *Earth Planet. Sci. Lett., 60*, 196–206, 1982.

Batiza, R., and D. Vanko, Petrology of young Pacific seamounts, *J. Geophys. Res., 89*, 11235–11260, 1984.

Bemis, K. G., and D. K. Smith, Production of small volcanoes in the Superswell region of the South Pacific, *Earth Planet. Sci., Lett., 118*, 251–262, 1993.

Bindschadler, D. L., and E. M. Parmentier, Mantle flow tectonics: The influence of a ductile lower crust and implications for the formation of topographic uplands on Venus, *J. Geophys. Res., 95*, 21329–21344, 1990.

Blake S., Volcanism and dynamics of open magma chambers, *Nature, 289*, 783–785, 1981.

Bowen, A., D. Fornari, J. Howland, and B. Walden, *The Woods Hole Oceanographic Institution's remotely-operated and towed vehicle facilities for deep ocean research*, Version 1.0, 26 pp., Woods Hole Oceanographic Institute, Woods Hole, MA, 1993.

Bridges, N. T., Submarine analogs to Venusian pancake domes, *Geophys. Res. Lett., 22*, 2781–2784, 1995.

Bridges, N. T., and J. H. Fink, Aspect ratios of lava domes on Earth, Moon, and Venus, *Lunar Planet. Sci., XXVIII*, 159–160, 1994.

Bulmer, M. H., and J. E. Guest, Modified volcanic domes and associated debris aprons on Venus, in *Volcano Instability on the Earth and Other Planets*, Geol. Soc. London Spec. Publ. No. 110, edited by McGuire *et al.*, pp. 349–372, The Geological Society, Bath, UK, 1996.

Bulmer, M. H., K. Beratan, G. Michaels, and S. Saunders, Debris avalanches and slumps on the margins of volcanic domes on Venus: Characteristics of deposits, *LPI Contrib., 789*, 14–15, 1992.

Bulmer, M. H., G. Michaels, and S. Saunders, Scalloped margin domes: What are the processes responsible and how do they operate? *Lunar Planet. Sci., XXIV*, 177–178, 1993.

Bussey, D. B. J., S. A. Sorensen, and J. E. Guest, Factors influencing the capability of lava to erode its substrate: Application to Venus, *J. Geophys. Res., 100* (E8), 16941–16948, 1995.

Campbell, B. A., and D. B. Campbell, Analysis of volcanic surface morphology on Venus with comparison of Arecibo, Magellan, and terrestrial airborne radar data, *J. Geophys. Res., 97*, 16293–16314, 1992.

Campbell, B. A., and M. K. Shepard, Effect of Venus surface illumination on photographic image texture, *Geophys. Res. Lett., 24*, 731–734, 1997.

Campbell, B. A., L. Glaze, and P. G. Rogers, Pyroclastic deposits on Venus: Remote-sensing evidence and modes of formation, Abstract #1810 (CD-ROM), *Lunar Planet. Sci., XXIX*, 1998.

Campbell, D. B., R. B. Dyce, and G. H. Pettengill, New radar image of Venus, *Science, 193*, 1123–1124, 1976.

Campbell, D. B., J. W. Head, A. A. Hine, J. K. Harmon, D. A. Senske, and P. C. Fisher, Styles of volcanism on Venus: New Arecibo high resolution radar data, *Science, 246*, 373–377, 1989.

Campbell, D. B., D. A. Senske, J. W. Head, A. A. Hine, and P. C. Fisher, Venus southern hemisphere: Character and age of terrains in the Themis–Alpha–Lada region, *Science, 251*, 180–183, 1991.

Cashman, K., H. Pinkerton, and J. Stephenson, Introduction to special section: Long lava flows, *J. Geophys. Res., 103* (B11), 27281–27289, 1998.

Chadwick, W. W., Jr., and M. R. Perfit, Magmatism at mid-ocean ridges: Constraints from volcanological and geochemical investigations, in *Faulting and Magmatism at Mid-Ocean Ridges*, edited by W. R. Buck, P. T. Delaney, J. A. Karson, and Y. Lagabrielle, *Geophys. Monogr. 106*, pp. 59–116, American Geophysical Union, Washington, DC, 1998.

Christensen, N. I., Seismic velocities, in *Handbook of Physical Properties of Rocks, Volume II*, edited by R. S. Carmichael, pp. 1–228, CRC Press, Boca Raton, FL, 1982.

Clague, D. A., R. T. Holcomb, J. M. Sinton, R. S. Detrick, and M. E. Torresan, Pliocene and Pleistocene alkalic flood basalts on the seafloor north of the Hawaiian islands, *Earth Planet. Sci. Lett., 98*, 175–191, 1990.

Crumpler, L. S., J. W. Head, and J. C. Aubele, Relation of major volcanic center concentration on Venus to global tectonic patterns, *Science, 261*, 591–598, 1993.

Crumpler, L. S., J. C. Aubele, D. A. Senske, S. W. Keddie, K. Magee, and J. W. Head, Volcanoes and centers of volcanism on Venus, in *Venus II*, edited by Bougher et al., pp. 697–756, University of Arizona Press, Tucson, 1997.

Davis, E. E., Evidence for extensive basalt flow on the sea floor, *Geol. Soc. Am. Bull., 93*, 1023–1029, 1982.

Ernst, R. E., J. W. Head, E. Parfitt, E. B. Grosfils, and L. Wilson, Giant radiating dyke swarms on Earth and Venus, *Earth Sci. Rev., 39*, 1–58, 1995.

Fagents, S. A., and L. Wilson, Explosive volcanism on Venus: Transient volcanic explosions as a mechanism for localized pyroclast dispersal, *J. Geophys. Res., 100*, 26327–26338, 1995.

Fink, J. H., and R. W. Griffiths, A laboratory analog study of the surface morphology of lava flows extruded from point and line sources, *J. Volcanol. Geotherm. Res., 54*, 19–32, 1992.

Ford, P. G., and G. H. Pettengill, Venus topography and kilometer-scale slopes, *J. Geophys. Res., 97*, 13103–13114, 1992.

Fornari, D. J., and R. W. Embley, Tectonic and volcanic controls on hydrothermal processes at the mid-ocean ridge: An overview based on near-bottom and submersible studies, in *Seafloor Hydrothermal Systems: Physical, Chemical, Biological, and Geological Interactions*, Geophys. Monogr., 91, 1–46, American Geophysical Union, Washington, DC, 1995.

Fornari, D. J., R. M. Haymon, M. R. Perfit, T. K. P. Gregg and M. H. Edwards, 1998, Axial summit trough of the East Pacific Rise (9°N to 10°N): Geological characteristics and evolution of the axial zone on fast-spreading mid-ocean ridges, *J. Geophys. Res., 103*:9827–9855.

Fornari, D. J., R. Batiza, and M. A. Luckman, Seamount abundances and distribution near the East Pacific Rise 0°–24°N based on SeaBeam data, in *Seamounts, Islands, and Atolls*, Geophys. Monogr., 43, 13–21, American Geophysical Union, Washington, DC, 1987.

Garvin, J. B., and R. S. Williams, Small domes on Venus: Probable analogs of Icelandic lava shields, *Geophys. Res. Lett., 17*, 1381–1384, 1990.

Glazner, A. F., and D. M. Miller, Late-stage sinking of plutons, *Geology, 25*, 1099–1102, 1997.

Glazner, A. F., and W. Ussler, Trapping of magma at midcrustal density discontinuities, *Geophys. Res. Lett., 15*, 673–675, 1988.

Goldstein, R. M., R. R. Green, and H. C. Rumsey, Venus radar brightness and altitude images, *Icarus, 36*, 334–352, 1978.

Gregg, T. K. P., and J. H. Fink, Quantification of submarine lava-flow morphology through analog experiments, *Geology, 23*, 73–76, 1995.

Gregg, T. K. P. and J. H. Fink, Quantification of extraterrestrial lava flow effusion rates through laboratory simulations, *J. Geophys. Res., 101* (E7), 16891–16900, 1996.

Gregg, T. K. P., and D. J. Fornari, Long submarine lava flows: Observations and results from numerical modeling, *J. Geophys. Res., 103*, 27517–27532, 1998.

Gregg, T. K. P., and R. Greeley, Formation of Venus canali: Considerations of lava types and their thermal behaviors, *J. Geophys. Res., 98* (E6), 10873–10882, 1993.

Gregg, T. K. P., and S. E. H. Sakimoto, Venusian lava flow morphologies: Variations on a basaltic theme, *Lunar Planet. Sci., XXVII*, 459–460, 1996.

Gregg, T. K. P., D. J. Fornari, M. R. Perfit. R. M. Haymon, and J. H. Fink, Rapid emplacement of a mid-ocean ridge lava flow on the East Pacific Rise at 9°46'–52'N, *Earth Planet. Sci. Lett., 144*, E1–E7, 1996.

Griffiths, R. W., and J. H. Fink, The morphology of lava flows in planetary environments: Predictions from analog experiments, *J. Geophys. Res., 97* (B13), 19739–19748, 1992.

Grimm, R. E., and P. C. Hess, The crust of Venus, in *Venus II*, edited by Bougher et al., pp. 1205–1244, University of Arizona Press, Tucson, 1997.

Grosfils, E. B., and J. W. Head, The global distribution of giant radiating dike swarms on Venus: Implications for the global stress state, *Geophys. Res. Lett., 21*, 701–704, 1994.

Grosfils, E. B., and J. W. Head, Radiating dike swarms on Venus: Evidence for emplacement at zones of neutral buoyancy, *Planet. Space Sci., 43*, 1555–1560, 1995.

Guest, J. E., K. Beratan, G. Michaels, K. Desmaris, and C. Weitz, Slope failure on the margins of volcanic domes on Venus: *EOS Trans. Am. Geophys. Union, 72*, 278–279, 1991.

Guest, J. E., et al., Small volcanic edifices and volcanism in the plains of Venus, *J. Geophys. Res., 97*, 15949–15966, 1992.

Hansen, V. L., J. J. Willis, and W. B. Banerdt, Tectonic overview and synthesis, in *Venus II*, edited by Bougher et al., pp. 797–845, University of Arizona Press, Tucson, 1997.

Haymon, R. M., D. J. Fornari, K. L. Von Damm, M. D. Lilley, J. M. Edmond, W. C. Shanks, III, R. A. Lutz, J. M. Grebmeier, S. Carbotte, D. Wright, E. McLaughlin, M. Smith, N. Beedle and E. Olson, Volcanic eruption of the mid-ocean ridge along the East Pacific Rise at 9°45–52′N: I. Direct submersible observation of seafloor phenomena associated with an eruption event in April, 1991, *Earth Planet. Sci. Lett., 119*, 85–101, 1993.

Head, J. W., and L. Wilson, Volcanic processes and landforms on Venus: Theory, predictions, and observations, *J. Geophys. Res., 91*, 9407–9446, 1986.

Head, J. W., and L. Wilson, Magma reservoirs and neutral buoyancy zones on Venus: Implications for the formation and evolution of volcanic landforms, *J. Geophys. Res., 97*, 3877–3903, 1992.

Head, J. W., D. B. Campbell, C. Elachi, J. E. Guest, D. P. McKenzie, R. S. Saunders, G. G. Schaber, and G. Schubert, Venus volcanism: Initial analysis from Magellan data, *Science, 252*, 276–288, 1991.

Head, J. W., L. S. Crumpler, J. C. Aubele, J. E. Guest, and R. S. Saunders, Venus volcanism: Classification of volcanic features and structures, associations, and global distributions from Magellan data, *J. Geophys. Res., 97*, 13153–13198, 1992.

Holliger, K., and A. Levander, Lower crustal reflectivity modeled by rheological controls on mafic intrusions, *Geology, 22*, 367–370, 1994.

Hooft, E. E., and R. S. Detrick, The role of density in the accumulation of basaltic melts at mid-ocean ridges, *Geophys. Res. Lett., 20*, 423–426, 1993.

Jaroslow, G. E., D. K. Smith, and B. E. Tucholke, Record of seamount production and off-axis evolution in the western North Atlantic Ocean 25 degrees 25′–27 degrees 10′N, submitted to *J. Geophys. Res.*, 2000, in press.

Keddie, S. T., and J. W. Head, Height and altitude distribution of large volcanoes on Venus, *Planet. Space Sci., 42*, 455–462, 1994.

Keszthelyi, L., A preliminary thermal budget for lava tubes on the Earth and planets, *J. Geophys. Res., 100* (B10), 20411–20420, 1995a.

Keszthelyi, L., Thermal constraints on the lengths of tube-fed lava flows on the Earth, Moon, Mars and Venus, *Lunar Planet. Sci., XXVI*, 739–740, 1995b.

Kleinrock, M. C., Capabilities of some systems used to survey the deep-sea floor, in *CRC Handbook of Geophysical Exploration at Sea, 2nd Edition, Hard Minerals*, edited by R. A. Geyer, pp. 35–86, CRC Press, Boca Raton, FL, 1992.

Komatsu, G., V. R. Baker, V. Gulick, and T. J. Parker, Venusian channels and valleys: Distribution and volcanological implications, *Icarus, 102*, 1–25, 1993.

Lide, D. R., *CRC Handbook of Chemistry and Physics*, 71st ed., CRC Press, Boca Raton, FL, 1990.

Mackwell, S. J., M. E. Zimmerman, and D. L. Kohlstedt, High-temperature deformation of dry diabase with application to tectonics of Venus, *J. Geophys. Res., 103*, 975–984, 1998.

Magde, L. S., and D. K. Smith, Seamount volcanism at the Reykjanes Ridge: Relationship to the Iceland hot spot, *J. Geophys. Res., 100*, 8449–8468, 1995.

Marsh, B. D., On the interpretation of crystal size distributions in magmatic systems, *J. Petrol. 39*, 553–599, 1998.

McGill, G. E., J. F. Warner, M. C. Malin, R. E. Arvidson, E. Eliason, S. Nozette, and R. D. Reasenberg, Topography, surface properties, and tectonic evolution, in *Venus*, edited by D. M. Hunten et al., pp. 69–130, University of Arizona Press, Tucson, 1983.

McKenzie, D., P. G. Ford, F. Liu, and G. H Pettengill, Pancake-like domes on Venus, *J. Geophys. Res., 97*, 15967–15976, 1992a.

McKenzie, D., J. M. McKenzie, and R. S. Saunders, Dike emplacement on Venus and on Earth, *J. Geophys. Res., 97*, 15977–15990, 1992b.

McNutt, M. K., and K. M. Fischer, The South Pacific Superswell, in *Seamounts, Islands and Atolls, Geophys. Monogr., 43*, 25–34, American Geophysical Union, Washington, DC, 1987.

Moore, H. J., J. J. Plaut, P. M. Schenk, and J. W. Head, An unusual volcano on Venus, *J. Geophys. Res., 97*, 13479–13494, 1992.

Parfitt, E. A., L. Wilson, and J. W. Head, Basaltic magma reservoirs: Factors controlling their rupture characteristics and evolution, *J. Volcanol. Geotherm. Res., 55*, 1–14, 1993.

Pettengill, G. H., E. Eliason, P. G. Ford, G. B. Loriot, H. Masursky, and G. E. McGill, Pioneer Venus radar results: altimetry and surface properties, *J. Geophys. Res., 85*, 8261–8270, 1980.

Perfit, M. R., D. J. Fornari, M. C. Smith, J. F. Bender, C. H. Langmuir, and R. M. Haymon, Small-scale spatial and temporal variations in mid-ocean ridge crest magmatic processes, *Geology, 22*, 375–379, 1994.

Phipps Morgan, J., and Y. J. Chen, The genesis of ocean crust: Magma injection, hydrothermal circulation, and crustal flow, *J. Geophys. Res., 98*, 6283–6297, 1993.

Purdy, G. M., L. S. L. Kong, G. L. Christenson, and S. C. Solomon, Relationship between spreading rate and the seismic structure of mid-ocean ridges, *Nature, 355*, 815–817, 1992.

Reidel, S. P., Emplacement of Columbia River Flood Basalt, *J. Geophys. Res., 103* (B11), 27393–27410, 1998.

Ristau, S., J. Sammons, E. Grosfils, L. Reinen, M. Gilmore, and S. Kozak, Distribution of intermediate volcanoes on Venus as a function of altitude, Abstract #1100 (CD-ROM), *Lunar Planet. Sci., Conf., XXIX*, 1998.

Roberts, K. M., J. E. Guest, J. W. Head, and M. G. Lancaster, Mylitta Fluctus, Venus: Rift-related, centralized volcanism and the emplacement of large-volume flow units, *J. Geophys. Res., 97* (E10), 15991–16015, 1992.

Rowland, S. K., and G. P. L. Walker, Pahoehoe and aa in Hawaii: Volumetric flow rate controls the lava structure, *Bull. Volcanol., 52*, 615–628, 1990.

Rubin, A. M., A comparison of rift-zone tectonics in Iceland and Hawaii, *Bull. Volcanol., 52*, 302–319, 1990.

Rubin, A. M., and D. D. Pollard, Origins of blade-like dikes in volcanic rift zones, *U.S. Geol. Surv. Prof. Pap., 1350*, 1449–1470, 1987.

Ryan, M. P., Neutral buoyancy and the mechanical evolution of magmatic systems, in *Magmatic Processes: Physiochemical Principles*, edited by B. O. Myser, Geochemical Society Special Publication No. 1, pp. 259–287, University Park, PA, 1987.

Ryan, M. P., Neutral buoyancy and the structure of mid-ocean ridge magma reservoirs, *J. Geophys. Res., 98*, 22321–22338, 1993.

Ryan, M. P., Neutral-buoyancy controlled magma transport and storage in mid-ocean ridge magma reservoirs and their sheeted dike complex: A summary of basic relationships, in *Magmatic Systems*, edited by M. P. Ryan, pp. 97–138, Academic Press, San Diego, 1994.

Ryan, M. P., and C. G. Sammis, The glass transition in basalt, *J. Geophys. Res., 86*, 9519–9535, 1981.

Sakimoto, S. E. H., Terrestrial basaltic counterparts for the Venus steep-sided or pancake domes, *Lunar Planet. Sci., XXV*, 1189–1190, 1994.

Sakimoto, S. E. H., and S. M. Baloga, Thermal controls on tube-fed planetary lava flow lengths, *Lunar Planet. Sci., XXVI*, 1217–1218, 1995.

Sakimoto, S. E. H., and M. T. Zuber, Flow and convective cooling in lava tubes, *J. Geophys. Res., 103*, 27465–27487, 1998.

Sakimoto, S. E. H., J. Crisp, and S. M. Baloga, Eruption constraints on tube-fed planetary lava flows, *J. Geophys. Res., 102* (E3), 6597–6613, 1997.

Schaber, G. G., Volcanism on Venus as inferred from the morphometry of large shields, *Proc. Lunar Planet. Sci. Conf., XXI, 3–11, 1991.*

Scheirer, D. S., and K. C. Macdonald, Near-axis seamounts on the flanks of the East Pacific Rise, 8°N to 17°N, *J. Geophys. Res., 100*, 2239–2259, 1995.

Searle, R. C., Submarine central volcanoes on the Nazca plate—high-resolution sonar observations, *Mar. Geol., 53*, 77–102, 1983.

Seiff, A., Thermal structure of the atmosphere of Venus, in *Venus*, edited by D. M. Hunten et al., 215–279, University of Arizona Press, Tucson, 1983.

Self, S., Th. Thordarson, L. P. Keszthelyi, G. P. L. Walker, K. Hon, M. T. Murphy, P. Long, and S. Finnemore, A new model for the emplacement of Colombia River basalts as large, inflated pahoehoe lava flow fields, *Geophys. Res. Lett., 23*, 2689–2692, 1996.

Senske, D. A., D. B. Campbell, J. W. Head, P. C. Fisher, A. A. Hine, A. deCharon, S. L. Frank, S. T. Keddie, K. M. Roberts, E. R. Stofan, J. C. Aubele, L. S. Crumpler, and N. Stacy, Geology and tectonics of the Themis Regio–Lavinia Planitia–Alpha Regio–Lada Terra Area, Venus: Results from Arecibo image data, *Earth Moon Planets, 55*, 97–161, 1991.

Simkin, T., and R. Batiza, Flattish summits, calderas and circumferential vents: A morphogenetic comparison of young EPR seamounts and Galapagos volcanoes, *EOS, 65*, 1080, 1984.

Sinton, J. M., and R. S. Detrick, Mid-ocean ridge magma chambers, *J. Geophys. Res., 97*, 197–216, 1992.

Smith, D. K., Shape analysis of Pacific seamounts, *Earth Planet. Sci. Lett., 90*, 457–466, 1988.

Smith, D. K., Comparison of the shapes and sizes of seafloor volcanoes on Earth and "pancake" domes on Venus, *J. Volcanol. Geotherm. Res., 73*, 47–64, 1996.

Smith, D. K., and J. R. Cann, The role of seamount volcanism in crustal construction at the Mid-Atlantic Ridge (24°–30°N), *J. Geophys. Res., 97*, 1645–1658, 1992.

Smith, D. K., and J. R. Cann, Building the crust at the Mid-Atlantic Ridge, *Nature, 365*, 707–715, 1993.

Smith, D. K., and J. R. Cann, Constructing the upper crust of the Mid-Atlantic Ridge: A reinterpretation based on the Puna Ridge, Kilauea Volcano, *J. Geophys. Res., 104*, 25379–25400, 1999.

Smith, D. K., and T. H. Jordan, Seamount statistics in the Pacific Ocean, *J. Geophys. Res., 93*, 2899–2918, 1988.

Smith, D. K., S. E. Humphris, and W. B. Bryan, A comparison of volcanic edifices at the Reykjanes Ridge and the Mid-Atlantic Ridge at 24°–30°N, *J. Geophys. Res., 100*, 22485–22498, 1995.

Smith, T. L., and R. Batiza, New field and laboratory evidence for the origin of hyaloclastite flows on seamount summits, *Bull. Volcanol., 51*, 96–114, 1989.

Smrekar, S. E., and S. C. Solomon, Gravitational spreading of high terrain in Ishtar Terra, Venus, *J. Geophys. Res., 97*, 16121–16148, 1992.

Smrekar, S., W. S. Kieffer, and E. R. Stofan, Large volcanic rises on Venus, in *Venus II*, edited by Bougher *et al.*, pp. 845–878, University of Arizona Press, Tucson, 1997.

Solomon, S. C., S. E. Smrekar, D. L. Bindschadler, R. E. Grimm, W. M. Kaula, G. E. McGill, R. J. Phillips, R. S. Saunders, G. Schubert, S. W. Squyres and E. R. Stofan, Venus tectonics: An overview of Magellan observations, *J. Geophys Res., 97*, 13199–13255, 1992.

Stephenson, P. J., and T. J. Griffin, Some long basaltic lava flows in North Queensland, in *Volcanism in Australasia*, edited by R. W. Johnson, pp. 41–51, Elsevier, Amsterdam, 1976.

Stofan, E. R., V. E. Hamilton, D. M. Janes, and S. E. Smrekar, Coronae on Venus: Morphology and origin, in *Venus II*, edited by Bougher *et al.*, pp. 931–965, University of Arizona Press, Tucson, 1997.

Tormey, D. R., T. L. Grove, and W. B. Bryan, Experimental petrology of normal MORB near the Kane fracture zone: 22°–25°N, Mid-Atlantic Ridge, *Contrib. Mineral. Petrol., 96*, 121–139, 1987.

Turner, J. S., and I. H. Campbell, Convection and mixing in magma chambers, *Earth Sci. Rev., 23*, 255–352, 1986.

von Zahn, U., S. Kumar, H. Niemann, and R. Prinn, Composition of the atmosphere of Venus: in *Venus*, edited by D. M. Hunten *et al.*, pp. 299–430, University of Arizona Press, Tucson, 1983.

Walker, G. P. L., Lengths of lava flows, *Philos. Trans. R. Soc. London Ser. A, 274*, 107–118, 1973.

Walker, G. P. L., Gravitational (density) controls on volcanism, magma chambers and intrusions, *Aust. J. Earth Sci., 36*, 149–165, 1989.

Watters, T. R., and D. M. Janes, Coronae on Venus and Mars: Implications for similar structures on Earth, *Geology, 23*, 200–204, 1995.

Weertman, J., Height of mountains on Venus and the creep properties of rock, *Phys. Earth Planet. Inter., 19*, 197–207, 1979.

Wenrich, M. L., and R. Greeley, Investigation of Venusian pyroclastic volcanism, *Lunar Planet. Sci., XXIII*, 1515–1516, 1992.

Wilson, L., J. W. Head, and E. A. Parfitt, The relationship between the height of a volcano and the depth to its magma source zone: A critical reexamination, *Geophys. Res. Lett., 19*, 1395–1398, 1992.

Wojcik, K. M., and R. W. Knapp, Stratigraphic control of the Hills Pond lamproite, Silver City Dome, southeastern Kansas, *Geology, 18*, 251–254, 1990.

Wood, J. A., Rock weathering on the surface of Venus, in *Venus II*, edited by Bougher *et al.*, pp. 637–664, University of Arizona Press, Tucson, 1997.

Zimbelman, J. R., Emplacement of long lava flows on planetary surfaces, *J. Geophys. Res., 103* (B11), 27503–27516, 1998.

Zuber, M. T., Constraints on the lithospheric structure of Venus with mechanical models and tectonic surface features, *J. Geophys. Res., 92*, 541–551, 1987.

6

Moon and Mercury

Volcanism in Early Planetary History

James W. Head III, Lionel Wilson, Mark Robinson,
Harald Hiesinger, Catherine Weitz, and Aileen Yingst

6.1. INTRODUCTION

The Moon and Mercury have a generally similar surface morphology, with their ancient landscapes characterized by heavily cratered terrain, impact basins, and expanses of smooth plains (Murray *et al.*, 1981; Taylor, 1982; Strom, 1987). In addition, volcanism is known (in the case of the Moon) and thought to have been (in the case of Mercury) an important part of their surface evolution. Volcanic processes on these bodies are likely to have shared some important attributes because of the ancient nature of the geologic record (e.g., early thermal evolution phases) and the lack of an atmosphere (e.g., the style of cooling of lava flows, influencing the distribution of volcanic ejecta). But what about the fundamental differences between the Moon and Mercury, and the lack of knowledge about their consequences? Should Mercury, with its very high surface temperatures, Mars-like surface gravity, and very large iron core, be characterized by volcanic activity that is similar to that of the Moon? Or do these variations dictate that volcanism should be manifested differentially on the two bodies?

It is generally thought that the Moon formed very early in solar system history when a Mars-sized object impacted Earth, ejecting crust and upper mantle material that reaccreted in Earth orbit. The energy associated with accretion caused large-scale melting, and this was accompanied by density segregation of the melt and formation of a low-density, plagioclase-rich crust. The solidification of the global crust was accompanied by, and succeeded for several hundred million years by, a massive influx of projectiles producing impact craters of many sizes and obscuring the record of any early volcanism, although some evidence of the eruptions in the latest part of heavy bombardment is described below. The formation of the globally continuous, low-density, buoyant crust apparently precluded the development of Earth-like plate tectonics early in the history of the Moon. This led to a dominance of conductive cooling through this continuous layer, producing a globally continuous lithosphere, in contrast to the multiple laterally moving and subducting plates on the Earth. Volcanism, or advective cooling, was minor compared with conduction through this globally continuous lithospheric layer, yet it produced surface deposits that make up some of the unique topography and albedo of the surface seen today.

Volcanic flooding of the surface of the Moon became evident only during the waning stages of heavy bombardment, about 3.7–3.8 billion years ago (Ga), following the formation of the youngest impact basins, Imbrium and Orientale (e.g., Wilhelms, 1987). After this time, lavas emplaced could retain their character and not be covered and modified by regional impact ejecta deposits. By about 2.5–3.0 Ga, basaltic lavas had covered approximately 17% of the lunar surface, preferentially filling in the nearside low-lying basin interiors to form the lunar maria (Figure 6.1). There is no evidence for major internally generated geologic activity on the Moon for the past 2.5 billion years. The Moon thus provides a picture of the first half of solar system history, characterized by impact bombardment and early volcanism, and serves as a cornerstone for the interpretation of the records preserved on other terrestrial planets.

The internal structure of the Moon is a very important component of our knowledge about volcanism. Density structure, source regions, and temperature gradients are all significant factors in the generation, ascent, and emplacement of magma. Remote sensing, surface seismic data, and Apollo and Luna samples all show that the Moon has been internally differentiated into a crust, mantle, and possibly a small core. The feldspar-rich crust is thinner on the central nearside (~ 55 km) but may reach thicknesses of 100 km on the farside. Seismic data and geologic mapping show that the lunar maria are relatively thin (maximum thickness of a few kilometers) and perched on a globally continuous feldspar-rich crust. The lunar highlands crust has been called a "primary" crust, derived from widespread melting associated with the energy of early impact cratering (Taylor, 1989).

The widespread melting that resulted in fractional crystallization and separation of a low-density plagioclase crust also produced a residual upper mantle layer. The residual layer

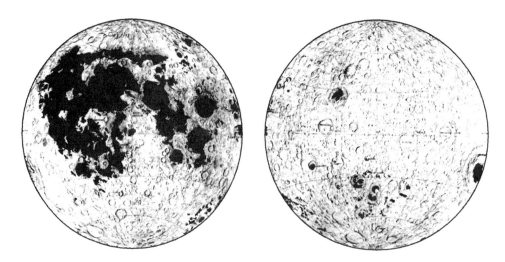

Figure 6.1. Distribution of the lunar maria.

below the low-density crust was denser than the underlying mantle, probably leading to gravitational collapse and sinking toward the interior, perhaps to form a core (e.g., Hess and Parmentier, 1995). During this time period, partial melting of the mantle led to the formation of magmas that were emplaced as secondary crust: the lunar maria, collecting in low-lying craters and basins. Secondary crust formation [e.g., crust derived by partial melting of the mantle (Taylor, 1989)] was apparently volumetrically minimal; the ascent of magma was inhibited by the thick, light crust causing the maria to form preferentially in basins on the thinner nearside highland crust (e.g., Solomon, 1975; Head and Wilson, 1992). The high surface area-to-volume ratio meant that the Moon underwent rapid cooling and little to no evidence of internal activity (e.g., magnetic field, surface volcanism) is seen in the last half of solar system history (e.g., Head and Solomon, 1981).

Mercury is about one-third the diameter, but about the same bulk density, as Earth. With a size near that of the Earth's Moon, and a density similar to that of the Earth, Mercury offers an opportunity to study the influence of size and internal structure on the geologic history and thermal evolution of planetary bodies, and the role that these factors might play in volcanic and magmatic processes. The Mariner 10 spacecraft returned images of about 35% of the planet's surface and revealed a lunarlike terrain, but detailed geologic mapping of the surface has shown that Mercury differs from the Moon in several important aspects related to volcanism and magmatism (e.g., Chapman, 1988; Spudis and Guest, 1988; Schubert *et al.*, 1988). Large areas of relatively ancient intercrater plains may indicate that more extensive volcanism accompanied the period of heavy cratering on Mercury than on the Moon. Large, extensive scarps on Mercury attest to episodes of regional shortening and perhaps even the global contraction that would result from a modest decrease in the planet's circumference during solidification (Melosh and McKinnon, 1988; Schubert *et al.*, 1988). This net compressive state of stress in the lithosphere could be an important factor in potentially inhibiting magma from reaching the surface. Areas of smooth plains have nearly the same albedo as that of heavily cratered regions, which has led to controversy over the origin of some of the smooth regions [e.g., volcanic or ponded impact ejecta (Strom *et al.*, 1975; Wilhelms, 1976)].

In this chapter, we assess the nature of the volcanic record on the Moon and Mercury and relationships between the two planetary bodies. We begin with the much better known record of the Moon, develop several themes that characterize volcanic activity there, and then consider the nature of the volcanic record on Mercury, concluding with several comparisons and delineations of some outstanding problems.

6.2. LUNAR VOLCANISM

6.2.1. Nature and Distribution

Lunar volcanic deposits are traditionally subdivided into two major types, mare (characterized by low-albedo plains and associated landforms) and nonmare or highland (characterized by intermediate- to high-albedo plains and a range of landforms suspected of being of volcanic origin) (e.g., Head, 1976). Mare deposits cover 17% of the lunar surface (Figure 6.1) (Head, 1975a,b). They occur preferentially in topographic lows on the nearside, and have a variety of modes of occurrence and styles of emplacement (Head, 1975a, 1976; Wilhelms, 1987). Mare deposits are concentrated in the interiors and margins of impact basins, and are particularly abundant in basins of Nectarian and post-Nectarian age (e.g.,

Nectaris, Humorum, Crisium, Serenitatis, and Imbrium). Relatively thin deposits are widespread and occur at the sites of older basins such as Tranquillitatis, Fecunditatis, and Procellarum. Farside maria are sparse and occur in the ancient South Pole–Aitken basin as patches, within other basins (such as Australe), and younger craters (such as Moscoviense and Tsiolkovsky).

The nearside/farside asymmetry in exposed mare deposits (Figure 6.1) is generally interpreted to be related to crustal thickness variations inferred from the observed offset center of figure/center of mass and from considerations of gravity and topography. The relevance of this global low-density crust and its thickness differences to mare volcanism was pointed out by Solomon (1975), who showed that dense mare basalt magmas should not erupt on the surface on the basis of buoyancy alone. Hydrostatic considerations suggest that melts may have risen to the surface preferentially in the deepest impact basins (regions of thinnest crust) which appear to be concentrated on the nearside. Solomon (1975) proposed that these considerations could account for the distribution of maria on the nearside (primarily within nearside basins) and the paucity of maria on the farside (enhanced crustal thickness there). We will return to this issue when we consider ascent and eruption of mare basalts.

6.2.2. Volume

Beginning with the known mare surface area (6.3×10^6 km^2; Head, 1976), several approaches have been used to estimate mare thickness and, thus, volume. A stratigraphic approach and morphometry of premare impact craters have been used to determine thicknesses of the nearside maria (e.g., DeHon, 1979). Values average less than about 400 m for most areas, with local regions in excess of 1–2 km. Other techniques include geochemical analyses of postmare impact craters, orbital sounding data, surface geophysical measurements, and topographic data from unfilled craters and basins. Together, these data show that the central parts of unmodified impact basins might contain lenses of basalt as thick as 6–8 km, but that most of the deposits in the basin shelves are generally less than about 2 km thick.

On the basis of stratigraphic evidence, the total volume of surface mare deposits is about 1×10^7 km^3, a volume that corresponds to less than 1% of the total volume of the lunar crust. If this melt volume represented the majority of the partial melting products generated in the zone between 100 and 300 km depth (assuming 10% partial melting), this would imply that less than $\sim 5\%$ of the zone underwent partial melting. Because of the density barrier of the lunar crust, more melts might be produced and solidify at depth than would reach the surface, so that the exact volume of melting is not known.

6.2.3. Duration

The duration of mare volcanism was thought to range from about 3.9 to 3.2 Ga on the basis of ages obtained from mare basalts sampled at the Apollo and Luna sites (Nyquist and Shih, 1992) and crater size-frequency distributions on exposed mare deposits (e.g., Taylor, 1982). Photogeologic studies suggested that volcanism may have continued until about 2.5 Ga in Mare Imbrium (Schaber, 1973). More recent information is available on both the dates of onset and termination of mare volcanism.

Ages of clasts in returned samples led Ryder and Spudis (1980) to suggest that the onset of mare basalt emplacement began prior to 3.9 Ga, perhaps as long as 4.2 Ga in the Apollo 14 region (Taylor *et al.*, 1983). A number of occurrences of intermediate-albedo highland plains have been proposed to represent the onset of mare volcanism prior to the termination of late heavy bombardment. Dark-haloed impact craters in some of these plains deposits (Schultz and Spudis, 1983; Bell and Hawke, 1984; Hawke *et al.*, 1990) and telescopic spectra that show the mare basalt affinities of the crater ejecta, support the idea that they represent impacts into buried mare deposits. These deposits are known as *cryptomaria* (Head and Wilson, 1992), and their recognition and distribution are discussed in a later section.

The duration of mare volcanism has been suggested to extend beyond the 2.5 Ga age proposed for the youngest Imbrium flows. Evidence for this includes possibly young flows dated using crater degradation ages, reinterpretation of the age of the youngest Imbrium flows [between 1.5 and 2.0 Ga (Schultz and Spudis, 1983)], and lava flows embaying Lichtenberg, a crater of Copernican age, and several apparently young mare deposits (Spudis, 1989). These latest deposits notwithstanding, lunar mare volcanism has apparently not been volumetrically significant surface process since the Imbrian (around the late Archean on Earth).

6.3. MORPHOLOGY

A rich and interesting variety of volcanic landforms have been observed in, and associated with, the maria (Figures 6.2–6.4), and spacecraft images documenting these, in addition to those reproduced here, can be found elsewhere (Head, 1976; Schultz 1976a; Head *et al.*, 1981; Greeley, 1985; Wilhelms, 1987; Christiansen and Hamblin, 1995). The presence, absence, and scale of these features can be used to develop an understanding of the nature of eruption conditions, which in turn reveals implications for the nature of source regions and modes of magma ascent. Here, we summarize the main features to address these issues.

6.3.1. Lava Flows and Sinuous Rilles

Flow fronts revealed at low lighting conditions in telescopic images were along the first evidence for the volcanic origin of lunar maria (e.g., Wilhelms, 1987), and the lack of similar evidence has been a factor in the debate about the origin of the mercurian plains. Lunar lava

Mare Imbrium

Rimae Prinz

Figure 6.2. Lunar lava flows and sinuous rilles. (Left) Oblique view of Eratosthenian-aged lava flows in Mare Imbrium; (right) sinuous rilles in the Rimae Prinz region east of the Aristarchus Plateau.

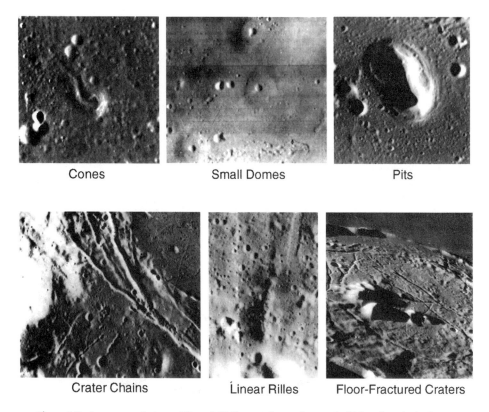

Figure 6.3. Lunar mare features. (Upper left) Cone at the southern end of Mare Serenitatis; (upper middle) series of small mare domes with summit pits; (upper right) pit crater in the lunar maria; (lower left) crater chains in the southwest part of Mare Serenitatis; (lower middle) cones aligned along Rima Parry V; (lower right) the floor-fractured crater Gassendi.

flow units (Figure 6.2) have been mapped on the basis of topographic, albedo, and color boundaries (e.g., Wilhelms, 1987), but distinctive, individual flow fronts are not common because flow thicknesses approach the thickness of the impact-generated regolith (5–10 m), and impact degradation has apparently obscured individual flow unit boundaries. The Eratosthenian-aged flows of Mare Imbrium are the most spectacular examples of flow units on the Moon, and occur in three phases, extending into the basin interior for distances of 1200, 600, and 400 km from the southwestern basin margin. The large volumes are comparable to some of those observed in the Columbia River basalts and this, together with individual flow lengths, led to the interpretation of high effusion rates (Schaber, 1973).

Sinuous rilles (Figure 6.2) are meandering channels that occur primarily in and along the edges of the lunar maria. They range in width up to about 3 km and in length from a few kilometers to more than 300 km. Individual sinuous rilles are commonly relatively constant in width along their length, and display a variety of planimetric and cross-sectioned forms. Sinuous rilles sometimes originate in circular or elongate depressions and some of these are located in the highlands adjacent to the maria. Sinuous rilles terminate in the maria, commonly shallowing imperceptibly until they are no longer visible.

The general morphologic similarity between sinuous rilles and terrestrial lava channels and tubes led to the interpretation that they were of similar origin (e.g., Greeley, 1971).

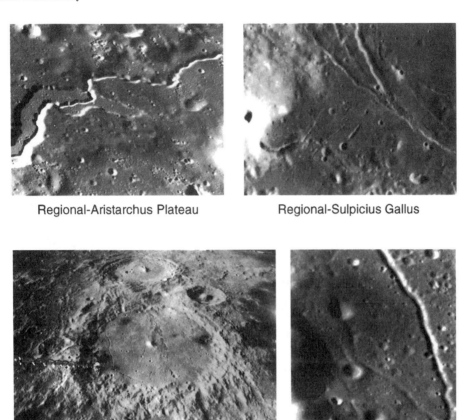

Figure 6.4. Dark mantle deposits. (Upper left) Regional deposit in the vicinity of Schroeters Valley in the Aristarchus Plateau (looking south); (upper right) regional deposits surrounding linear rilles in the southwest part of Mare Serenitatis; (lower left) oblique view of the crater Alphonsus (looking south) showing the dark-halo craters along linear rilles at the edges of the crater floors; (lower right) closeup view of one of the dark-halo craters on the Alphonsus crater floor.

However, sinuous rilles are generally an order of magnitude larger and often much more highly sinuous than terrestrial lava channels. Many of the characteristics of lunar sinuous rilles unexplained by simple lava channel/tube models can be accounted for by thermal erosion (Hulme, 1973, 1982; Carr, 1973). Thermal erosion processes presently appear to best explain the majority of characteristics of larger sinuous rilles. Lava channels and tubes do exist on the Moon, however, and some thermal erosion may have been initiated or enhanced by initially covered lava channels, or controlled by preexisting structures.

The length, width, depth, and the nature of rille source regions provide important information on eruption conditions associated with large sinuous rilles. Hulme (1973) calculated that a 50-km-long rille in the Marius Hills had an effusion rate of 4×10^4 m^3 s^{-1}, an eruption duration of about 1 year, and a total volume of about 1200 km^3. Wilson and Head (1980) and Head and Wilson (1980) analyzed the characteristics of source depression of sinuous rilles, which led to independent evidence for extremely high effusion rate eruptions of long duration. Together, these studies suggest that key factors in the formation of sinuous rilles by thermal erosion are twofold: (1) turbulent flow (requiring high effusion rates and aided by low yield strength) and (2) sustained flow (implying very long eruptions and thus very high eruption volumes to cause the continued downcutting to the observed rille depths). This lies in dramatic contrast to typical eruption volumes for shield-related flows on Earth, which are much less than $1\,km^3$ (Peterson and Moore, 1987), with the largest historic lava flow (Laki, Iceland) being about $12\,km^3$!

6.3.2. Shields and Calderas

Some of the most prominent volcanic features on the Earth, Mars, and Venus are large (>50 km diameter) shield volcanoes and associated calderas. These mark the accumulation of thousands of individual small lava flows from shallow magma reservoirs fed over long periods of time by deeper sources. No shield volcanoes larger than about 20 km have been observed on the Moon (Guest and Murray, 1976). Instead, lunar shields (Figure 6.2) are represented by about 50 small structures (3–20 km in diameter) similar to those seen on Earth in Iceland and Hawaii (Head and Gifford, 1980). The small lunar shield volcanoes (domes) have summit pits ($\sim$1–3 km diameter) but there is little evidence for the presence of large circular calderalike structures as seen on the Earth, Mars, and Venus. Some smooth-rimmed craters in the size range of 20–40 km have been interpreted as volcanic in origin (DeHon, 1971), but these types of features are much more likely to be modified impact craters than shieldlike structures or calderas (e.g., Wilhelms, 1987).

How do we account for the absence of major shield volcanoes on the Moon? Shield volcanoes are commonly built up of large numbers of flows whose lengths average less than the basal diameter of the edifice. These flow are derived from a shallow magma reservoir (commonly underlying or associated with a caldera) where magma stalled because it reached a neutral buoyancy zone (e.g., Ryan, 1987; Wilson and Head, 1990). Volume changes and excess pressures in shallow reservoirs lead to summit and flank eruptions, caldera formation, and edifice construction. Thus, the presence of shield volcanoes and calderas strongly implies the presence of shallow neutral buoyancy zones, the stalling and evolution of magma there, the production of numerous eruptions of relatively small volume and duration, and associated shallow magma migration to cause caldera collapse. The apparent absence of large shield volcanoes and calderas on the Moon implies that shallow neutral buoyancy zones are not common there, and that magma has not been extruded in continuing sequences of flows of relatively short duration and low volume or shallow source regions (e.g., Head and Wilson, 1991). In some cases, however, magma may stall in the near-surface environment (see discussion below). For example, floor-fractured craters have been interpreted to be the location of shallow sill-like intrusions (Schultz, 1976b).

6.3.3. Volcanic Complexes

Several areas on the Moon show unusual concentrations of volcanic features and most of these occur in Oceanus Procellarum. Two of the most significant are the Marius Hills ($\sim$35,000 km^2), which display 20 sinuous rilles and over 100 domes and cones (Weitz and Head, 1999) (Figure 6.3), and the Aristarchus Plateau/Rima Prinz region ($\sim$40,000 km^2), which is dominated by 36 sinuous rilles (Guest and Murray, 1976; Whitford-Stark and Head, 1977) (Figure 6.4). The large number of sinuous rilles in these two locations suggests that the complexes are the site of multiple high-effusion-rate, high-volume eruptions, which may be the source of much of the lava in Oceanus Procellarum emplaced during the Imbrian and Eratosthenian Periods (Whitford-Stark and Head, 1980).

6.4. PYROCLASTIC VOLCANISM

Regions of very low albedo compared with the mare and surrounding highlands have been noted on the Moon since the telescopic observation phase of exploration (e.g., Wilhelms, 1987). These deposits (Figure 6.4) consist of two types: (1) regional dark mantle deposits in excess of about 2500 km^2 that are scattered across the nearside (e.g., Taurus-Littrow, Sulpicius Gallus, Rima Bode, and Aristarchus Plateau) and (2) many hundreds of smaller localized dark mantle deposits that consist of dark-halo craters and related deposits. Dark mantling deposits attracted attention because they appear to drape and mantle surrounding highland terrain. Thus, in contrast to mare lavas emplaced as fluid flows to form flat plains, dark mantle deposits were interpreted to form from pyroclastic eruptions. In addition, their very low albedo and the tendency of Copernican-aged crater rays to disappear at the edges of these deposits led to the incorrect interpretation that they might represent very young volcanic activity on the Moon. Thus, early in the Apollo lunar exploration program, these deposits figured prominently in the development of the exploration and sampling strategy (see Wilhelms, 1987), and the Taurus-Littrow dark mantle deposits were explored during the Apollo 17 mission.

In early analyses, dark-halo craters found on the ejecta deposits of Copernican-aged craters (e.g., Copernicus H) were thought to be the sites of very young volcanic eruptions, but further analyses showed that these represented impact craters that excavated darker, mare-rich material from below (Schultz and Spudis, 1979; Bell and Hawke, 1984). However, a large number of small dark-halo craters are located along fractures and in large crater floors, and do not have the typical appearance of impact craters; these are thought to represent volcanic eruptions. Analysis of this latter type of dark-halo crater in Alphonsus (Figure 6.4) shows that they most plausibly result from vulcanian-style eruptions (Head and Wilson, 1979; Coombs *et al.*, 1990a). Hawke *et al.* (1989a) and Coombs *et al.* (1990b) mapped similar deposits in many other parts of the Moon. On the other hand, regional dark mantle deposits tend to occur at impact basin margins and in association with large vents and sinuous rilles that are candidates for large-volume sustained eruptions. Deposits range from patchy regional units (the margins of the Taurus-Littrow deposit) to thick deposits perhaps reaching depths of many tens of meters (Aristarchus Plateau; Zisk *et al.*, 1977; Lucey *et al.*, 1986; McEwen *et al.*, 1994). Recent analyses of Clementine data suggest that additional dark mantle deposits may occur on the lunar farside (Craddock *et al.*, 1997; Hawke *et al.*, 1997).

The Apollo samples provide important information about the nature of dark mantle deposits. The volcanic beads seen in all of the Apollo samples indicate that gas-rich eruptions occurred on the Moon, even though the Apollo samples revealed that the Moon is volatile-poor and the rocks lack water. The source of the gas driving the pyroclastic eruptions is thought to be CO derived from graphite oxidation in the shallow portions of the crust (e.g., Sato, 1979; Fogel and Rutherford, 1995). Gas exsolution and bubble coalescence in the lunar near-surface environment is accompanied by extremely rapid expansion into the lunar surface vacuum, breakup of vesicular magma into beads, and the widespread dispersal of these beads as a result of low gravity and lack of an atmosphere (Wilson and Head, 1981). The submillimeter-size volcanic beads returned in Apollo and Luna samples consisted of two types: glasses, which appear to have quenched rapidly, and slightly larger crystallized beads, which are interpreted to have experienced longer cooling times (e.g., Heiken et al., 1974). Laboratory cooling experiments have shown that the crystallized beads experienced cooling durations longer than blackbody cooling in a vacuum (e.g., Arndt and von Engelhardt, 1987), suggesting that they were erupted into an environment characterized by an optically thick hot plume that insulated them from simple radiative cooling in a vacuum. Assessment of pyroclastic eruption environments (e.g., Head and Wilson, 1989) suggests that the inner parts of a pyroclastic fountain would be optically thick and could provide such an environment. Thus, crystallized beads are interpreted as having been erupted as part of a gas/pyroclast fountain that was optically thick enough to inhibit cooling, and the glass beads formed by more rapid cooling in the marginal optically thin part of the cloud (Figure 6.5) or cooled more rapidly because of their small sizes (Weitz et al., 1999).

Dark mantle deposits composed of these volcanic beads have been studied telescopically (Gaddis et al., 1985; Hawke et al., 1989), and using Galileo (Greeley et al., 1993) and Clementine UV/VIS data (Weitz et al., 1998). Clementine data have sufficient spectral and spatial resolution to determine if the deposits are dominated by crystallized beads or glasses. Weitz et al. (1998) studied seven regional dark mantle deposits using Clementine data and assessed the degree of crystallinity and the nature and distribution of their deposits to constrain eruption conditions. Using the ratio of 415/750 and 750/950 nm values, estimates were made of the ratio of glass beads to crystallized beads. One extreme is represented by patches of dark mantle in the Sinus Aestuum region, which appear to be dominated by crystallized beads. The other extreme is represented by the Aristarchus Plateau dark mantle deposit, which is dominated by glass beads. Other regional dark mantle deposits (Taurus-Littrow, Sulpicius Gallus, Rima Bode, and Mare Vaporum) lie between these extremes, and represent mixtures of glass, crystallized beads, and other soils. Based on the distribution and crsytallinity of the beads within the deposits, Weitz et al. (1998) inferred the most plausible locations of the source vents and the likely optical density of the volcanic plumes.

An unusual dark ring about 180 km in diameter has been noted in the Orientale region and interpreted to be of pyroclastic origin (Greeley et al., 1993). Head et al. (1997) identified an elongate vent located central to the ring and interpreted the ring to be produced by a pyroclastic eruption from this vent. Analysis showed that the plume that produced the dark mantle ring was probably umbrella-shaped and gas-rich with a height of 20 km and characterized by ejection velocities of 320 m s^{-1} (Weitz et al., 1998). In contrast, other regional deposits are interpreted to have been produced by Hawaiian-style fire-fountain eruptions that formed both mare lavas and submillimeter beads.

Assessment of the influence of the lunar environment on the types of pyroclastic eruptions described above has therefore led to a better understanding of the mode of emplacement of the regional dark mantling deposits (Wilson and Head, 1981; Weitz et al., 1998). Magmas containing even very small amounts of volatiles will undergo near-surface gas exsolution and marked decompression, and vertical and horizontal gas expansion, guaranteeing that pyroclasts are dispersed away from the vent to produce deposits of pyroclastic material (Wilson and Head, 1983). The wide variety of glass beads of probable pyroclastic origin in Apollo and Luna samples (e.g., Heiken et al., 1974; Delano, 1986) and the extensive deposits surrounding individual vents (hundreds to thousands of square kilometers) (e.g., Weitz et al., 1998) attest to their wide dispersal. A variety of compositions are also inferred from remote sensing data (Gaddis et al., 1985). These regional dark mantling deposits are interpreted to be the result of sustained effusive eruptions, in which continuous gas exsolution in the lunar environment causes 'Hawaiian-like" fountaining. Pyroclasts are spread to ranges of tens to hundreds of kilometers and eruptions continue for a relatively long

Moon and Mercury

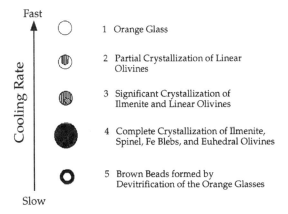

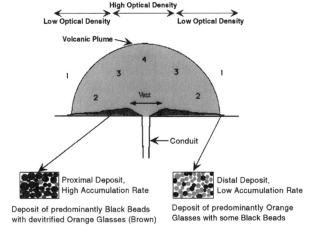

Figure 6.5. Dark mantle pyroclastic bead types and interpreted cooling environments. (Top) Types of volcanic beads identified in the Apollo 17 74001/2 core, described as a function of interpreted cooling rate. Orange glasses form at the fastest cooling rate while slower rates allow larger and more numerous crystals to develop in the beads. Brown beads are interpreted to have formed by devitrification of the orange glass. (Bottom) Interpreted position of the different bead types in an eruption and in the resulting deposits. Orange glasses form in the outer portions of the plume where the cooling rates are high because of the low optical density of the plume. The optical density increases further inward and beads are able to cool more slowly and to develop crystals. The deposit is dominated by orange glasses at a greater distance from the vent while black beads are concentrated more in the proximal part of the deposit.

period of time to build up the observed deposit thicknesses (Wilson and Head, 1981, 1983). The common association of these deposits with sinuous rilles and their sources suggests that the rilles are among the sites of the sustained eruptions that produced the mantling deposits. Regional dark mantling deposits appear to mark the locations of high-effusion-rate, long-duration eruptions. Despite the similarity between the Moon and Mercury in terms of the lack of an atmosphere, lunar pyroclastic deposits show several distinctive characteristics relative to the origin of their volatiles and the role of low gravity in their dispersal.

6.5. NONMARE VOLCANISM

6.5.1. Lunar Light Plains

During the telescopic phase of the mapping of the Moon (see Wilhelms, 1987) and in the early days of the Apollo missions, the Cayley Formation (intermediate- to high-albedo "light" plains in the highlands, and around the margins of some maria) was thought to be of volcanic origin because of its smooth nature, and the fact that it fills highland craters. Its stratigraphic position and absolute age based on crater counts, place the Cayley Formation between the latest basin-scale impact events, such as Imbrium, and the earliest mare. Thus, it was thought that a significant phase of highland (i.e., high-albedo) volcanism characterized the early Imbrian history of the Moon. The Apollo 16 mission was targeted to explore light plains units in the Descartes region of the central highlands and their possible volcanic source regions. Instead, pervasive impact-generated breccias were found at the Apollo 16 site, causing this model to fall into disfavor. Subsequently, interpretation of the light plains at the site centered on two major concepts. Some investigators thought they originated as primary basin ejecta blankets, either from Imbrium or from Orientale (e.g., Moore *et al.*, 1974). Others considered that they formed through the mixing of ejecta from basin impacts with local material (e.g., Head, 1972; Oberbeck *et al.*, 1974). Additional field and laboratory study of the impact cratering process led to the conclusion that light plains can be produced by ballistic erosion and sedimentation processes (Oberbeck *et al.*, 1974; Oberbeck, 1975). Large cratering event ejecta reimpacts and mixes with local target material, obscuring preimpact topography and resulting in the formation of relatively smooth light plains deposits in low-lying regions. Light plains formed in this way do not represent an ejecta "blanket" laid down over the surface, but rather an ejecta deposit that consists of a dynamic mixture of primary ejecta and local material, where the proportion of primary ejecta material to local target material decreases exponentially with radial distance from the primary crater (e.g., Oberbeck *et al.*, 1974; Head, 1974). As we will discuss later, this changing paradigm evolved at about the same time as the Mariner 10 data revealed the presence of Cayley-like plains on Mercury.

Light plains display a wide range of crater ages, leading some investigators to continue to support a volcanic origin for many light plains deposits (e.g., Neukum, 1977). This, together with the possibility that in some cases, cryptomaria might exist below a floor of impact breccias forming smooth plains, has caused many workers to assess smooth plains origins on a case-by-case basis, using a host of methods to determine their origins (e.g., Antonenko *et al.*, 1995). In addition, some light plains and related nonmare activity may be associated with possible KREEP-related volcanism (e.g., Spudis, 1978; Hawke and Head, 1978).

6.5.2. Nonmare Volcanic Domes and Cones

In addition to smooth plains, features that have been attributed to highland volcanism include domes, mounds, and crater chains. Although landforms at the Apollo 16 site originally attributed to highland volcanic constructional features have since been reinterpreted as impact-related, there are several features that have albedo and spectral characteristics different than maria, and are morphologically unlike mare volcanic features. The most prominent example of these are the Gruithuisen and Marian domes and cones, located in the northwest part of the lunar nearside. Some of these features (Figure 6.6) are very large (typically 20 km in diameter and over 1 km in height), have shapes that are suggestive of both

Figure 6.6. Gruithuisen domes. Oblique view of the Gruithuisen domes showing their distinctive morphology and superposition on the highlands surface.

extrusion of viscous magma (the distinctive Gruithuisen domes) and explosive volcanism (the Marian cone), and have several unusual associated features (Head and McCord, 1978; Chevrel *et al.*, 1999). Their spectral signature has a distinctive downturn in the ultraviolet (Head and McCord, 1978; Chevrel *et al.*, 1999), and are similar spectrally to "red spots" mapped elsewhere (e.g., Wood and Head, 1975; Bruno *et al.*, 1991).

These deposits are apparently contemporaneous with maria emplacement. Ejecta from the post-Imbrian basin Iridium crater predates the Gruithuisen domes (e.g., Wagner *et al.*, 1996). No known material from the lunar sample collection fits the spectral characteristics of this unit, and thus its petrogenetic affinities are unknown. Contemporaneity with maria suggests that there may be petrogenetic linkages; one possibility is that mare basalt diapirs stalled at the base of the lunar crust might partially melt and remobilize basal crustal layers, leading to localized extrusion of magmas more viscous than the maria (e.g., Malin, 1974). Interestingly, this process might also be applicable to Mercury and indeed, the Gruithuisen domes have been cited as possible analogues to some features there (Malin, 1978).

6.6. RECOGNITION AND ASSESSMENT OF CRYPTOMARIA

Cryptomaria are marelike deposits obscured from view by younger deposits of higher-albedo material (Head and Wilson, 1992), commonly ejecta from impact craters and basins. Cryptomare deposits have been identified through analysis of dark-halo craters (Schultz and Spudis, 1979, 1983; Hawke and Bell, 1981), which penetrate through higher-albedo ejecta material to mare material below and emplace low-albedo material in a halo around the crater. Cryptomare deposits have also been revealed in multispectral images of the Moon (e.g., Head *et al.*, 1993; Greeley *et al.*, 1993; Mustard and Head, 1996). The light plains emplacement process incorporates local material in basic ejecta deposits, which can be recognized by mixing analysis of spectral endmembers (e.g., Mustard and Head, 1996). Using these and related data, the number of candidate cryptomaria (Figure 6.7) is such that the total area covered by mare deposits may now exceed 20% of the lunar surface area (Antonenko *et al.*, 1995).

The time of initiation of extrusive mare volcanism is presently not known. The distribution of candidate cryptomaria clearly indicates the presence of regional deposits of volcanism, in some cases with significant depths (Mustard and Head, 1996), emplaced prior to the formation of Orientale and several other major impact basins. If earlier cryptomaria exist, they are likely to be increasingly obliterated by superposed large impacts and, because of the small thicknesses of deposits relative to crater excavation depths, the percentage of basalt clasts in impact breccias is also likely to be small. These two factors conspire to obscure evidence for mare volcanism and related extrusive volcanic activity in earliest lunar history. These observations are also important for the detection and discrimination of plains of volcanic origin on Mercury.

6.7. LAVA PONDS

In the large continuous nearside maria the analysis of the characteristics and interpretation of lunar eruptive events making up the stratigraphic sequence is complicated by the presence of multiple flows, burial of source vents, and possible variation of source regions within basins. To address these problems, Yingst and Head (1997, 1998, 1999) have examined clusters of discrete mare deposits that occur in isolated patches in the highlands adjacent to the major maria. Following previous workers, these ponds are interpreted to represent single eruptive phases (e.g., Whitford-Stark, 1982). Evidence for this comes from deposit homogeneity in albedo, multispectral image characteristics, and crater density, as well as their self-contained nature and lack of other evidence for multiple eruptive events.

For example, Yingst and Head (1997) examined the lateral distribution and topographic occurrence of mare deposits in the pre-Nectarian South Pole–Aitken basin, and the post-Imbrian Orientale basin and its environs (Figure 6.7). The total area covered by 52 lava ponds in the South Pole–Aitken basin is approximately 290,000 km^2, less than 5% of the total surface area covered by lunar maria. Extension of this analysis to the global distribution of lava ponds (Figure 6.7) reveals a total of 305 lava ponds considered to be candidates for individual eruptive episodes. Although a wide range of volumes are observed ($\sim$10–3860 km^3, more typical values are about 350–400 km^3 (Figure 6.8).

These data have important implications for lunar eruption conditions, magma transport phenomena, and the filling of the major mare basins. Typical volumes for lava ponds lie between those for single eruptive phases for Hawaiian shallow magma reservoirs (e.g., <1 km^3 per eruption (Peterson and Moore, 1987) and the largest estimates for terrestrial flood basalts (e.g., $\sim$1200 km^3 for the Roza Member of the Columbia River Basalt; Tolan *et al.*, 1989), but well within the terrestrial flood basalt range. These values suggest that the individual eruptive phases that were the sources of the lava ponds were more analogous to terrestrial flood basalts, rather than being fed from a shallow reservoir, as in the Hawaiian example. This interpretation is consistent with the presence of sinuous rilles in some of the lava ponds (Yingst and Head, 1997) and the lack of associated shield volcanoes and calderas here and elsewhere on the Moon that might constitute evidence for large magma reservoirs in the shallow lunar crust. These volumes are consistent with magma transport mechanisms that call for overpressurized reservoirs at the base of the crust, a neutral buoyancy zone between mantle melts and the lower-density primary lunar crust. Propagation of magma-filled cracks (dikes) from depth through the lunar crust would inhibit the development of shallow

Moon and Mercury

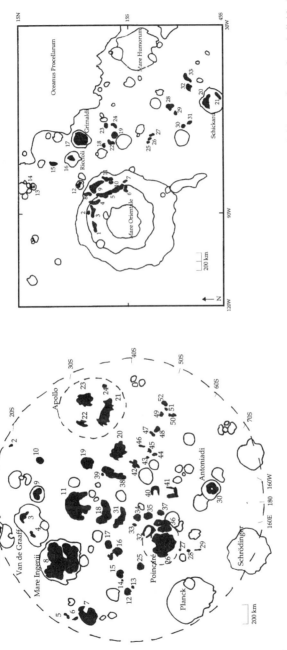

Figure 6.7. Areal distribution of lava ponds (shown in black) in South Pole–Aitken (left) and in the vicinity of the Orientale basin (right). Numbers refer to individual ponds listed and described in Yingst and Head (1997).

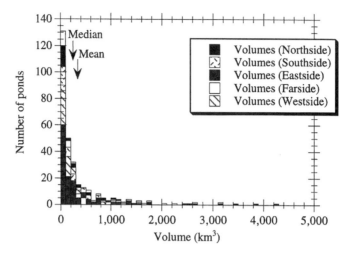

Figure 6.8. Global frequency distribution of lava pond volumes for those ponds considered the best candidates for single eruptive phases. From Yingst and Head (1997).

reservoirs, and produce a range of erupted volumes tending toward those typical of flood basalts (e.g., Head and Wilson, 1992). In addition, the generally inverse correlation of pond occurrence and crustal thickness is consistent with a model in which an overpressurized reservoir emplaces dikes in areas of thicker crust and extrudes magma onto the surface in areas of thinner crust (Figure 6.9).

Typical volumes of lava ponds can also be used to estimate reservoir volumes. Accounting for the dike volume between the source and the surface, and using the relationships for volumes propagated out of a reservoir in individual overpressurization events (Blake, 1981), the typical volumes of individual reservoirs have been estimated to represent a sphere with a diameter of slightly less than 100 km (Head and Wilson, 1992; Yingst and Head, 1997). If mare pond source reservoirs represent portions of rising mantle diapirs, these values may provide estimates of diapir size, and indeed the values lie within the range estimated for some lunar mantle diapirs (e.g., Hess, 1991). If lava pond separation distances are related to distances between individual reservoirs and thus between plumes, the density of rising plumes was high, at least in the areas examined in these analyses.

Typical lava pond volumes ($\sim$350–400 km^3) can also be used to estimate the number of eruptive events in the contiguous lunar maria under the assumption that eruptive dynamics and source regions are the same. If the total volume of lunar maria ($\sim 1 \times 10^7$ km^3; Head, 1975b) were emplaced in such events, the total number of eruptive events characterizing the emplacement of the contiguous lunar maria would be $\sim$27,000. Larger basins (such as Imbrium) would be filled by many thousands of flow units, while smaller basins (such as Humorum) would be characterized by hundreds of eruptive phases. Although these data provide some interesting guidelines, they only represent average values, and individual geologic mapping as well as consideration of magma eruption dynamics in evolving crustal and lithospheric environments must be used in individual basins. For example, Schaber (1973) has estimated that the three young flows in the Imbrium basin have a combined total volume of 4×10^4 km^3, considerably higher than the average values used here. In addition,

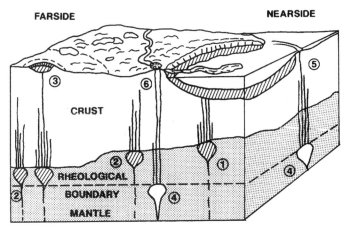

Figure 6.9. Magma transport from neutral buoyancy zones. Diagrammatic representation of the emplacement of secondary crust on the Moon. At 1, early basaltic magmas rise diapirically to the density trap at the base of the crust. Those below topographic lows (thin crust) associated with lunar impact basins are in a favorable environment for dike propagation and extrusion of lavas to fill the basin interior. Diapirs reaching the base of the thicker crust on the farside and parts of the nearside (2) at the same time, stall and propagate dikes into the crust, most of which stall and solidify in the crust and do not reach the surface. Variations in regional and local compensation produce a favorable setting for emplacement of some lavas in craters and the largest basins on the farside (3). With time, the lithosphere thickens and ascending diapirs stall at a rheologic boundary, (4) building up excess pressure to propagate dikes toward the surface. At the same time, loading and flexure of the earlier mare deposits creates a stress environment that favors extrusion at the basin edge (5); lavas will preferentially emerge there, flowing into the subsiding basin interior. The latest eruptions are deepest and require high stress buildups and large volumes in order to reach the surface; thus, these tend to be characterized by high volume flows and sinuous rilles (6). Ultimate deepening of source regions and cooling of the Moon causes activity to diminish and eventually to cease. From Head and Wilson (1992).

changes in topography as mare basins fill will also influence the style of emplacement and the mode of occurrence of the deposits (Head and Wilson, 1992).

6.8. CONTIGUOUS MARIA

In contrast to most of the individual lava ponds discussed above, mare deposits in the contiguous maria (Figure 6.1) are known from remote sensing data to be heterogeneous in composition (in terms of both mineralogy and elemental abundances (e.g., Metzger *et al.*, 1977; Adler and Trombka, 1977; Pieters, 1978; Adams *et al.*, 1981; Head *et al.*, 1981; Wilhelms, 1987; Johnson *et al.*, 1991; Lucey *et al.*, 1995; Staid *et al.*, 1996). There is also evidence for systematic variations of basalt types in space and time (e.g., Head, 1976; Pieters, 1978; Taylor, 1982; Hiesinger *et al.*, 1998a,b). Sequences and modes of emplacement of basaltic lavas began to be clarified by studies of mare stratigraphy in individual basins (e.g., Pieters *et al.*, 1975; and Lucey *et al.*, 1991, Mare Humorum; Pieters *et al.*, 1980, Flamsteed region; Head *et al.*, 1978, Mare Crisium; Whitford-Stark and Head, 1980, Oceanus Procellarum; Whitford-Stark, 1979, Mare Australe; Whitford-Stark, 1990, Mare Frigoris; Staid *et al.*, 1996, Mare Tranquillitatis; Hiesinger *et al.*, 1998a,b, several nearside basins).

These data contribute to the emerging picture of the distribution of mare basalts in space and time, data crucial to the understanding of mantle heterogeneity and basalt peterogenesis. Together, these data show that for the circular maria, large quantities of lavas were emplaced in the late Imbrian Period (Figure 6.10) in the center of the basins, causing basin loading and subsidence. Later lava emplacement was less voluminous, often originated from source vents along the edge of the basin (e.g., the Imbrium flows; Schaber, 1973), and flowed downslope to continue to infill the sagging interior (e.g., Solomon and Head, 1980). In the noncircular maria, lava volumes were less, subsidence less extreme, and flows tended to cover larger areas throughout the emplacement of the maria (e.g., Pieters *et al.*, 1980; Whitford-Stark and Head, 1980; Whitford-Stark, 1990).

Recent analyses of the maria in Australe, Tranquillitatis, Humboldtianum, Serenitatis, and Imbrium have been undertaken using unit boundaries defined by multispectral images. Crater size-frequency distribution ages for these units were obtained and correlations of TiO_2 content and unit volumes in space and time could be derived (Hiesinger *et al.*, 1998a,b). These data support the general conclusion that mare volcanism represented by exposed mare deposits was at its peak during the Late Imbrium Period, and declined drastically during the Eratosthenian. Volcanism lasted longer in the western basins and in the younger basins. In contrast to earlier models derived from returned samples, in which mare volcanism begins with high TiO_2 content and shows a decrease with time (e.g., Taylor, 1982), these data show no distinctive correlation between deposit age and corresponding TiO_2 content in any of the

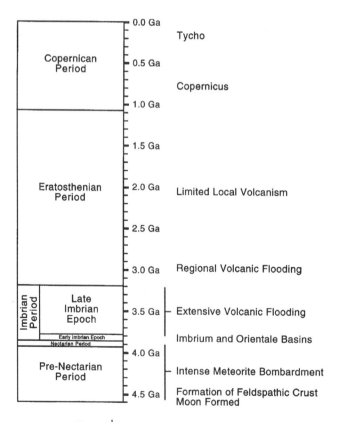

Figure 6.10. Chronology of lunar volcanism.

basins. Rather, within each basin, TiO_2 content appears to vary independently with time, and generally simultaneous eruptions of TiO_2-rich and TiO_2-poor basalts occur. If basin stratigraphies are compared, however, basalts with the highest TiO_2 content appear to be emplaced at approximately the same time in most of the basins.

Although studies such as these have resulted in an increased understanding of the styles, volumes, and modes of emplacement of mare lavas, spectral units are defined by the characteristics of mature soils, and a close correlation of fresh basalt types in the Apollo and Luna collections to these units remains to be completely determined. The situation is even more acute for basalt types identified by remote sensing techniques for areas away from the landing sites. The wide variety of basalt types identified by remote sensing techniques (e.g., Pieters, 1978, and subsequent studies) suggests that perhaps as many as two-thirds of the mare types on the Moon remain unsampled. High-resolution Clementine multispectral data provide hope for the identification of fresh basalt types in crater walls, thereby linking these both to the sample collection and to regional stratigraphic units (e.g., Staid and Pieters, 1998).

Attempts have been made to estimate the general mare volcanic flux on the Moon by measuring surface areas and thicknesses of mare units of different ages in the contiguous maria (Hartmann et al., 1981; Wilhelms, 1987; Head and Wilson, 1992). The estimated volume of Upper Imbrian deposits is $\sim 9.3 \times 10^6$ km^3 and if this was emplaced evenly over the Late Imbrian (Figure 6.10) it would imply an average extrusion rate of 0.015 km^3 a^{-1} during this time. Similar approaches yield an average flux for Eratosthenian mare deposits of $\sim 1.3 \times 10^{-4}$ km^3 a^{-1} and for the candidate Copernican mare deposits of less than 2.4×10^{-6} km^3 a^{-1} (Head and Wilson, 1992). These values are miniscule compared with the current rate of formation of Earth's oceanic crust (intrusive and extrusive) of about 20 km^3 a^{-1}, and the current terrestrial global volcanic output of 3.7–4.1 km^3 a^{-1}. At peak average flux, the lunar global volcanic output rate was about 10^{-2} km^3 a^{-1}, which is about the same as recent output rates for single volcanoes on Earth such as Vesuvius (1.07×10^{-2} km^3 a^{-1}) and Kilauea, Hawaii (1.7×10^{-2} km^3 a^{-1}) (Crisp, 1984). Of course, mare deposits were not emplaced evenly throughout the Eratosthenian and Copernican periods and thus local effusion rates were certainly much higher. For example, some sinuous rilles (Hulme, 1973) have been estimated to represent emplacement of ~ 1000; km^3 of lava in a yearlong eruption. This value is the equivalent of $\sim 70,000$ years of the average flux!

Lunar mare volcanic eruptions peaked in the Late Imbrium Period, and were relatively infrequent, but characterized by large volumes of lava in individual eruptions, yielding a net very low average volcanic flux. Eruptions have been extremely rare over the last 2.5–3 Gyr, but the age of the youngest eruptions remains to be determined with confidence.

6.9. EFFECTS OF A GLOBAL LOW-DENSITY CRUST

Petrologic and geochemical evidence indicates that source zones for the mafic mare basalt magmas lay within the lunar mantle, that melts were less dense than the mantle and so would have been naturally buoyant, but were denser than the anorthositic crust, making their eruption to the surface difficult (e.g., Solomon, 1975; Taylor, 1982). Early work (Solomon, 1975; Wilson and Head, 1981) showed how the positive buoyancy of these melts within the mantle could compensate for their negative buoyancy within the crust provided that each eruption took place through a dike that was continuously open to the surface from a source zone at a sufficiently great depth in the mantle. This kind of analysis was extended (Head and

Wilson, 1992) to take into account the fact that magmas collecting in reservoirs are subject to a fluid pressure greater than the lithostatic pressure in the surrounding country rocks. This excess pressure is commonly inherited from the volume increase during the partial melting process. The excess pressure in the melt allows upward dike propagation to occur into overlying rocks within which the melt is negatively buoyant.

The excess pressure required to drive dikes to the surface (Head and Wilson, 1992) from a magma reservoir at the base of the lunar crust is on the order of 15 MPa for the nearside crustal thicknesses of ~ 64 km, and ~ 20 MPa for typical farside crustal thicknesses of ~ 86 km. Although these excess pressures would allow dikes to propagate to the surface, they would not be large enough to force magma to the surface through the available fractures; slightly higher excess pressure (~ 21 and 28 MPa) would be needed to accomplish this. The above conditions correspond to dikes with mean widths of a few to several hundred meters. The relatively small differences between the excess pressures in magma reservoirs that either would or would not ensure eruptions taking place underline the fact that the conditions for surface eruptions of lunar basalts were finely balanced, and that the crustal thickness reductions provided by the presence of impact basins were a major factor in ensuring that eruptions took place preferentially into such basins (e.g., Head and Wilson, 1992; Yingst and Head, 1997).

6.10. LUNAR ERUPTION CONDITIONS

On the basis of these considerations, it seems likely that throughout much of lunar history, combinations of magma pressure and source depth must have existed that would have allowed magma-filled dikes to penetrate close to, but not reach, the surface. Using models of the stress fields around fluid-filled elastic cracks, Head and Wilson, (1993) showed that dikes with mean thicknesses on the order found above (hundreds of meters) propagating to shallow (~ 1–2 km) depths in the lunar crust were capable of producing surface stress fields leading to the development of graben with widths of ~ 1–3 km. This process was identified as the likely cause of the development of two linear rilles (Rima Sirsalis and Rima Parry V).

The surface manifestation of a dike that does not actually reach the surface can take a range of forms (Figure 6.11) (Head and Wilson, 1996, 1998). If the dike stalls at a sufficiently great depth, there will be some undetectably small amount of surface extension and uplift. If it penetrates to shallower depths, there may still be no noticeable topographic effects at the scale of available images, but incipient failure or activation of preexisting fractures may generate pathways along which gas (probably mainly CO; Sato, 1979) released by magma in the shallowest parts of the dike can reach the surface. Still shallower penetration will lead to a large volume of melt being exposed to the relatively low pressure environment near the surface and will encourage the generation of a greater mass of CO from the magma since the chemical reaction producing it is pressure-dependent. Any development of convective motions in the magma in the dike while it is cooling (particularly likely in a wide dike) will also cycle melt from depth through the low-pressure zone and add to the gas generation process (e.g., Head et al., 1997). Subsequent loss of this gas, coupled with a magma volume decrease on cooling, may lead to collapse features (or event explosion craters) forming on the surface above the dike.

Sufficiently close approach of the dike tip to the surface will cause new fractures to form and allow significant movements along parallel faults to occur. As the dike tip further approaches the surface, the main effect will be for the graben to become progressively deeper

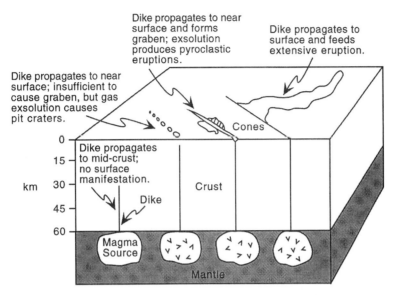

Figure 6.11. Models for the geometry of dike emplacement in the lunar crust and its surface manifestation.

as more strain is accommodated. Very shallow intrusion may lead to further fractures developing on the floor of the graben and will encourage the formation of small secondary intrusions and possible eruptions. This type of activity will eventually produce small cones and localized dark mantle deposits, as in the case of Rima Parry V (Head and Wilson, 1993). Solidification of lava in the tip of the dike and subsequent buildup of pressure to levels that explosively fragment the cap and adjacent country rock can produce vulcanian eruptions similar to those thought responsible for the Alphonsus dark-halo craters (Head and Wilson, 1979) and that form many other small dark mantle deposits (e.g., Hawke *et al.*, 1989; Coombs *et al.*, 1990a,b). Propagation of the dike directly to the surface will produce a variety of landforms. Those dikes just reaching the surface should produce small shields and cones, such as those seen in the Marius Hills and elsewhere on the Moon (Weitz and Head, 1999). Alternatively, dikes that overshoot the surface (Figure 6.11) should produce large-volume, areally extensive flows (Head and Wilson, 2000) such as those seen in central Mare Imbrium. Gas exsolution processes associated with long-duration, high-effusion-rate, Hawaiian-style eruptions are the best candidates for the mode of emplacement of regional dark mantle deposits (e.g., Wilson and Head, 1981; Weitz *et al.*, 1998).

6.11. SYNTHESIS OF LUNAR VOLCANISM

Lunar volcanism consists of mare basalt deposits and nonmare, or highland, volcanism. Mare basalt deposits have a total volume estimated at $\sim 1 \times 10^7$ km^3, and occur preferentially in topographic lows on the nearside. Returned sample, photogeologic, and remote sensing studies show that mare volcanism began prior to the end of heavy bombardment in pre-Nectarian times, during the period of cryptomare formation. Although potentially significant deposits were emplaced during this time, their volumes and affinities to mare or other mafic volcanism have not been firmly established. Mare volcanism continued until the Copernican Period, a total duration approaching 3.5–4 Ga, but the actual age of the last mare volcanism is not well constrained. Stratigraphic analyses show that the flux of mare basalt lavas was not constant, but peaked in nearly lunar history, during the latter part of the Imbrian Period (which spans the period 3.85–3.2 Ga). During this peak period the average volcanic output rate was about 10^{-2} km^3 a^{-1}, a very low relative to the present global terrestrial volcanic output rate and comparable to the present local output rates for individual volcanoes on Earth such as Vesuvius, Italy, and Kilauea, Hawaii.

The morphology of volcanic landforms indicates that many eruptions were of high volume and long duration. Some eruptions associated with sinuous rilles may have lasted about 1 year, during which time about 10^3 km^3 of lava was emplaced. Such an eruption would represent the equivalent in 1 year of about 70,000 years at the average flux. On the basis of the absence of large shield volcanoes and associated calderas, shallow magma chambers are interpreted to have been uncommon.

The observed nearside–farside mare deposit asymmetry can be readily explained by differences in crustal thickness. Magma ascending from the mantle or from a buoyancy trap at the base of the crust should preferentially extrude to the surface on the nearside, but should generally stall and cool in dikes in the farside crust, extruding only in the deepest basins. Dikes that establish pathways to the surface on the nearside should have very high volumes, comparable to the volumes associated with many observed flows and sinuous rilles. An abundance of dikes should exist in the lower crust of the Moon, many more than those feeding surface eruptions. As pointed out by Head and Wilson, (1992), the presence and abundance of such dike swarms have important implications for the interpretation of the average composition of the lunar crust and the composition of basin and crater ejecta.

Mare volcanism is intimately linked to lunar thermal evolution. The interplay between thermal contraction and differentiation leads to net cooling and ultimate contraction of the outer portions of the Moon, resulting in a regional horizontal compressive stress acting on the lunar crust (Solomon and Head, 1980). In addition, overall cooling and deepening of sources require the production of ever larger volumes of magma in order to reach the surface. With time, stresses became sufficiently high that few dikes could open to the surface, causing eruptive activity to be severely diminished in the Eratosthenian, and perhaps to cease in the Copernican Period (e.g., Head and Wilson, 1992). Lower stress levels are required to terminate eruptive activity on the lunar farside, consistent with the Imbrian age of the farside maria and the nearside location of the youngest maria.

Highland (nonmare) volcanism is poorly understood and may be linked to early stages of plutonic and volcanic activity related to initial crustal differentiation (e.g., Taylor, 1982). Some recognizable nonmare volcanic activity persisted into the period of mare volcanism and includes possible KREEP-related plains (e.g., Spudis, 1978; Hawke and Head, 1978) and steep-sided domes that may be related to more viscous magma mobilized during mare basalt emplacement (e.g., the Gruithuisen domes).

Lunar mare deposits provide an example of the transition from primary crusts (the highland crust) to secondary crusts (e.g., Taylor, 1989). Their formation illustrates the significance of several factors in the evolution of secondary crust, such as crustal density, variations in crustal thickness, presence of impact basins, state and magnitude of stress in the lithosphere, and general thermal evolution. These factors are also responsible for the extremely low volcanic flux, even during periods of peak extrusion. Lunar pyroclastic activity was driven primarily by shallow crustal formation of CO, and predominantly occurs during vulcanian, strombolian, and Hawaiian-style eruptions. The low gravity and lack of an atmosphere caused the wide dispersal and unique characteristics of the pyroclasts.

Major strikes have taken place in terms of mapping the volumes and heterogeneity of mare basalts in space and time. Can this heterogeneity be linked to vertical and/horizontal diversity in the mantle source regions? At least four issues must be addressed before this is more confidently known: (1) Superposition of younger maria makes the establishment of early mare basalt volumes and ages difficult, and more refined estimates are required. More sophisticated mixing models and higher-resolution multispectral imaging and spectrometer data will aid in these determinations. (2) The low-density lunar crust likely acted as a density barrier to the buoyant ascent of mare basalt magmas and this was apparently responsible for much of the areal difference in distribution of mare basalt deposits, most notably in the nearside–farside asymmetry, but also in the volumes of individual deposits. The effects of this filter must be understood before models of mantle heterogeneity based on the distribution of volcanic deposits in space and time can be confidently developed. (3) The style of emplacement of mare deposits (ranging from shallow, near-surface dike emplacement to flood basalts) differs sufficiently from typical eruptions from shallow magma reservoirs that volumes of individual eruption events vary widely depending on this factor, rather than on the significance of melting in the source region. More detailed models are required to assess quantitatively these two factors. (4) Thermal evolution is a very significant overprint on the volumes of mare basalts generated, the depth of melting, and the state of stress in the lithosphere and thus the ability of magma to reach the surface. Successful interpretation of the surface record of mare basalt volcanism in terms of petrogenesis must also factor in the influence of this factor.

Using this summary as a background, we now discuss the surface of Mercury for a review of the recognition, mode of occurrence, and style of emplacement of volcanism there.

6.12. VOLCANISM ON MERCURY

Did Mercury, like the Moon, experience crustal intrusion and surface extrusion, resulting in volcanic deposits and landforms during its early thermal evolution? Did secondary crustal formation occupy a significant part of the resurfacing history of Mercury as it did for the Moon? How did the environmental parameters associated with Mercury modify and change the geologic record and eruption styles there, relative to the Moon? And how can the better-known lunar record be used to address these issues?

Such questions may seem trivial after geologic exploration of the Earth, Moon, Mars, and Venus. The dominant endogenic geologic process on these bodies is volcanism, mostly characterized by extrusions of basaltic lavas (e.g., Basaltic Volcanism Study Project, 1981). Yet to this day the question remains as to whether there are any signs of volcanism visible on Mercury's surface. The root of this debate stems from the initial analyses of the Mariner 10 black and white images (~ 1 km pixel^{-1} over about half the planet). Workers identified the existence of widespread plains deposits, occurring as relatively smooth surfaces between craters, and as apparently ponded material, somewhat analogous to lunar mare units. It was proposed that at least some of these units were volcanic in origin (Murray, 1975; Murray *et al.*, 1975; Trask and Guest, 1975; Strom *et al.*, 1975; Strom, 1977; Dzurisin, 1978). Others argued that the plains deposits must be considered as basin ejecta, similar to those found at the lunar Apollo 16 landing site (Wilhelms, 1986; Oberbeck *et al.*, 1977), raising the possibility that there are no identifiable volcanic units on Mercury. Detailed crater counts of

Caloris basin ejecta facies and smooth plains deposits, however, indicate that the smooth plains were emplaced well after the Caloris basin forming vent (Spudis and Guest, 1988). A detailed summary of this controversy can be found in Spudis and Guest (1988).

Initial studies utilizing Mariner 10 color images to delimit units on Mercury led to three broad conclusions: (1) crater rays and ejecta blankets are bluer (higher UV/orange ratio) than average Mercury; (2) color boundaries often do not correspond to photogeologic units; and (3) no low-albedo, blue materials are found that are analogous to titanium-rich lunar mare deposits (Hapke *et al.*, 1980; Rava and Hapke, 1987). Significantly, little correlation was found between color units (Hapke *et al.*, 1980; Rava and Hapke, 1987) and mapped plains boundaries (Figure 6.12), further weakening the case for volcanic plains on Mercury. However, the calibration employed in these earlier studies did not adequately remove vidicon blemishes and radiometric residuals. These artifacts were sufficiently severe that the authors were forced to present an interpretive color unit map overlain on monochromatic mosaics, while publishing only a subset of the color ratio coverage of Mercury (Hapke *et al.*, 1975, 1980; Hapke, 1977; Rava and Hapke, 1987). Newly calibrated UV (375 nm) and orange (575 nm) Mariner 10 mosaics with significantly increased signal-to-noise ratio have been interpreted to indicate that indeed color units correspond to mapped plains units on Mercury, and further that some color units are the result of compositional heterogeneities in the mercurian crust (Robinson *et al.*, 1992, 1997; Robinson and Lucey, 1997).

The newly calibrated Mariner 10 color data were interpreted in terms of the color-reflectance paradigm for Mercury and the Moon articulated by Hapke and others (Hapke *et*

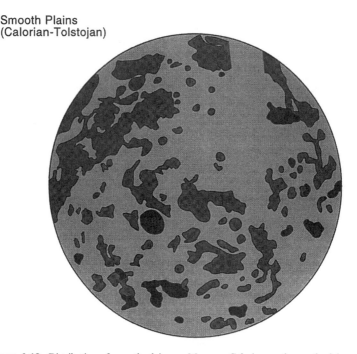

Figure 6.12. Distribution of smooth plains on Mercury. Calorian-aged smooth plains are shown in dark gray and Calorian and/or Tolstojan are shown in black. Together, these smooth plains cover about 10.4×10^6 km^2 or about 40% of the part of Mercury imaged by Mariner 10. Lambert equal-area projection centered on 0, 100, with north at top. From Spudis and Guest (1988). Copyright, Arizona Board of Regents.

al., 1980; Rava and Hapke, 1987; Cintala, 1992; Lucey *et al.*, 1995, 1998). This view holds that ferrous iron lowers the albedo and reddens (relative decrease in the UV/visible ratio) a lunar or mercurian soil. Soil maturation has a similar effect. with increasing maturity, soils darken and redden with the addition of submicroscopic iron metal (Figure 6.13). In contrast, addition of spectrally neutral opaque minerals (i.e., ilmenite) results in a trend that is nearly perpendicular to that of iron and maturity: Opaque minerals lower the albedo and *increase* the UV/visible ratio (Hapke *et al.*, 1980; Rava and Hapke, 1987; Lucey *et al.*, 1998). In lunar data, the orthogonal effects of opaques and iron-plus-maturity are manifested in two trends on a plot of reflectance and UV ratio (Figure 6.13): one comprising the mare basalts which vary in ilmenite along the trend, and the other the highlands which vary both in iron and in maturity along the perpendicular trend (Robinson and Lucey, 1997; Lucey *et al.*, 1998).

Transforming the Mariner 10 color data (UV and orange mosaics) using the lunar coordinate rotation allows the separation of the two perpendicular trends (opaque mineral abundance from iron plus maturity) into two separate image (Robinson and Lucey, 1997). The rotation clearly distinguishes maturity effects from compositional units on Mercury (Figures 6.14 and 6.15). From these new data, color units can be identified that correspond with previously mapped smooth plains deposits. The three best examples are the plains associated with Rudaki crater (2°S, 55°W), Tolstoj basin (16°S, 163°W), and Degas crater (33°N, 133°W), each distinguished by their low opaque index relative to their corresponding basement materials (Robinson and Lucey, 1997; Robinson *et al.*, 1997, 1998). In all three cases, the basement material is bluer (higher UV/orange ratio), and enriched in opaques: None show a distinct unit boundary in the iron-plus-maturity image that corresponds with the morphologic plains boundary. It is noteworthy that the iron-plus-maturity data do not exhibit

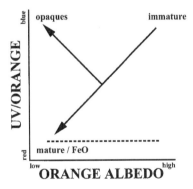

Figure 6.13. Trends in the visible color of the lunar surface. The visible color the lunar surface can be described by two perpendicular trends (opaque mineral concentration and iron-plus-maturity). The addition of ferrous iron to an iron-free silicate material (i.e., anorthosite) reddens the visible slope and lowers the albedo (a translation down the iron–maturity line: upper right to lower left). Color changes in lunar soil during maturation mimic the iron trend, as soils mature they redden (UV brightness/orange brightness) and their albedo decreases (orange brightness): also a translation down the iron–maturity line, upper right to lower left. Adding spectrally neutral opaque minerals, such as ilmenite, results in a color trend that is nearly perpendicular to the iron–maturity line: Opaques lower the albedo, but *decrease* the relative redness (an increase in the UV/orange ratio) of lunar soils. These two trends can be used to map the distribution of opaques (opaque index) and the iron-plus-maturity parameter through a coordinate rotation such that their perpendicular axes become parallel with the x and y axes of the color–albedo plot (Robinson and Lucey, 1997; Lucey *et al.*, 1998). The dotted line indicates the position of the iron–maturity line after rotation. Adapted from Robinson and Lucey (1997).

unit boundaries corresponding to the morphologic plains boundaries, thus indicating that there are no large differences in iron content between the overlying plains materials and basement material (see below). In the case of Tolstoj basin (Figure 6.16) (Robinson et al., 1998), a distinct mappable opaque index unit corresponds with the asymmetric NE–SW trending ejecta pattern of the basin [the Goya Formation (Schaber and McCauley, 1980; Spudis and Guest, 1988)]. This stratigraphic relation implies that formation of the Tolstoj basin (~ 550 km diameter) resulted in excavation of anomalously opaque-rich material from within the crust. The Goya Formation is not a mappable unit in the iron-plus-maturity image, indicating that its FeO content does not differ significantly from the local (and hemispheric) average.

The dark blue material (Figure 6.15) associated with the crater Homer exhibits diffuse boundaries consistent with ballistically emplaced material, either from an impact event or from pyroclastic activity. However, there is no impact crater central to this deposit, and it straddles a linear feature on the southwestern margin of Homer. Therefore, a pyroclastic origin is favored. A similar deposit is seen northwest of crater Lermontov; examination of the iron–maturity parameter and opaque index images (Figures 6.15 and 6.16) reveals that the darkest and bluest material in this deposit is not associated with an ejecta pattern. The relatively blue color, high opaque index, and low albedo of these materials (for both areas) are consistent with a more mafic material, possibly analogous to a basaltic or gabbroic composition. Sprague et al. (1994) reported a tentative identification of basaltlike material in this hemisphere with Earth-based thermal IR measurements, while later microwave measurements were interpreted to indicate a total lock of areally significant basaltic materials on Mercury (Jeanloz et al., 1995). From the data currently available it is not possible to make an identification of basaltic material or of any rock type; however, the spectral parameters, stratigraphic relations, and morphology are consistent with volcanically emplaced materials. Regardless of the mode of emplacement, these materials found around the craters Homer and Lermontov, and the plains units identified above (Figures 6.14–6.16) argue that significant compositional units occur within the mercurian crust and that at least some of them were emplaced during volcanic eruptions.

The key observation from the image data is that the mercurian crust is not homogeneous in the color image (UV/orange) and the opaque index image (Figure 6.14). The exact composition of the units seen in these data can only be inferred from lunar analogy and geologic common sense as no samples or other relevant compositional data exist for Mercury. Regarding the mercurian smooth plains discussed here (Robinson and Lucey, 1997), the observation that they do show boundaries in the opaque index image strongly supports the previous morphologic interpretation that some mercurian plains units are indeed volcanic in origin (Trask and Guest, 1975; Strom, 1977; Keiffer and Murray, 1987; Spudis and Guest, 1988; and others) and not basin ejecta emplaced during the Caloris basin (or any other basin) forming event (Wilhelms, 1976). Importantly, these volcanic units are not identifiable in the iron-plus-maturity image, indicating that they have similar FeO contents as the rest of the mercurian crust. From this interpretation one can speculate on the nature of the mercurian upper mantle magma source regions.

The FeO abundance of mantle source regions corresponds, to a first order, to the FeO content of the erupted magma. This follows from the observation that the partition coefficient for FeO is close to 1 for mafic magmas (Longhi et al., 1992). The observation that volcanics identified on Mercury do not have FeO abundances differing from the hemispheric average indicates that the mercurian mantle source of these volcanics is not enriched in FeO relative to the crust, or conversely that the ancient crust is not depleted in FeO relative to the upper

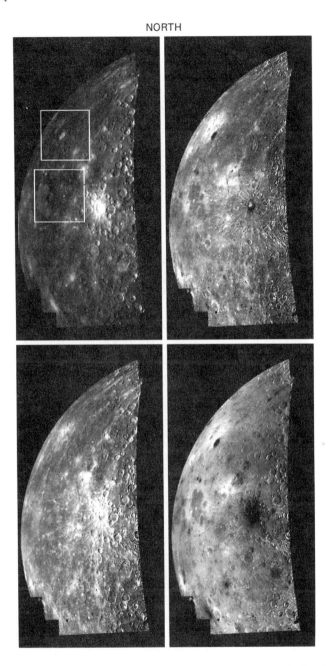

Figure 6.14. Spectral parameters for Mariner 10 data. Essential spectral parameters for the Mariner 10 incoming hemisphere: (Upper left) Orange (575 nm) albedo, boxes indicate areas enlarged in Figure 15 (top is B, bottom is A). (Upper right) Relative color (UV/orange), brighter tones indicate increasing blueness. (Lower left) Parameter 1—iron–maturity parameter, brighter tones indicate decreasing maturity and/or decreasing FeO content; (lower right) parameter 2—opaque index, brighter tones indicate increasing opaque mineral content. The relatively bright feature in the center right of the albedo image is the Kuiper-Muraski crater complex centered at latitutde 12°S, longitude 13°E. Adapted from Robinson and Lucey (1997).

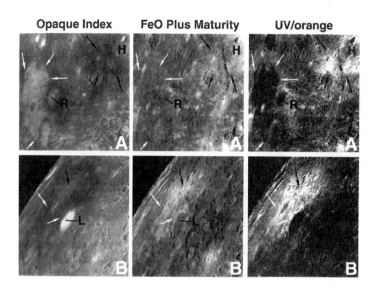

Figure 6.15. Color ratio images of portions of Mercury. Enlargement of areas found on the Mariner 10 incoming hemisphere, keying on color units indicative of volcanically emplaced materials near the craters Rudaki (Top Row–A), Homer (Top Row–A), and Lermontov (Bottom Row–B). Red is formed from the inverse of the opaque index (increasing redness indicates decreasing opaque mineralogy—Figure 14, lower right), green component is the iron–maturity parameter (Figure 14, lower left), and blue shows the relative color (UV/orange ratio—Figure 14, upper right). The plains unit seen west and south, and filling the interior of the crater Rudaki (R—120 km diameter; Top Row–A) exhibits embaying boundaries (white arrows) indicative of material emplaced as a flow and it has a distinct color signature relative to its surroundings. The blue material on the southwest margin of the crater Homer (H—320 km diameter; Top Row–A) exhibits diffuse boundaries, is insensitive to local topographic undulations (black arrows), and is aligned along a linear segment of a Homer basin ring. A portion of the blue material seen northwest of crater Lermontov (L—160 km diameter; Bottom Row–B) is somewhat concentric to a small impact crater (black arrow) and may represent material excavated from below during the impact. However, examination of the iron–maturity parameter and opaque index images (Bottom Row) suggests that the darkest and bluest material (white arrows) in this deposit is not associated with an impact ejecta pattern, rather the anomalous lighter blue ejecta is composed of the darker material, though less mature and possibly with an admixture of basement material, overlying the darker blue portions of the deposit. Adapted from Robinson and Lucey (1997).

mantle (Robinson *et al.*, 1997, 1998). If the plains deposits had a significant increase (or decrease) in FeO relative to the basement rock that they overlie (ancient crust), then they would appear as a mappable unit in the iron-plus-maturity image. In contrast, mare deposits found on the Moon (mare lavas versus anorthositic crust) have a significant contrast in FeO content relative to the ancient anorthositic crust they overlie (typically >15 wt% versus <6 wt%, respectively; see Lucey *et al.*, 1998). The global crustal abundance of FeO on Mercury has been estimated to be less than 6 wt% from remote sensing data (McCord and Adams, 1972; Vilas and McCord, 1975; Vilas *et al.*, 1984; Vilas, 1985, 1988; Veverka *et al.*, 1988; Sprague *et al.*, 1994; Blewett *et al.*, 1997). The lack of structure corresponding to the plains units in the iron-plus-maturity image are consistent with mantle magma source regions roughly sharing the crustal FeO composition, and so support the idea that Mercury is highly reduced and most of its iron is sequestered in a metallic core.

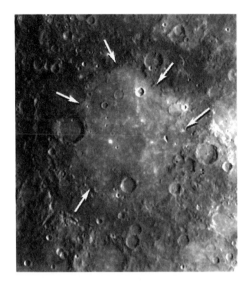 

Figure 6.16. A visual comparison of putative mercurian flood lavas found on the floor of Tolstoj basin (left—17°S, 164°W) with Mare Humboldtianum on the Moon (right—56°N, 80°E). Both data sets acquired by Mariner 10 with similar resolution (~ 1 km pixel^{-1}; each image ~ 625 km across) and viewing geometries (incidence angle = 65° for Mercury, incidence angle = 55° for the Moon). The most obvious distinguishing characteristic of the lunar mare deposit is its albedo contrast with the underlying highlands (right); not true for Mercury (left). The key morphologic properties used to identify flood lavas on the Moon (other than albedo) are embaying contacts and ponding in topographic lows (usually basins; see arrows on both images). For the Moon, classic flow features such as flow fronts, spatter ramparts, vents, and so on are for the most part simply not visible at a scale of 1 km (see review and references in Spudis and Guest, 1988).

The inferred relative uniformity of iron abundance in the areas seen by Mariner 10 begs the question, "what are volcanics composed of on Mercury?" Low-iron mafic lavas are found on the Moon in the Apennine Bench Formation (Spudis and Hawke, 1986). Komatiitic volcanics are found on the Earth with FeO abundances of under 5 wt%. However, identification of any specific rock types on Mercury will have to wait for a return mission to Mercury, preferably an orbiting spacecraft with both spectral imaging and compositional mapping (i.e., X-ray and gamma-ray spectrometer) capabilities. Mercury is the last unexplored terrestrial planet, and a detailed examination of its geologic history is overdue. Superficially it looks like the Moon; Mariner 10 and terrestrial remote sensing data hint that it shares some lunar spectral affinities but that it also must be very different in many respects.

6.13. CONCLUSION: ENVIRONMENTAL INFLUENCES, SYNTHESIS, AND OUTSTANDING PROBLEMS

Perhaps the most pervasive factor influencing volcanic processes on any planetary body is the acceleration related to gravity. The buoyancy force driving convective flows in the mantle is directly proportional to the gravity (as well as to thermally driven density contrasts, of course), implying that rates of melt formation and extraction may be significantly higher within Mercury than the Moon. However, the relative widths of dikes transferring magma from a given depth to the surface scale inversely with, approximately, the cube root of the gravity, so that individual dikes feeding eruptions on Mercury may be narrower than those on the Moon. If magma rise to the surface from a given depth is driven by excess pressure, individual eruption rates may therefore be systematically smaller on Mercury than the Moon; in contrast, if buoyancy controls the force driving magma to the surface, the eruption rates will be a little higher on Mercury. In neither case, however, is the influence of gravity likely to be as great as the depth from which the magma must be driven to the surface: If shallow magma reservoirs existed on Mercury, then the dikes connecting them to the surface will have been much narrower than those that were required to penetrate the lunar crust and the eruption rates will have been much more like those on the Earth now than the high rates inferred for the mare lava flows and sinuous rilles.

Surface volcanic deposits are also strongly influenced by gravity, but other factors assume more importance at the surface. The velocity of ejection of pyroclasts is governed by the amount of gas released from the magma and the amount by which the gas expands. The latter is normally strongly influenced by atmospheric pressure, but in the absence of an atmosphere on both Mercury and the Moon, only the magma volatile content is relevant. Lunar magmas were very deficient in volatiles relative to the Earth, but in view of the lack of insight into likely volatile amounts and species on Mercury it is hard to predict which of Mercury and the Moon might have had the more vigorous explosive eruptions. What is clear, however, is that for the same magma volatile content, the ranges of deposits on Mercury would be less than on the Moon because of the higher gravity. This will allow less time for radiative cooling (there being no other mechanism for heat loss in the absence of atmospheres) and ensure that pyroclasts land systematically hotter on Mercury than the Moon, therefore being more prone to produce welded deposits.

When lava flows form on the surface, whether from direct effusion from the vent or as a result of accumulation of hot pyroclasts around the vent, their thicknesses, widths, and flow speeds depend on the local surface slope and the gravity as well as the lava rheology and effusion rate. For any given magma composition and eruption temperature, and hence rheology, the thickness and speed of a channelized flow will be larger if the gravity and surface slope are smaller. However, the main observable property of the flow will be its length. On very shallow slopes, local topography variations, and processes such as ponding and overflow, will strongly control the final length and shape. On steeper slopes, however, the lengths of individual flow units will be limited by cooling, and will then be somewhat longer if the slope is greater or the gravity is less or the heat loss rate is less. At first sight this might imply that we should expect systematically shorter flows under the higher gravity of Mercury, perhaps partly compensated by the higher surface temperature, than the Moon unless eruptions commonly occurred on slopes steeper than those of the mare basins. However, the strongest dependence of flow length is on the effusion rate from the vent, and so the lengths to be expected for flows on Mercury hinge almost entirely on the depths of the reservoirs from which eruptions commonly took place.

The Moon and Mercury share several attributes in terms of their lack of an atmosphere, well-preserved ancient heavily cratered surfaces, and abundance of impact basins and related deposits. They differ, however, in several fundamental ways. First, the position of the bodies in the solar system is very different, with Mercury forming nearest the Sun, where temperature–pressure gradients will considerably influence the starting materials and conditions of accretion. Second, the Moon may have formed from the impact of a Mars-sized body into the proto-Earth in early Earth history. Although Mercury may have been influenced by similar large impacts early in its history, the differences in their formational environments are fundamental. Third, the internal structure of the bodies is quite different, with Mercury characterized by an iron core approximately the size of the Moon itself. Fourth, the large core means that the density, and thus the surface gravity, is larger on Mercury than on the Moon, and this affects a arrange of processes, including volcanism and impact cratering. Fifth, it is clear from the internal structure of Mercury, and the surface mineralogic characteristics described previously, that petrogenetic processes operate on different starting material on Mercury than on the Moon. Sixth, the differences in in internal structure and starting conditions mean that the thermal evolution will not be the same and that the manifestation of this into heat loss processes (e.g., tectonism and volcanism) will also be distinct. Despite these fundamental variations, and even because of them, the comparison of the Moon and Mercury is, and will continue to be, very instructive.

The difficulty in these comparisons at present, however, is that data availability is so asymmetric that it is premature to address many of the questions raised in the introduction. For Mercury, Earth-based observations and Mariner 10 flyby data constitute the data set. For the Moon, humans have deployed geophysical instruments, made surface observations, and returned samples from six sites. Automated spacecraft have traversed large regions of the lunar surface and returned samples from three sties. Orbiting spacecraft have obtained global geochemical, mineralogic, topographic, gravity, magnetic, and image data. Thirty scientific conferences have been dedicated to the analysis of the lunar data, and countless thousands of pages of journal articles attest to the success of these endeavors.

Did Mercury, like the Moon, form a primary crust that served as a barrier to, and filter for, magma ascent? Did secondary crustal formation (e.g., the lunar maria) occupy a significant part of the resurfacing history of Mercury? Did basaltic volcanism contribute to the resurfacing history of Mercury or does the formation of the iron core so alter the mantle geochemistry that other rock types dominate? Does the tectonic history of Mercury (significant global compressional deformation) mean that extrusive volcanism is inhibited or precluded by the state of stress in the lithosphere in its early to intermediate history? Will high resolution and global image data reveal presently lacking evidence for vents and landforms associated with extrusive volcanism? As outlined in this chapter and the cited references, the lunar volcanic record offers a rich and detailed paradigm with which to address these questions. Scientists currently await the time when spacecraft return to Mercury to advance our knowledge of that planet beyond the level we had for the Moon over 30 years ago.

ACKNOWLEDGMENTS

We gratefully acknowledge the help of Anne Cote in manuscript formatting and editing and Peter Neivert in figure preparation. Authors were assisted by grants from the Planetary Geology and Geophysics Program of the National Aeronautics and Space Exploration, which are gratefully acknowledged.

6.14. REFERENCES

Adams, J. B., C. M. Pieters, A. E. Metzger, I. Adler, T. B. McCord, C. R. Chapman, T. V. Johnson, and M. J. Bielefeld, Remote sensing of basalts in the solar system, in *Basaltic Volcanism on the Terrestrial Planets*, pp. 439–490, Pergamon Press, New York, 1981.

Adler, I., and J. Trombka, Orbital chemistry: Lunar surface analysis from the X-ray and gamma-ray remote sensing experiments, *Phys. Chem. Earth,* 10, 17–43, 1977.

Antonenko, I., J. W. Head, J. F. Mustard, and B. R. Hawke, Criteria for the detection of lunar cryptomaria, *Earth Moon Planets,* 69, 141–172, 1995.

Arndt, J., and W. von Engelhardt, Formation of Apollo 17 orange and black glass beads, *Proc. 17th Lunar Planet, Sci. Conf., J. Geophys. Res. Suppl., 92*, E372–E376, 1987.

Basaltic Volcanism Study Project, *Basaltic Volcanism on the Terrestrial Planets*, 1286 pp., Pergamon Press, New York, 1981.

Bell. J. F., and B. R. Hawke, Lunar dark-haloed impact craters: Origin and implications for early mare volcanism, *J. Geophys. Res., 89*, 6899–6910, 1984.

Blake, S., Volcanism and the dynamics of open magma chambers, *Nature, 289*, 783–785, 1981.

Blewett, D. T., P. G. Lucey, B. R. Hawke, G. G. Ling, and M. S. Robinson, A comparison of mercurian reflectance and spectral quantities with those of the Moon, *Icarus, 129*, 217–231, 1997.

Bruno, B. C., P. G. Lucey, and B. R. Hawke, High-resolution UV-visible spectroscopy of lunar red spots, *Proc. 21st Lunar Planet Sci. Conf.*, 405–415, 1991.

Carr, M. H., The role of lava erosion in the formation of lunar rilles and Martian channels, *Icarus, 22*, 1–22, 1973.

Chapman, C. R., Mercury: Introduction to an end-member planet, in *Mercury*, edited by F. Vilas, C. R. Chapman, and M. S. Matthews, pp 1–23, University of Arizona Press, Tucson, 1988.

Chevrel, S. D., P. C. Pinet, and J. W. Head, Gruithuisen domes region: A candidate for an extended nonmare volcanism unit on the Moon, *J. Geophys. Res., 104*, 16515–16529, 1999.

Christiansen, E. H., and W. K. Hamblin, *Exploring the Planets*, Prentice–Hall, Englewood Cliffs, NJ, 1995.

Cintala, M. J., Impact-induced thermal effects in the lunar and mercurian regoliths, *J. Geophys. Res., 97*, 947–973, 1992.

Coombs, C. R., B. R. Hawke, P. G. Lucey, P. D. Owensby, and S. Zisk, The Alphonsus region: A geological and remote-sensing perspective, *Proc. 20th Lunar Planet. Sci. Conf.*, 161–174, 1990a.

Coombs, C. R., D. S. McKay, and B. R. Hawke, The violent side of mare volcanism, *LPI-LAPST Workshop on Mare Volcanism and Basalt Petrogenesis: Astounding Fundamental Concepts (AFC) Developed Over the Last Fifteen Years*, 1–2, 1990b.

Craddock, R. A., M. S. Robinson, B. R. Hawke, and A. S. McEwen, Clementine-based geology of the Moscoviense Basin, lunar farside, *Lunar Planet. Sci. Conf., 28*, 265–266, 1997.

Crisp, J. A., Rates of magma emplacement and volcanic output, *J. Volcanol. Geotherm. Res., 20*, 177–211, 1984.

DeHon, R. A., Cauldron subsidence in lunar craters, Ritter and Sabine, *J. Geophys. Res., 76*, 5712–5718, 1971.

DeHon, R. A., Thickness of the western mare basalts, *Proc. 10th Lunar Planet. Sci. Conf.*, 2935–2955, 1979.

Delano, J. W., Pristine lunar glasses: Criteria, data and implications, *Proc. 16th Lunar Planet. Sci. Conf.*, D201–D213, 1986.

Dzurisin, D., The tectonic and volcanic history of Mercury as inferred from studies of scarps, ridges, troughs, and other lineaments, *J. Geophys. Res., 83*, 4883–4906, 1978.

Fogel, R. A., and M. J. Rutherford, Magmatic volatiles in primitive lunar glasses: I. FTIR and EPMA analyses of Apollo 15 green and yellow glasses and revision of the volatile-assisted fire-fountain theory, *Geochim. Cosmochim. Acta, 59*, 201–215, 1995.

Gaddis, L. R., C. M. Pieters, and B. R. Hawke, Remote sensing of lunar pyroclastic mantling deposits, *Icarus, 61*, 461–489, 1985.

Greeley, R., Observations of actively forming lava tubes and associated structures, Hawaii, *Mod. Geol., 2*, 207–233, 1971.

Greeley, R., *Planetary Landscapes*, Allen & Unwin, London, 1985.

Greeley, R., S. D. Kadel, D. A. Williams, L. R. Gaddis, J. W. Head, A. S. McEwen, S. L. Murchie, E. Nagel, G. Neukum, C. M. Pieters, J. M. Sunshine, R. Wagner, and M. J. S. Belton, Galileo imaging observation of lunar maria and related deposits, *J. Geophys. Res., 98*, 17813–17205, 1993.

Guest, J. E., and J. B. Murray, Volcanic features of the nearside equatorial lunar maria, *J. Geol. Soc., 132*, 251–258, 1976.

Hapke, B., Interpretation of optical observations of Mercury and the Moon, *Phys. Earth Planet. Inter., 15*, 264–274, 1977.

Hapke, B., G. Danielson, K. Klaasen, and L. Wilson, Photometric observations of Mercury from Mariner 10, *J. Geophys. Res., 80*, 2431–2443, 1975.

Hapke, B., C. Christman, B. Rava, and J. Mosher, A color-ratio map of Mercury, *Proc. 11th Lunar Planet. Sci. Conf.*, 817–822, 1980.

Hartmann, W. K., R. G. Strom. S. J. Weidenschilling, K. R. Blasius, A. Woronow, M. R. Dence, R. A. F. Grieve, J. Diaz, C. R. Chapman, E. M. Shoemaker, and K. L. Jones, Chronology of planetary volcanism by comparative studies of planetary cratering, in *Basaltic Volcanism on the Terrestrial Planets*, pp. 1049–1127, Pergamon Press, New York, 1981.

Hawke, B. R., and J. F. Bell, Remote sensing studies of lunar dark-halo impact craters: Preliminary results and implications for early volcanism, *Proc. 12th Lunar Planet. Sci. Conf.*, 665–678, 1981.

Hawke, B. R., and J. W. Head, Lunar KREEP volcanism: Geologic evidence for history and mode of emplacement, *Proc. 9th Lunar Planet. Sci. Conf.*, 3285–3309, 1978.

Hawke, B. R., C. R. Coombs, L. R. Gaddis, P. G. Lucey, and P. D. Owensby, Remote sensing and geologic studies of localized dark mantle deposits on the Moon, *Proc. 19th Lunar Planet. Sci. Conf.*, 255–268, 1989.

Hawke, B. R., P. G. Lucey, J. F. Bell, and P. D. Spudis, Ancient mare volcanism, *LPI-LAPST Workshop on Mare Volcanism and Basalt Petrogenesis: Astounding Fundamental Concepts (AFC) Developed Over the Last Fifteen Years*, 5–6, 1990.

Hawke, B. R., C. R. Coombs, L. R. Gaddis, P. G. Lucey, C. A. Peterson, M. S. Robinson, G. A. Smith, and P. D. Spudis, Remote sensing studies of geologic units in the Eastern Nectaris Region of the Moon, *Lunar Planet. Sci. Conf.*, 28, 529–530, 1997.

Head, J. W., Small-scale analogs of the Cayley Formation and Descartes Mountains in impact-associated deposits, in Apollo 16 Preliminary Science Report, pp. 29-16–29-20, NASA, Washington, DC, 1972.

Head, J. W. Stratigraphy of the Descartes region (Apollo 16): Implications for the origin of samples, *Moon*, 11, 77–99, 1974.

Head, J. W., Mode of occurrence and style of emplacement of lunar mare deposits, in *Origins of Mare Basalts*, 61–65, 1975a.

Head, J. W., Lunar mare deposits: Areas, volumes, sequence, and implication for melting in source areas, in *Origins of Mare Basalts*, 66–69, 1975b.

Head, J. W., Lunar volcanism in space and time, *Rev. Geophys. Space Phys.*, 14, 265–300, 1976.

Head, J. W., and A. Gifford, Lunar mare domes: Classification and mode or origin, *Moon Planets*, 22, 235–258, 1980.

Head, J. W., and T. B. McCord, Imbrian-age highland volcanism on the Moon: The Gruithuisen and Marian domes, *Science*, 199, 1433–1436, 1978.

Head, J. W., and S. C. Solomon, Tectonic evolution of the terrestrial planets, *Science*, 213, 62–76, 1981.

Head, J. W., and L. Wilson, Alphonsus-type dark halo craters: Morphology, morphometry and eruption conditions, *Proc. 10th Lunar Planet. Sci. Conf.*, 2861–2897, 1979.

Head, J. W., and L. Wilson, The formation of eroded depressions around the sources of lunar sinuous rilles: Observations, *Lunar Planet. Sci. Conf.*, 11, 426–428, 1980.

Head. J. W., and L. Wilson, Basaltic pyroclastic eruptions: Influence of gas-release patterns and volume fluxes on fountain structure, and the formation of cinder cones, spatter cones, rootless flows, lava ponds and lava flows, *J. Volcanol. Geotherm. Res.*, 37, 261–271, 1989.

Head, J. W., and L. Wilson, The lack of shield volcanoes on the Moon: Consequence of magma transport phenomena, *Geophys. Res. Lett.*, 18, 2121–2124, 1991.

Head, J. W., and L. Wilson, Lunar mare volcanism: Stratigraphy, eruption conditions, and the evolution of secondary crusts, *Geochim. Cosmochim. Acta*, 56, 2155–2175, 1992.

Head, J. W., and L. Wilson, Lunar graben formation due to near-surface deformation accompanying dike emplacement, *Planet. Space Sci.*, 41, 719–727, 1993.

Head, J. W., and L. Wilson, Lunar linear rilles as surface manifestations of dikes: Predictions and observations, *Lunar Planet. Sci. Conf.*, 27, 519–520, 1996.

Head, J. W., and L. Wilson, Near-surface emplacement in the lunar crust: Emplacement dynamics and associated structure and morphology, in preparation, 2000.

Head, J. W., J. B. Adams, T. B. McCord, C. M. Pieters, and S. Zisk, Regional stratigraphy and geologic history of Mare Crisium, *Geochim. Cosmochim. Acta Suppl.*, 9, 43–74, 1978.

Head, J. W., W. B. Bryan, R. Greeley, J. E. Guest, P. H. Schultz, R. S. J. Sparks, G. P. L. Walker, J. L. Whitford-Stark, C. A. Wood, and M. H. Carr, Distribution and morphology of basalt deposits on planets, in *Basaltic Volcanism on the Terrestrial Planets*, pp. 701–800, Pergamon Press, New York, 1981.

Head, J. W., S. M. Murchie, J. F. Mustard, C. M. Pieters, G. Neukum, A. S. McEwen, R. F. Greeley, E. Nagel, and M. J. S. Belton, Lunar impact basin: New data for the western limb and far side (Orientale and South Pole–Aitken Basins) from the first Galileo flyby, *J. Geophys. Res.*, 98, 17149–17181, 1993.

Head, J. W., L. Wilson, and C. M. Weitz, The dark ring in southwestern Lunar Orientale Basin: Origin as a single pyroclastic eruption, *Lunar Planet. Sci. Conf.*, 28, 543–544, 1997.

Heiken, G. H., D. S. McKay, and R. W. Brown, Lunar deposits of possible pyroclastic origin, *Geochim. Cosmochim. Acta*, 38, 1703–1718, 1974.

Hess, P. C., Diapirism and the origin of high TiO_2 mare glasses, *Geophys. Res. Lett.*, 18, 2069–2072, 1991.

Hess, P. C., and E. M. Parmentier, A model for the thermal and chemical evolution of the Moon's interior: Implications for the onset of mare volcanism, *Earth Planet. Sci. Lett.*, 134, 501–514, 1995.

Hiesinger, H., R. Jaumann, G. Neukum, and J. W. Head, Ages of lunar mare basalts, *Lunar Planet. Sci. Conf., 29*, CD-ROM #1242, 1998a.

Hiesinger, H., G. Neukum, and J. W. Head, On the relation of age and titanium content of lunar mare basalts, *Lunar Planet. Sci. Conf., 29*, CD-ROM #1243, 1998b.

Hulme, G., Turbulent lava flow and the formation of lunar sinuous rilles, *Mod. Geol. 4*, 107–117, 1973.

Hulme, G., A review of lava flows processes related to the formation of lunar sinuous rilles, *Geophsy. Surv., 5* 245–279, 1982.

Jeanloz, R., D. L. Mitchell, A. L. Sprague, and I. de Pater, Evidence for a basalt-free surface on Mercury and implications for internal heat, *Science, 268*, 1455–1457, 1995.

Johnson, J., S. M. Larson, and R. Singer, Remote sensing of potential lunar resources. I. Near side compositional properties, *J. Geophys. Res., 96*, 18861–18882, 1991.

Keiffer, W. S., and B. C. Murray, The formation of Mercury's smooth plains, *Icarus, 72*, 477–491, 1987.

Longhi, J., E. Knittle, J. R. Holloway, and H. Wanke, The bulk chemical composition, mineralogy and internal structure of Mars, in *Mars*, edited by H. Kieffer, B. Jakosky, C. Snyder, and M. Mathews, pp. 184–208, University of Arizona Press, Tucson, 1992.

Lucey, P. G., B. R. Hawke, C. M. Pieters, J. W. Head, and T. B. McCord, A compositional study of the Aristarchus region of the Moon using near-infrared reflectance spectroscopy, *J. Geophys. Res., 91*, D344–D354, 1986.

Lucey, P. G., B. C. Bruno, and B. R. Hawke, Preliminary results of imaging spectroscopy of the Humorum Basin region of the Moon, *Proc. 21st Lunar Planet. Sci.*, 391–403, 1991.

Lucey, P. G., G. J. Taylor, and E. Malaret, Abundance and distribution of iron on the Moon, *Science, 268*, 1150–1153, 1995.

Lucey, P. G., D. T. Blewett, and B. R. Hawke, Mapping the FeO and TiO_2 content of the lunar surface with multispectral imagery, *J. Geophys. Res., 103*, 3679–3699, 1998.

Malin, M., Lunar red spots: Possible pre-mare material, *Earth Planet. Sci. Lett., 21*, 331–341, 1974.

Malin, M. C., Surfaces of Mercury and the Moon: Effects of resolution and lighting conditions on the discrimination of volcanic features, *Proc. 9th Lunar Planet. Sci. Conf.*, 3395–3409, 1978.

McCord, T. B., and J. B. Adams, Mercury: Interpretation of optical observations, *Icarus, 17*, 585–588, 1972.

McEwen, A. S., M. S. Robinson, E. M. Eliason, P. G. Lucey, T. C. Duxbury, and P. D. Spudis, Clementine observations of the Aristarchus region of the Moon, *Science, 266*, 1858–1862, 1994.

Melosh, H. J., and W. B. McKinnon, The tectonics of Mercury, in *Mercury*, edited by F. Vilas, C. R. Chapman, and M. S. Matthews, pp. 374–400, University of Arizona Press, Tucson, 1988.

Metzger, A. E., E. L. Haines, R. E. Parker, and R. G. Radocinshi, Thorium concentrations in the lunar surface, I: Regional values and crustal content, *Proc. 8th Lunar Sci. Conf.*, 949–999, 1977.

Moore, H. J., C. A. Hodges, and D. H. Scott, Multiring basins: Illustrated by Orientale and associated features, *Proc. 5th Lunar Conf.*, 71–100, 1974.

Murray, B. C., The Mariner 10 pictures of Mercury: An overview, *J. Geophys. Res., 80*, 2342–2344, 1975.

Murray, B. C., M. J. S. Belton, G. E. Danielson, M. E. Davies, D. E. Gault, B. Hapke, B. O'Leary, R. G. Strom, V. Suomi, and N. Trask, Mercury's surface: Preliminary description and interpretation from Mariner 10 pictures, *Science, 185*, 169–179, 1975.

Murray, B., M. C. Malin, and R. Greeley, *Earthlike Planets: Surfaces of Mercury, Venus, Earth, Moon, Mars*, Freeman, San Francisco, 1981.

Mustard, J. F., and J. W. Head, Buried stratigraphic relationships along the southwestern shores of Oceanus Procellarum: Implications for early lunar volcanism, *J. Geophys. Res., 101*, 18913–18925, 1996.

Neukum, G., Different ages of lunar light plains, *Moon, 17*, 383–393, 1977.

Nyquist, L. E., and C.-Y. Shih, The isotopic record of lunar volcanism, *Geochim. Cosmochim. Acta, 56*, 2213–2234, 1992.

Oberbeck, V. R., The role of ballistic erosion and sedimentation in lunar stratigraphy, *Rev. Geophys. Space Phys., 13*, 337–362, 1975.

Oberbeck, V. R., R. H. Morrison, F. Hörz, W. L. Quaide, and D. E. Gault, Smooth plains and continuous deposits of craters and basins, *Proc. 5th Lunar Sci. Conf.*, 111–136, 1974.

Oberbeck, V. R., W. L. Quaide, R. E. Arvidson, and H. R. Aggarwal, Comparative studies of lunar, mantle, and mercurian craters and plains, *J. Geophys. Res., 82*, 1681–1698, 1977.

Peterson, D. W., and R. B. Moore, Geologic history and evolution of geologic concepts, Island of Hawaii, *U. S. Geol. Surv. Prof. Pap 1350*, 149–189, 1987.

Pieters, C. M., Mare basalt types on the front side of the Moon: A summary of spectral reflectance data, *Proc. 9th Lunar Planet. Sci. Conf.*, 2825–2849, 1978.

Pieters, C. M., J. W. Head, T. B. McCord, J. B. Adams, and S. H. Zisk, Geochemical and geological units of Mare Humorum: Definition using remote sensing and lunar sample information, *Proc. 6th Lunar Sci. Conf.*, 2689–2710, 1975.

Pieters, C. M., J. W. Head, J. B. Adams, T. B. McCord, S. H. Zisk, and J. L. Whitford-Stark, Late high titanium basalts of the western maria: Geology of the Flamsteed region of Oceanus Procellarum, *J. Geophys. Res., 85*, 3913–3938, 1980.

Rava, B., and B. Hapke, An analysis of the Mariner 10 color ratio map of Mercury, *Icarus, 71*, 397–429, 1987.

Robinson, M. S., and P. G. Lucey, Recalibrated Mariner 10 color mosaics: Implications for mercurian volcanism, *Science, 275*, 197–200, 1997.

Robinson, M. S., B. R. Hawke, and P. G. Lucey, Mariner 10 multispectral images of the eastern limb and farside of the Moon, *J. Geophys. Res., 97*, 18265–18274, 1992.

Robinson, M. S., B. R. Hawke, P. G. Lucey, G. J. Taylor, and P. D. Spudis, The color of Mercury, *Lunar Planet. Sci., 28*, 1189–1190, 1997.

Robinson, M. S., B. R. Hawke, P. G. Lucey, G. J. Taylor, and P. D. Spudis, Compositional heterogeneity of Mercury's crust, *Lunar Planet. Sci., 29*, CD-ROM #1860, 1998.

Ryan, M. R., Elasticity and contractancy of Hawaiian olivine tholeiitte and its role in the stability and structural evolution of subcaldera magma reservoirs and rift systems, in *Volcanism in Hawaii*, edited by R. W. Decker, T. L. Wright, and P. H. Stauffer, Vol. 2, pp. 1395–1447, *U. S. Geol. Surv. Prof. Pap. 1350*, 1987.

Ryder, G., and P. D. Spudis, Volcanic rocks in the lunar highlands, Lunar and Planetary Institute, Compiler, *Conference on the Lunar Highlands Crust, Houston, TX, 1979, Proceedings*, Pergamon, *Geochim. Cosmochim. Acta Suppl., 12*, 353–375, 1980.

Sato, M., The driving mechanism of lunar pyroclastic eruptions infered from the oxygen fugacity behavior of Apollo 17 orange glass, *Proc. 10th Lunar Planet. Sci. Conf.*, 311–325, 1979.

Schaber, G., Lava flows in Mare Imbrium: Geologic evaluation from Apollo orbital photography, *Proc. 4th Lunar Sci. Conf.*, 73–92, 1973.

Schaber, G. G., and J. F. McCauley, Geological map of the Tolstoj quadrangle of Mercury (H-8), *U.S. Geol. Surv. Map, I-1199*, 1980.

Schubert, G., M. N. Ross, D. J. Stevenson, and T. Spohn, Mercury's thermal history and the generation of its magnetic field, in *Mercury*, edited by F. Vilas, C. R. Chapman, and M. S. Matthews, pp. 429–460, University of Arizona Press, Tucson, 1988.

Schultz, P. H., *Moon Morphology*, University of Texas Press, Austin, 1976a.

Schultz, P. H., Floor-fractured lunar craters, *Moon, 15*, 241–273, 1976b.

Schultz, P. H., and P. Spudis, Evidence for ancient mare volcanism, *Proc. 10th Lunar Planet. Sci. Conf.*, 2899–2918, 1979.

Schultz, P. H., and P. Spudis, The beginning and end of lunar mare volcanism, *Nature, 302*, 233–236, 1983.

Schultz, S. C., Mare volcanism and lunar crustal structure, *Proc. 6th Lunar Sci. Conf.*, 1021–1042, 1975.

Solomon, S. C., and J. W. Head, Lunar mascon basins: Lava filling, tectonics, and evolution of the lithosphere, *Rev. Geophys. Space Phys., 18*, 107–141, 1980.

Sprague, A. L., R. W. H. Kozlowski, F. C. Witteborn, D. P. Cruikshank, and D. H. Wooden, Mercury: Evidence for anorthosite and basalt from mid-infrared (7.4–11.4 μm) spectroscopy, *Icarus, 109*, 156–167, 1994.

Spudis, P. D., Composition and origin of the Apennine Bench Formation, *Proc. 9th Lunar Planet. Sci. Conf.*, 3379–3394, 1978.

Spudis, P. D., Young dark mantle deposits on the Moon, *NASA Tech. Memo. 4210*, 406–407, 1989.

Spudis, P. D., and J. E. Guest, Stratigraphy and geologic history of Mercury, in *Mercury*, edited by F. Vilas, C. R. Chapman, and M. S. Matthews, pp. 118–164, University of Arizona Press, Tucson, 1988.

Spudis, P. D., and B. R. Hawke, The Apennine Bench Formation revisited, in *Workshop on the Geology and Petrology of the Apollo 15 Landing Site, LPI Tech. Rep. 86-03*, pp. 105–107, Lunar and Planetary Institute, Houston, 1986.

Staid, M., and C. M. Pieters, Reevaluation of lunar basalt types through spectral analyses of fresh mare craters, *Lunar Planet. Sci. Conf., 29*, 1853, 1998.

Staid, M. I., C. M. Pieters, and J. W. Head, Mare Tranquillitatis: Basalt emplacement history and relation to lunar samples, *J. Geophys. Res., 101*, 23213–23228, 1996.

Strom, R. G., Origin and relative age of lunar and mercurian intercrater plains, *Phys. Earth Planet. Inter.*, 156–172, 1977.

Strom, R. G., *Mercury: The Elusive Planet*, Smithsonian Institution Press, Washington, DC, 1987.

Strom, R. G., N. J. Trask, and J. E. Guest, Tectonism and volcanism on Mercury, *J. Geophys. Res., 80*, 2478–2507, 1975.

Taylor, L. A., J. W. Shervais, R. H. Hunter, and J. C. Laul, Ancient (4.2 AE) highlands volcanism: The gabbronorite connection? *Lunar Planet. Sci. Conf., 14*, 777–778, 1983.

Taylor, S. R., *Planetary Science: A Lunar Perspective*, Lunar and Planetary Institute, Houston, 1982.

Taylor, S. R., Growth of planetary crust, *Tectonophysics, 161*, 147–156, 1989.

Tolan, T. L., S. Reidel, J. L. Anderson, M. H. Beeson, K. R. Fecht, and D. A. Swansen, Revisions to the estimates of the areal extent and volume of the Columbia River Basalt Group, *Geol. Soc. Am. Spec. Publ. SP 239*, 1–20, 1989.

Trask, N. J., and J. E. Guest, Preliminary geologic terrain map of Mercury, *J. Geophys. Res., 80*, 2462–2477, 1975.

Veverka, J., P. Helfenstein, B. Hapke, and J. D. Goguen, Photometry and polarimetry of Mercury, in *Mercury*, edited by F. Vilas, C. Chapman, and M. S. Mathews, pp. 37–58, University of Arizona Press, Tucson, 1988.

Vilas, F., Mercury: Absence of crystalline Fe^{2+} in the regolith, *Icarus, 64*, 133–138, 1985.

Vilas, F., Surface composition of Mercury from reflectance spectrophotometry, in *Mercury*, edited by F. Vilas, C. Chapman, and M. S. Mathews, pp. 59–76, University of Arizona Press, Tucson, 1988.

Vilas, F., and T. B. McCord, Mercury: Spectral reflectance measurements (0.33–1.06 µm) 1974/75, *Icarus, 28*, 593–599, 1976.

Vilas, F., M. A. Leake, and W. W. Mendell, The dependence of reflectance spectra of Mercury on surface terrain, *Icarus, 59*, 60–68, 1984.

Wagner, R. J., J. W. Head, U. Wolf, and G. Neukum, Age relations of geologic units in the Gruithuisen region of the Moon based on crater size-frequency measurements, *Lunar Planet. Sci. Conf., 27*, 1367–1368, 1996.

Weitz, C. M., and J. W. Head, Diversity of lunar eruptions at the Marius Hills volcanic complex: Implications for the formation of domes and cones, *J. Geophys. Res., 104*, 18933–18956, 1999.

Weitz, C. M., J. W. Head, and C. M. Pieters, Lunar regional dark mantle deposits: Geologic, multispectral, and modeling studies, *J. Geophys. Res., 103*, 22725–22759, 1998.

Weitz, C. M., M. J. Rutherford, J. W. Head, and D. S. McKay, Ascent and eruption of a lunar high-Ti magma as inferred from the petrology of the 74001/2 core, *Meteor. Planet. Sci., 34*, 527–540, 1999.

Whitford-Stark, J. L., Charting the southern seas: The evolution of the lunar Mare Australe, *Proc. 10th Lunar Planet. Sci. Conf.*, 2975–2994, 1979.

Whitford-Stark, J. L., A preliminary analysis of lunar extra-mare basalts: Distribution, compositions, ages, volumes, and eruption styles, *Moon Planets, 26*, 323–338, 1982.

Whitford-Stark, J. L., The volcanotectonic evolution of Mare Frigoris, *Proc. 20th Lunar Planet. Sci. Conf.*, 175–186, 1990.

Whitford-Stark, J. L., and J. W. Head, The Procellarum volcanic complexes: Contrasting styles of volcanism, *Proc. 8th Lunar Sci. Conf.*, 2705–2724, 1977.

Whitford-Stark, J. L., and J. W. Head, Stratigraphy of Oceanus Procellarum basalts: Sources and styles of emplacement, *J. Geophys. Res., 85*, 6579–6609, 1980.

Wilhelms, D. E., Mercurian volcanism questioned, *Icarus, 28*, 551–558, 1976.

Wilhelms, D. E., The geologic history of the Moon, *U. S. Geol. Surv. Prof. Pap. 1348*, 1987.

Wilson, L., and J. W. Head, The formation of eroded depressions around the sources of lunar sinuous rilles: Theory, *Lunar Planet. Sci. Conf., 11*, 1260–1262, 1980.

Wilson, L., and J. W. Head, Ascent and eruption of basaltic magma on the Earth and Moon, *J. Geophys. Res., 86*, 2971–3001, 1981.

Wilson, L., and J. W. Head, A comparison of volcanic eruption processes on Earth, Moon, Mars, Io and Venus, *Nature, 302*, 663–669, 1983.

Wilson, L., and J. W. Head, Factors controlling the structures of magma chambers in basaltic volcanoes, *Lunar Planet. Sci. Conf., 21*, 1343–1344, 1990.

Wood, C. A., and J. W. Head, Geologic setting and provenance of spectrally distinct pre-mare material of possible volcanic origin, in *Papers Presented to the Conference on Origins of Mare Basalts*, The Lunar Science Institute, pp. 189–192, 1975.

Yingst, R. A., and J. W. Head, Volumes of lunar lava ponds in South Pole–Aitken and Orientale Basins: Implications for eruption conditions, transport mechanisms, and magma source regions, *J. Geophys. Res., 102*, 10909–10931, 1997.

Yingst, R. A., and J. W. Head, Characteristics of lunar mare deposits in Smythii and Marginis basins: Implications for magma transport mechanisms, *J. Geophys. Res., 103*, 11135–11158, 1998.

Yingst, R. A., and J. W. Head, Geology of mare deposits in South-Pole Aitken basin as seen by Clementine, UV/VIS data, *J. Geophys. Res., 104*, 18957–18979, 1999.

Zisk, S. H., C. A. Hodges, H. J. Moore, R. W. Shorthill, T. W. Thompson, E. A. Whitaker, and D. A. Wilhelms, The Aristarchus–Harbinger region of the Moon: Surface geology and remote sensing from recent remote sensing observations, *Moon, 17*, 59–99, 1977.

7

Extreme Volcanism on Jupiter's Moon Io

Alfred S. McEwen, Rosaly Lopes-Gautier, Laszlo Keszthelyi,
and Susan W. Kieffer

7.1. INTRODUCTION AND BACKGROUND

7.1.1. Basics

Io (Figure 7.1) is the innermost of the four large Galilean satellites of Jupiter discovered by Galileo Galilei in 1610. Io's mean radius (1821 km) and bulk density (3.53 g cm^{-3}) are comparable to those of the Moon. However, long before the Voyager spacecraft encounters, it was apparent from Earth-based observations that Io is very different from the Moon: It has an unusual spectral reflectance and anomalous thermal properties, and it is surrounded by immense clouds of ions and neutral atoms. Following discovery of a thermal anomaly (Witteborn *et al.*, 1979) and the prediction of intense tidal heating (Peale *et al.*, 1979), the Voyager spacecraft revealed a world covered by volcanoes, many of them active (Smith *et al.*, 1979a).

Gravity measurements from tracking of the Galileo spacecraft indicate that Io has a large iron or iron/iron sulfide core, comprising ~20% of the satellite's mass (Anderson *et al.*, 1996). The bulk density of Io, its rugged topography, and models of satellite origin suggest that the bulk composition of the crust and mantle is silicate. However, the surface composition has been profoundly altered by volcanic resurfacing and outgassing (Nash *et al.*, 1986). Sulfur dioxide (SO_2) frost is present over most of the surface and SO_2 gas is the major atmospheric component. Elemental sulfur is considered a likely surface component on the basis of (1) the similarity of Io's ultraviolet through near-infrared spectral reflectance to that of powdered sulfur and (2) the detection of sulfur ions in Io's plasma torus (a doughnut-shaped region of charged particles surrounding Io's orbit) (Spencer and Schneider, 1996). The detections of clouds of Na and K around Io have led to proposals that alkali sulfides, polysulfides, or sulfates may be present on the surface (Fanale *et al.*, 1982). The atmospheric pressure is ~10^{-4} Pa (~10^{-9} bar), but varies spatially and temporally, and is strongly influenced by volcanic activity (Lellouch, 1996).

Because of Io's sulfur-rich surface environment, many Voyager-era investigators thought that the active volcanism was dominated by sulfurous eruptions rather than the silicate

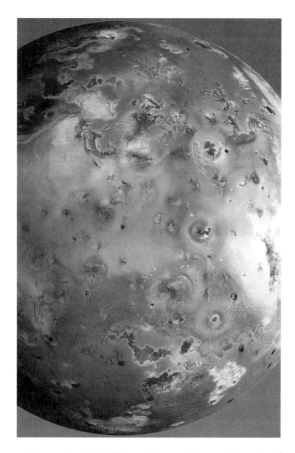

Figure 7.1. Image of Io acquired by Galileo, with a resolution of 3 km pixel^{-1}, seen in front of Jupiter's atmosphere. Several of the most active volcanic centers are shown, including Zamama, Prometheus, and Culann. The bright areas probably consist of nearly pure coarse-grained SO_2 frost, the localized dark regions are probably silicates, and the intermediate gray areas (red, yellow, and green when seen in color) probably consist of a variety of sulfur-rich materials mixed with SO_2 (Geissler *et al.*, 1999). For color version of this image and Figures 7.4. and 7.5, see http://photojournal.jpl.nasa.gov. Photo courtesy NASA/JPL.

volcanism seen in the inner solar system (see reviews by Nash *et al.*, 1986, and McEwen *et al.*, 1989). More recent observations have shown that the volcanism on Io is dominated by high-temperature eruptions of silicate lava, confirming the view of Carr (1986). Some measured temperatures are high enough to suggest ultramafic compositions (McEwen *et al.*, 1998a; see Chapter 8).

Basic facts about Io are summarized in Table 7.1. Much fundamental information is not yet available. Topographic measurements are very spotty, extracted from the images in places via stereo, shape-from-shading, and limb measurements. We have little direct information on the composition of Io's interior or erupting lavas. The best images have a resolution of no better than ~ 0.5 km pixel^{-1}, although Galileo should improve on that late in 1999.

The environment of Io, which influences the style of volcanism, is unique in comparison with other planets and satellites in our solar system (see Chapter 1). The low gravitational acceleration and atmospheric pressure are similar to conditions on the Moon, but Io is rich in

Table 7.1. Basic Facts about Io

Orbital period	42.456 h
Surface gravity	1.8 m s^{-2}
Triaxial shape	$a = 1830$ km, $b = 1819$ km, $c = 1815$ km (Thomas et al., 1998)
Bulk density	3530 kg m^{-3}
Core	Fe (radius 656 km) or Fe–FeS (radius 947 km)
Global average heat flow	>2.5 W m^{-2}
Local topographic relief	Up to 15 km
Active hot spots	>60, ~uniform global distribution, persistent hot spots at low latitudes
Hot spot temperatures	Up to at least 1600 K
Active plumes observed	15, concentrated at low latitudes
Surface composition	Ubiquitous SO$_2$ frost, other unidentified components
Typcial surface temperatures	85 K (night) to 125 K (day)
Atmospheric pressure	10^{-4} Pa (10^{-9} bar) or less, higher under plumes
Corona and neutral cloud species	Na, K, S, O

volatiles. The combination of low atmospheric pressure, volatile-rich environment, and extensive volcanic activity results in the large and spectacular plumes. Ionian volatiles are primarily SO$_2$ rather than H$_2$O or CO$_2$, which are more abundant on the terrestrial planets. The high heat flow and ultramafic volcanism of Io may be analogous to conditions on early Earth (see Chapter 8). Voluminous outpourings of lava on Io may be comparable to rapidly emplaced flood lavas on the Moon and Mars (see Chapters 4 and 6). Interactions between lava and SO$_2$ could be similar to lava/ground ice interactions on Earth and Mars (see Chapter 3).

7.1.2. Heat Flow and Tidal-Heating Mechanisms

Io's average heat flow is similar to that of the most volcanically active spots on Earth, such as Kilauea. The current global average heat flow on Io is at least 2.5 W m^{-2} (Veeder et al., 1994). Shallow conducted heat flow from recent igneous activity could double this estimate (Stevenson and McNamara, 1988), but it is unclear how much of this heat flow is already included in the Veeder et al. (1994) estimate, because their result is sensitive to model parameters. The total energy flow from Io must also include contributions from heat conducted through the base of the lithosphere and kinetic energy from the active plumes, but these quantities probably do not exceed 0.5 W m^{-2} (McEwen et al., 1989). Therefore, the total energy loss from Io is probably in the range of 2.5 to 5 W m^{-2}, about two orders of magnitude greater than the average heat loss from Earth (0.09 W m^{-2}) or from the Moon (0.02 W m^{-2}).

The importance of tidal heating in powering the enhanced heat flow and active volcanism of Io is widely accepted (reviewed by Schubert et al., 1986). All plausible nontidal heat sources are about two orders of magnitude less energetic than the power output of Io. The basic tidal-heating mechanism involves deformation of Io into a triaxial ellipsoid by Jupiter's gravitational field (Greenberg, 1982). Because Io's orbit is forced to be eccentric by orbital resonances among Io, Europa, and Ganymede, the satellite undergoes a periodic deformation. The total rate of tidal dissipation within Io depends inversely on its dissipation factor, which is a function of the internal structure and material properties of the satellite. Segatz et al. (1988) have shown that reasonable parameters for mantle rheology can produce dissipation rates as high as 3×10^{15} W, well above Io's current output (1–2×10^{14} W). However, the transfer of orbital energy from Jupiter to Io is a consequence of the bulge raised on Jupiter by Io, and the rate of such transfer depends inversely on Jupiter's dissipation factor (Q_J). This value is poorly known, but giving Q_J the lowest possible value averaged over 4.5×10^9 years results in an upper limit to Io's average energy dissipation rate of 3.3×10^{13} W (Peale, 1986; Greenberg, 1989). [Note that the upper limit of 9×10^{13} W cited by Schubert *et al.* (1986) and others has been corrected and revised to 3.3×10^{13} W.] Io's power output ($1-2 \times 10^{14}$ W) clearly exceeds this value in recent decades. We consider five hypotheses to resolve this discrepancy in Section 7.6.1.

The mechanism of heat dissipation is intimately tied to Io's internal structure. The initial model of Peale *et al.* (1979), with "runaway" melting of the interior leading to tidal energy dissipation within a thin, elastic lithosphere, now seems unlikely. Dissipation of all of the tidal energy within the lithosphere requires that it be 8–18 km thick, and such a thin lithosphere is inconsistent with the presence of mountains more than 10 km high. We discuss models for Io's internal structure in greater detail in Section 7.6.2.

7.1.3. Geomorphology

The morphology and surface markings on Io are largely the result of volcanic processes, with the possible exception of the rugged topographic features (high mountains, plateaus, and scarps). Impact craters, the dominant landform on most solid planetary bodies, have not been observed on Io, despite the comet flux concentration by Jupiter's gravity. Recent estimates of the impact cratering rates on the Galilean satellites show that a crater 20 km in diameter should form on Io every 2.3 Myr (Zahnle *et al.*, 1999), so the absence of any large (>20 km) impact craters (which would be easily recognized) indicates that no large terrains are more than a few tens of million years old. The surface features can be grouped into three general categories: volcanic centers (including calderas), mountains, and plains (Figure 7.2).

Io is spotted by local dark markings that probably consist of recently erupted silicate lavas and pyroclastics at volcanic centers. All of the hot spots and plumes emanate from these local dark areas, which cover a few percent of the surface (Figure 7.1). Many of these markings are seen to lie within and around depressions that resemble terrestrial calderas. Calderas have steep, scalloped walls, flat floors, and smooth rims; they form by collapse over shallow magma chambers following rapid removal of volcanic materials. Dark markings may cover all or part of a caldera floor and adjacent regions. The calderas range from about 20 to 200 km in diameter and are as deep as ~ 2 km. Most calderas are not associated with an obvious edifice, but some low shields with summit caldera are present. Steep caldera walls and association with high-temperature hot spots are inconsistent with the concept of a thick sulfur crust. Io's upper crust is probably composed of silicate rocks, and the sulfurous materials form a relatively thin surface veneer as well as ground fluids at depth.

The caldera morphology does not provide a good match to that of either basaltic shield volcanoes or silicic ash-flow calderas (Wood, 1984). Ionian calderas have scalloped margins and multiple areas of collapse, similar to calderas on terrestrial and Martian shield volcanoes (see Chapter 4), but they are larger and generally lie on flat plains or very low shields. The large size and low relief are similar to ash-flow calderas (Schaber, 1982), but the scalloped margins and absence of resurgence argue against this interpretation, if a terrestrial analogy is valid. More directly, the high-temperature hot spots indicate that the current activity is associated with mafic to ultramafic effusive eruptions. The absence of high shield volcanoes may be related to the low viscosity of ultramafic flows, which retards the accumulation of material near the vent (see Chapter 8).

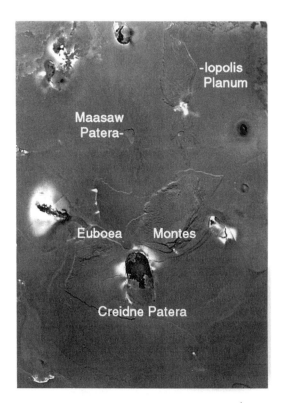

Figure 7.2. Mosaic of Voyager images of Io at a scale of 1 km pixel^{-1}. The landforms include tall mountains (montes), a plateau (planum), calderas (paterae), and layered plains. There are also albedo markings suggestive of lava flows and pyroclastic deposits. Orthographic map projection centered at latitude −45°, longitude 339°. Image scene is 720 km wide. Photo courtesy NASA/JPL.

Dark elongated features with sinuous margins, probably lava flows, can be seen at many volcanic centers (Figures 7.1–7.3). Many dark markings are sources of high-temperature hot spots consistent with ongoing eruptions of silicate lava. Thus, the dark areas are probably all sites of ongoing or very recent activity, and there are about 100 such locations on Io (McEwen et al., 1985; Lopes-Gautier et al., 1999). The spectral reflectances of the dark materials are consistent with silicates rich in orthopyroxene (a magnesium-rich silicate mineral), and with varying degrees of sulfurous coatings (Geissler et al., 1999). Diffuse bright white, yellow, and red markings in and around volcanic centers are common and may be the result of sulfurous fumarolic and plume activity.

The second major class of surface feature is mountains, which typically have irregular outlines and rugged surfaces that appear to be tectonically disrupted. They may be as large as 600 km in basal diameter and 15 km high (Schaber, 1982; Schenk et al., 1997). The origin of these high mountains is not resolved. They may be degraded volcanoes or uplifted crustal blocks. Several mountains show possible layering and have sharp angular boundaries suggesting that they are tilted crustal blocks (Schenk and Bulmer, 1998; Carr et al., 1998). The mountains are often considered to be the oldest materials exposed on Io's surface, but higher-resolution images are required to confirm this hypothesis. There has been no clear indication of active volcanism associated with any mountain on Io. Plateaus and mesas are

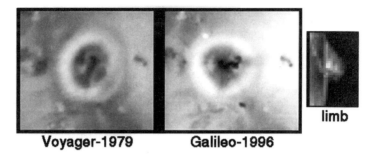

Figure 7.3. Three views of the active plume Prometheus. The left image shows the Prometheus region as seen by Voyager, and the middle view shows the same region seen by Galileo. There are new dark lava flows and the plume vent has moved about 70 km to the west. The right view shows the plume on the limb, against black sky, as seen by Galileo. Plume's height is 75 km, and bright ring has a diameter of about 300 km. Photo courtesy NASA/JPL.

present as well, often surrounding or grouped with mountains. The detailed relations are unclear, but some form of mass wasting seems likely. Some of the plateaus may also be uplifted crustal blocks, but with lesser degrees of uplift (or downdropping of adjacent terrains). It is possible that mountains subside over 10^4–10^6 years as a result of thermal erosion at the base of the lithosphere (Webb and Stevenson, 1987).

Most of Io's surface is covered by plains with little apparent relief. They have complex albedo and color patterns, and, where relief is discernible, they are commonly seen to be layered. The plains and their surface patterns must have formed by some combination of lava flows, mass wasting, deposition from plumes, condensation of volatiles, and possibly pyroclastic flows.

7.1.4. Surface Changes

About 30 large-scale (tens of kilometers) changes in color and albedo, covering $\sim 5\%$ of the surface, were obvious from comparison of the Galileo images obtained in 1996 with those acquired by Voyager in 1979 (McEwen et al., 1998b; Figure 7.3). These include new pyroclastic deposits of several colors, bright and dark flows, and caldera-floor materials. There have also been significant surface changes on Io during the Galileo mission, such as a new 400-km-diameter dark pyroclastic deposit around Pillan Patera (Figure 7.4). While these surface changes are impressive, the number of large-scale changes observed in the 4 months between the Voyager 1 and Voyager 2 flybys in 1979 suggested that over 17 years the cumulative changes would have been much more impressive. There are two reason why this was not actually the case. First, the most widespread plume deposits are ephemeral, and seem to fade away within a few years. Second, it appears that a large fraction of the volcanic activity is confined to repeated resurfacing of dark calderas and flow fields that cover only a few percent of Io's surface. No topographic changes have been detected, but only changes of a kilometer or more in relief would be detectable given the Voyager and Galileo resolution and coverage limits.

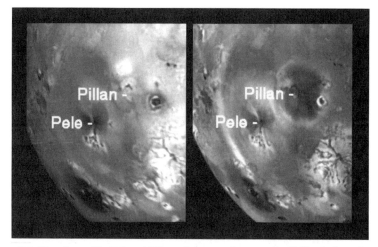

Figure 7.4. Two views of the Pele and Pillan volcanoes on Io, as seen by Galileo in the spring of 1997 (left) and fall of 1997 (right). The large ring is the fallout deposit of the 460-km-high Pele plume. The eruption near Pillan Patera northeast of Pele produced the new dark deposits seen in the right-hand image, with a diameter of 400 km. Pillan also produced a very high temperature hot spot, with lava temperatures exceeding 1700 K, probably related to the eruption of ultramafic lava from Io's mantle. Photo courtesy NASA/JPL.

7.2. VOLCANIC HOT SPOTS

7.2.1. Early Observations

Evidence for thermal anomalies on Io was acquired years before the Voyager encounters, but it was not understood. Thermal infrared observations had revealed two curious relations. First, when Io is eclipsed by Jupiter (Figure 7.5), its temperature falls as expected for a surface with low thermal inertia, but is minimum temperature is too high. Second, the brightness temperature of Io is significantly higher at shorter wavelengths. After the Voyager 1 encounter and the discovery of active volcanism, these observations were understood as being the result of hot spots covering a few percent of the surface (Matson *et al*, 1981).

The term *hot spot* is used to indicate resolved regions of thermal emission significantly above the local background temperature. Background temperatures range from ~ 65 K at the poles to ~ 140 K for low-albedo patches near the subsolar point. The hot spots and observations of tall plumes served as definitive evidence for active volcanism. Since then, analysis of Voyager 1 data from the Infrared Interferometer Spectrometer (IRIS) showed at least 22 active hot spots over the one-third of the surface that was observed by this instrument (Lopes-Gautier *et al.*, 1999). There were hot spots near all of the Voyager plume sites observed by IRIS. Fits to the IRIS spectra indicated maximum temperatures for the hot spots of about 650 K (Pearl and Sinton, 1982). These relatively low temperatures were initially interpreted as possible evidence for sulfur rather than silicate volcanism. However, IRIS could not detect small areas at higher temperatures because of limitations in sensitivity and wavelength coverage, and the data were also consistent with silicate volcanism (Carr, 1986).

Between the Voyager observations in 1979 and the Galileo observations that started in 1996, several of Io's hot spots were observed from ground-based telescopes; some observations revealed high temperatures. For example, Johnson *et al.* (1988) reported a temperature

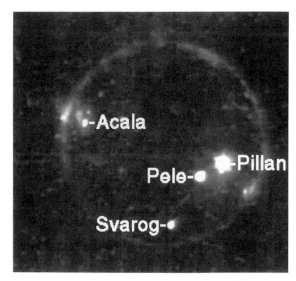

Figure 7.5. Image of Io in eclipse (when Io is in Jupiter's shadow). Small bright spots are high-temperature hot spots. The two brightest spots are Pele and Pillan (see Figure 7.4). Diffuse glows from electronic excitation of gases highlight Io's limb and active plumes. Photo courtesy NASA/JPL.

of at least 900 K, Veeder *et al.* (1994) reported 1225 and 1500 K, and Stansberry *et al.* (1997) reported two events at 1400 K or hotter. These measurements are consistent with silicate but not with sulfur volcanism. Sulfur boils vigorously on Io's surface at ~ 500 K and although sulfur compounds such as sodium polysulfides could be several hundred degrees hotter (Lunine and Stevenson, 1985), temperatures exceeding 1000 K strongly suggest silicate lavas.

Voyager and ground-based measurements have revealed that Loki is the single most powerful source of thermal emission on Io and that it undergoes major variations in thermal flux. Voyager observed Loki while it was in a relatively quiet state, yet the IRIS spectra of Loki were consistent with a power of more than 10^{13} W, exceeding the summed output of all of Earth's active volcanoes. Ground-based measurements showed that Loki brightenings typically last several months before fading (Spencer and Schneider, 1996). The most dramatic brightenings are called *outbursts*, defined as events that at least double the 5-μm flux from Io, which is normally dominated by reflected sunlight. The onset of an outburst can occur in 2 h or less, can show significant evolution in 3 h, and can fade in 7 days or less. A particularly well observed outburst at Loki in 1990 is discussed in Section 7.2.3.

7.2.2. Galileo Observations

The Galileo mission has provided an unprecedented opportunity to monitor the activity of Io's hot sots. The key instruments are the Near-Infrared Mapping Spectrometer (NIMS) and the Solid State Imaging System (SSI). The NIMS wavelength range, 0.7 to 5.2 μm, enables it to detect hot spots from the highest temperatures down to 180 K (if a pixel is filled) (Smythe *et al.*, 1995). The spatial resolution of SSI is 40 times better than that of NIMS, but it is only sensitive from about 0.4 to 1.0 μm, enabling it to detect smaller hot spots provided that the temperatures are higher than ~ 800 K. Between June 28, 1996, and September 20, 1997 (Galileo orbits G1 to C10), NIMS and SSI detected 37 hot spots o Io, including 22 that were not known from Voyager or ground-based measurements (Lopes-Gautier *et al.*, 1999). At three of these locations, SSI detected multiple hot spots, corresponding to two or three areas that could be different vents or flows in the same volcanic region. Other areas that are interpreted as active volcanic centers on Io include hot spots and plume sites observed by Voyager but not by Galileo, areas identified as hot spots from ground-based observations, and areas with Voyager-to-Galileo surface changes (Figure 7.6).

The locations of active centers correspond to calderas (paterae), flows (flucti), and other features interpreted to be volcanic. Galileo results have confirmed the suggestion from Voyager observations (Pearl and Sinton, 1982; McEwen *et al.*, 1985) that hot spots are almost always associated with low-albedo materials, and have shown that the hot spots lie directly on the dark materials.

The monitoring of Io's hot spots has shown that they can be persistently active over time scales of months to years and perhaps decades (Table 7.2). Analysis of NIMS data showed that 22 hot spots were active for longer than 1 year and several had also been seen by telescopic observers or by Voyager nearly two decades earlier. Apparently short-lived or sporadic events also occur, such as the 5 hot spots detected only once by Galileo and the 5 others detected once by Spencer *et al.* (1997b) from ground-based observations during 1995–1997. These events could represent short-lived activity or activity that falls to levels below the detection limits. Persistent and sporadic types of hot spots may represent different type of volcanism on Io, perhaps controlled by the rate of magma supply or eruption mechanism. Plumes seen above Io's surface are generally, or perhaps always, associated with persistent hot spots, although plumes have not been detected at many other persistent hot spots. Surface changes suggest the existence of short-lived plumes, which may correspond to sporadic hot spots.

One of the most surprising findings about Io's volcanism was the discovery of very high temperatures, hotter than any active volcanism on Earth (McEwen *et al.*, 1998a). This seems

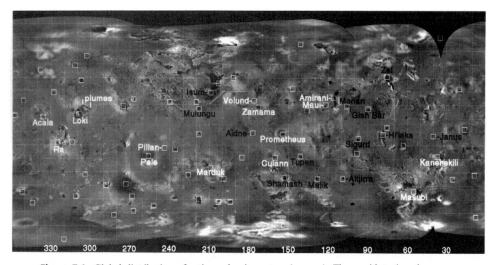

Figure 7.6. Global distribution of active volcanic centers (squares). Those with active plumes are labeled in white, and other persistent hot spots are labeled in black. Simple cylindrical global mosaic of Galileo images; 10° latitude–longitude grid. The sub-Jupiter hemisphere is centered at longitude 0° (left and right sides) and anti-Jupiter hemisphere at longitude 180 (middle). Photo courtesy NASA/JPL.

Table 7.2. Persistent Hot Spots on Io

Hot spot	% observations detected[a]	Latitude (deg.)	Plume?
Pele	100	−18	yes
Mulungu	100	+17	no
Marduk	100	−27	yes
Isum	100	+31	no
Amirani	100	+25	yes
Hi'iaka	100	−2	no
Kanehekili	100	−16	yes
Pillan	89	−10	yes
Loki	88	+11	yes
Malik	88	−35	no
Prometheus	88	−2	yes
Janus	83	−4	no
Zal	83	+41	no
Tupan	75	−18	no
Culann	75	−19	probably
Altjirra	56	−35	no
Zamana	56	+18	yes
Aidne	56	−1	no
Gish Bar	44	+17	no
Sigurd	44	−5	no
Monan	44	+19	no
Shamash	38	−34	no

[a] Percentage of NIMS observations covering each region in which the hot spot was detected (Lopes-Gautier et al., 1999).

especially surpsing following the Voyager-era models favoring low-temperature sulfur volcanism. Lava temperatures probably exceed the typical temperature of basalts (∼1500 K) in about a dozen hot spots, although the measurement uncertainties are significant. The eruption of Pillan Patera in the summer of 1997 provided an unambiguous example of extremely hot lava (Figures 7.4 and 7.5). SSI and NIMS observed Pillan in eclipse within a few minutes of each other, and results from both instruments indicate temperatures from 1600 to 2500 K. It is possible that such temperatures characterize the lavas at most of the active volcanic centers on Io, but are only detectable from distant observations during especially active eruptive episodes.

7.2.3. Models of Effusive Volcanism

Interpreting the data from these hot spots has posed a unique challenge. While the dark flowlike features associated with the hot spots are assumed to be lava, we have not been able to image the effusion of lava onto the surface of Io and can only infer the processes via extrapolation from our low-spatial-resolution data, modeling of the infrared spectra, and comparison with terrestrial volcanoes. The low image resolution (0.5 to 10 km pixel^{-1}) does not allow us to identify clearly features such as individual lava domes, channels, levees, spatter ramparts, or lava lakes. Thus, many of the techniques used to study volcanoes on the Earth and other terrestrial planets are not currently applicable to Io. For example, the bulk rheology of a flow could be estimated given measurements of flow thickness and width, levee widths, and slopes, but such data are not all available for any location on Io. However, the fact that the volcanism on Io is currently active enables other techniques. In particular, observations of the thermal emissions from active eruptions have been extremely valuable.

It is crucial to remember that the temperature models described in Section 7.2.2 are derived by averaging the emissions for the entire hot spot, corresponding to volcanic centers tens to hundreds of kilometers wide. Only a tiny fraction of the emitting surface will be at the temperature of the liquid lava. The surface of a lava flow will typically cool 200 to 400 K in a few seconds, making it essentially impossible to measure directly the temperature of the liquid lava via remote sensing (Keszthelyi and McEwen, 1997a). Even cracks in the surface of flow typically only break through to the brittle–ductile transition of the crust (slightly below the solidus of the lava) (Hon *et al.*, 1994). The great majority of the thermal emission must come from cooling crusts on recently exposed lava.

The effect of the emissions from the cooling crust is minimized by using observations at short wavelengths. SSI data in the region from 0.7 to 1.0 µm cannot detect temperatures below ~ 800 K, but are very sensitive to temperatures greater than 1000 K. The NIMS instrument, sensitive from about 1 to 5 µm, is best suited to detect blackbody temperatures in the 300–1500 K region. The IRIS instrument on Voyager, working at longer wavelengths (~ 5–50 µm), detects temperatures from about 50 to 650 K. The areas of each hot spot reported by the different instruments are directly related to the temperature range to which they are sensitive. SSI sees small areas (typically on the order of 10^{-2} km^2) at temperatures of 900–1700 K; single-temperature fits to NIMS data suggest areas on the order of a few square kilometers at 400 to 600 K, and two or three temperature fits to the Voyager IRIS data result in areas of tens of square kilometers at temperatures generally under ~ 400 K. Yet these instruments were observing identical (or very similar) hot surfaces. Clearly, using a limited wavelength range and fitting to a small number of temperatures can result in misleading interpretations.

Interpreting these studies of data requires modeling the full range of surface temperatures expected across an entire volcano. Needless to say, such models require a host of simplifying assumptions, the validity of which is difficult to test directly with the current data sets. There are currently a handful of models attempting to predict the infrared flux from effusive eruptions on Io. Since the observations do not resolve the vents and flows, these models examine each part of the eruption separately, then sum their infrared emissions.

All of the models to date have assumed a roughly basaltic composition, an assumption that is probably violated at least in the case of the 1997 eruption at Pillan. However, increasing the initial temperature of the lava and adjusting thermophysical properties such as density and heat capacity will not fundamentally change the models. In the near vacuum of Io, heat loss is dominated by thermal radiation from the surface (e.g., Howell, 1997). It is important to note that the porosity (i.e., bubble content) of the lava plays a key role in the cooling of the surface (Keszthelyi and McEwen, 1997a), but this parameter is difficult to estimate remotely. Lava porosities on Io may be very high because of the combination of low atmospheric pressure and abundant volatiles. The models also contain a description of the advance of the lava in order to predict the surface area at a given temperature as a function of time. This is done by combining a model for the rate of cooling of the lava with a model of the rate of advance of the flow, adding a whole suite of assumptions about the rheology of the lava and topography that the flow traverses. (Davies, 1996; Davies *et al.*, 1999) also explicitly included the effects of fountaining lava at the vent and cracks in the surface of the flow.

Keszthelyi and McEwen (1997a) attempted to condense all of these assumptions into a single parameter, the maximum age of the hot surfaces (tau). Any eruption will have surfaces ranging in age from zero to tau, but tau should normally be significantly less than the duration of the entire eruption. Estimates of tau for different styles of volcanism can be made based on terrestrial observations. In the case of fountains and rapid channels, tau is on the order of seconds to tens of seconds. For lava lakes, the surface typically survives for minutes; and for slower lava flows, the surface typically survives for hours to weeks before being covered again. Turbulent, ultramafic terrestrial lava flows have never been witnessed by humans, but we expect tau to be on the order of a few seconds.

Figure 7.7 shows the expected response of SSI as a function of tau and melt temperature for a lava with 50% porosity on Io. We have found the ratio of the clear (CLR) and 1-μm (1MC) filters to be the most useful parameter for translating SSI data into temperatures. The 1MC filter effectively cuts out the visible and shortest infrared photons while the CLR filter is sensitive to the same near-IR flux as the 1MC, as well as shorter wavelengths. Figure 7.7 suggests that active basaltic eruptions (liquids at ~1400 K) would be expected to produce CLR/1MC ratios of about 7–8, and perhaps as high as 12 for very active fountains and rapid channels. CLR/1MC ratios of 12–17 have been repeatedly observed on Io. These results require lavas hotter than typical basalts and could be consistent with lavas approaching 2000 K. If the lavas have porosities greater than 50%, which is likely in Io's low-pressure and volatile-rich environment, then even higher initial temperatures (or extremely short tau) would be needed to explain the observations. These extremely hot lavas suggest ultramafic (magnesium-rich) compositions. Nonthermal explanations for the emissions seen by SSI and NIMS are inconsistent with the observations over a broad range of wavelengths.

Could the high temperatures be the result of superheated lava of basaltic composition? Magmas rising rapidly from depth have the potential to be superheated (i.e., have a temperature greater than the equilibrium melting temperature). This is because, as a function of pressure, the adiabatic temperature gradient is often less steep than the liquidus. Thus, as the liquid rises into a lower-pressure region, the temperature of the liquid drops more slowly than the melting temperature, producing magma hotter than can be produced in equilibrium conditions. This process would allow mafic lavas produced in the mantle to reach the surface at temperatures typical of ultramafic melts if they rise rapidly from a depth of 300–800 km. Although melts from such depths have not been found on Earth, we cannot rule them out on Io.

Another process that could produce superheated lavas is tidal heating focused within large magma chambers. Tidal heating is expected to be maximized at rheologic boundaries

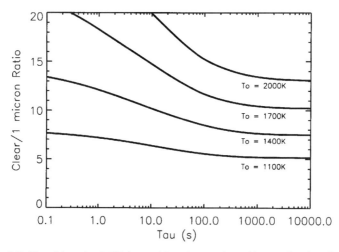

Figure 7.7. Plot of the ratio of SSI clear and 1 μm hot spot intensities as a function of tau (maximum lava age) and T_0 (initial lava temperature). See text for discussion.

(where strain can be large) such as the interface between solid rock and liquid mama. At this time we do not have quantitative comparisons of the rate at which heat could be added to the magma via tides versus the rate of heat loss via conduction, melting of the wall rocks, and escape of the magma.

Therefore, we must conclude for now that the high temperatures alone do not prove ultramafic compositions. However, additional support for the ultramafic interpretation comes form SSI 6-color data, which shows an absorption band consistent with magnesium-rich orthopyroxene in the low-albedo materials (Geissler et al., 1999), and from the morphology of the calderas (see Section 7.1.3). An additional possibility is that the lavas are both ultramafic and superheated. This could become the preferred interpretation if we see evidence for lavas hotter than ~ 2100 K.

Some of the most energetic thermal events on Io have been interpreted as the result of very high mass eruption rates. The well-observed outburst at Loki in 1990 has been modeled as reflecting a basaltic eruption rate of 10^5 to 10^6 m^3 s^{-1} (Blaney et al., 1995; Davies, 1996). The eruption near Pillan Patera in 1997 has been modeled with a comparably large eruption rate (Davies et al., 1998). These eruption rates are huge by terrestrial standards. The largest observed eruption rate was $\sim 10^4$ m^3 s^{-1} at Laki, Iceland (Thordarson and Self, 1993). Such large eruption rates indicate a fundamental difference between Io and Earth, perhaps simply reflecting the larger volumes of melt available within Io.

There are additional processes that must be occurring on Io that are not included in any of the thermal models to date. In particular, there must be intense interactions between the SO_2-rich surface deposits and the advancing lava flows. In many ways, these interactions should be similar to lava flows entering the ocean on the Earth (see Chapter 5), except the "steam" is blasting into a near vacuum instead of the thick atmosphere of the Earth.

7.3. PLUMES

7.3.1. Background and Descriptions

The most visually spectacular phenomena at Io are the active volcanic plumes. Nine eruption plumes were observed during the Voyager 1 encounter; eight of these were also observed 4 months later by Voyager 2. Galileo has observed a total of 10 plumes as of mid-1998, 4 of which appear related to Voyager-era plumes, so there are a total of 15 volcanic centers with observed plumes (McEwen et al., 1998b). Ring-shaped surface deposits suggest that many other plumes have been recently active as well. Many of the plumes, such as that of Prometheus (Figure 7.3), are 50–150 km high, long lived (years or decades), deposit bright white material (probably SO_2 frost) and/or bright red material (probably a bright frost containing short-chain sulfur) and are associated with high-temperature hot spots. Prometheus and other plumes show signs of lateral migrations over time of up to 100 km (or these are separate but nearby plumes). The presence of dark flows suggests that the migrations could be related to advance of the lava through SO_2-rich surface deposits.

Pele's plume is very different from Prometheus-type plumes, as it is very faint at visible wavelengths, up to 460 km high, and deposits a large ring of bright red material that is probably rich in short-chain sulfur (Spencer et al., 1997c). The eruptions that deposited large red rings around two other centers in 1979 may have been similar to Pele (McEwen and Soderblom, 1983). The Pele hot spot is also unusual because it maintains an intense high-temperature component at surprisingly constant levels over long periods of time, perhaps reflecting a vigorously churning or fountaining lava lake. No convincing and complete hypothesis for Pele's characteristics and uniqueness has been presented.

The dominant volatiles driving explosive volcanism on Earth, H_2O and CO_2, seem to be highly depleted on Io. Discernible water or carbonate absorption bands are absent or extremely weak, and elemental carbon has not been detected in Jupiter's magnetosphere near Io. Hydrogen Lyman-α emission from Io's polar regions was recently detected by the Hubble Space Telescope (Woodward *et al.*, 1998), and there have been suggestions of H_2S (Nash and Howell, 1989), so small amounts of magmatic H_2S or H_2O cannot be ruled out. In contrast, SO_2 is ubiquitous over the surface (Carlson *et al.*, 1997) and is widely believed to be the dominant atmospheric component (Lellouch, 1996). Hence, SO_2 (and maybe elemental sulfur) probably drives the explosive volcanism (Kieffer, 1982).

7.3.2. Plume Dynamics

The dynamics of Ionian volcanism can be separated into two topics for purposes of discussion, namely, thermodynamics and fluid dynamics, although these two subjects intertwine in a complex way. Possible thermodynamic conditions for Io were described by Kieffer (1982) and are summarized in Figure 7.8. These conditions range from unusually cold to unusually hot as compared with terrestrial volcanism. These conditions have been simplified and generalized to six different initial states in the reservoir, corresponding to various phases of SO_2 or S, and to various temperatures. Although the reservoirs were initially derived with reference to the Smith *et al.* (1979b) model for a sulfur-rich crust on Io, they are not restricted to that specific model, and are equally applicable to a silicate crust that contains SO_2 and/or sulfur as ground fluids.

The coolest reasonable volcanism on Io resembles geyser volcanism on the Earth (Reservoir I, Figure 7.8a, liquid SO_2 with a temperature below the liquids of sulfur, 393 K). Liquid SO_2 in a shallow crustal reservoir begins ascending because of buoyancy, and starts boiling at higher levels in the crust. As it erupts, a plume of vapor and ice condensate forms as the fluid emerges into the cold, low-pressure atmosphere of Io. At slightly higher temperatures on the liquidus of sulfur, the SO_2 boils in the reservoir and on ascent (Reservoir II, Figure 7.8b). This fluid also forms a mixture of vapor and ice in the plume. These two reservoir conditions are termed *low-entropy volcanism* because the initial entropy is less than the critical-point entropy. The characteristic phase change sequence on eruption is from liquid to vapor to ice.

SO_2 can coexist with sulfur at the sulfur liquidus conditions in a different state—as a vapor—if there is sufficient entropy in the system (Reservoir III, Figure 7.8b). On eruption, this vapor can condense into a liquid–vapor mixture, and then into an ice–vapor plume. Reservoirs III, IV, and V (described below), are termed *high-entropy volcanism* because the entropy is greater than the critical-point entropy. The Prometheus-type plumes most likely originate from conditions like Reservoirs II, III, or IV, as they have scattering properties consistent with the presence of small particles of condensing SO_2.

Finally, either SO_2 or S can exist in superheated vapor states whose temperature is determined by other conditions in the reservoir. The most plausible governor of temperature is proximity of a silicate magma. Temperatures as high as 2000 K may be present (see Section 7.2.3). Reservoirs IV, V, and V' (Figure 7.8c) represent variations on these conditions. Reservoir V' contains a pseudogas (gas with entrained solids) whereas Reservoir V is pure vapor. All-vapor plumes from reservoirs V may be difficult to detect via scattered light, and

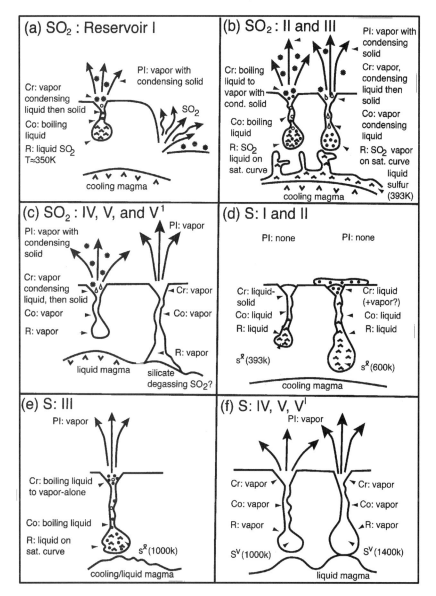

Figure 7.8. Models of Ionian plumes, modified from Kieffer (1982). In each figure, a plausible position for the reservoir is shown within a silicate crust containing sulfur and SO_2 fluids. The annotation shows the composition of erupting fluid in the reservoir (R), conduit (Co), crater (Cr), and plume (Pl). Liquid phase is indicated by the wavy pattern, boiling flow by circles, condensing flow by drops, and condensation of solid phase by snowflakes. Vapor is unpatterned.

have been called *stealth plumes* (Johnson *et al.*, 1995). The Acala plume, seen only via diffuse emissions while in eclipse, is the best candidate for a stealth plume on Io. Pele may be an example of a plume from Reservoir V', in which entrained solids keep the surface deposits sufficiently warm to prevent condensation of SO_2 (McEwen *et al.*, 1998b).

A similar sequence of reservoirs can be considered but with sulfur rather than SO_2 as the working fluid (Figure 7.8d–f). Reservoirs I and II do not produce plumes, but may produce sulfur flows. Sulfur Reservoir III is similar to SO_2 Reservoir II but at higher temperatures. Reservoirs III, IV, and V all result in plumes of sulfur vapor, and V' erupts gas with entrained solids. Plumes driven by a mixture of SO_2 and sulfur are also plausible.

To the extent that the fluid dynamics of Ionian volcanism can be modeled, the only initial conditions that are important are temperature, mass fraction of gas versus particle, and initial pressure. It is important to emphasize that the dynamics of the plumes give only limited information about reservoir conditions. At the present state of observational constraint and modeling capability, this information is restricted to temperature, plausible gas/particle ratios, particle sizes, and—possibly—inital pressure. However, initial pressure conditions may not be distinguishable from initial kinetic energy because the transformation of pressure–energy into kinetic energy causes these two variables to be intertwined.

Early models for volcanism on Io considered transformation of thermal energy to velocity to plume height. Thus, we expect plume height to increase from Reservoir I to Reservoir V (Figure 7.8). These models suggested that some plume shapes—those that are relatively symmetric umbrellas, such as Prometheus—could be explained by ballistic-trajectory models (Strom *et al.*, 1981), whereas others probably reflect complex pressure conditions and/or vent geometries. Almost nothing could be stated about the plume dynamics.

Subsequently, numerical models of the plume dynamics have been initiated (S. W. Kieffer *et al.*, manuscript in preparation). A preliminary result is shown in Figure 7.9. Because of computer limitations, the calculations to date have been restricted to small plumes and, therefore, to low-temperature conditions. The plume illustrated in Figure 7.9 is assumed to have originated from a reservoir at 400 K. It has 15% volatiles, a vent radius of 1 km, and an initial velocity at the vent of $90\,\text{m s}^{-1}$. This is the sonic velocity for a mixture of this composition and temperature, i.e., the plume is assumed to emerge from the conduit into the atmosphere at Mach 1 conditions. The mass discharge rate is $3.6 \times 10^9\,\text{kg s}^{-1}$. Atmospheric pressure is taken to be initially 10^{-4} Pa (10^{-9} bar). The parameter plotted in Figure 7.9 is log of the void fraction of particles, and so the snapshots correspond crudely to visual cross sections. For calibration, an observer viewing such a plume from outside might be able to see in only as far as the contours of -6.

Several phenomena are revealed by the simulations that were not apparent in earlier simpler models. First, the plume rises to about 10 km within a minute. It is supersonic because of the high ratio between vent pressure ad atmospheric pressure, and a supersonic shock structure is formed near the base. Material decelerates through the shock structure, reaching an altitude of 12 km over the first few minutes of an eruption. At about 400 s, the ascending plume collapses to form a lower fountain. The material reimpacts the ground at a radius of about 12 km. This reimpacted material feeds an inwardly directed recirculating eddy, and an outwardly directed pyroclastic flow.

Examination of temperature profiles through the plume (not shown) reveals that the plume remains relatively hot because of the mass loading (e.g., the collapsed fountain material and pyroclastic flow are >180 K). The plume interacts in a complex way with the preeruption atmosphere, generating large mixing eddies and changing the pressure conditions

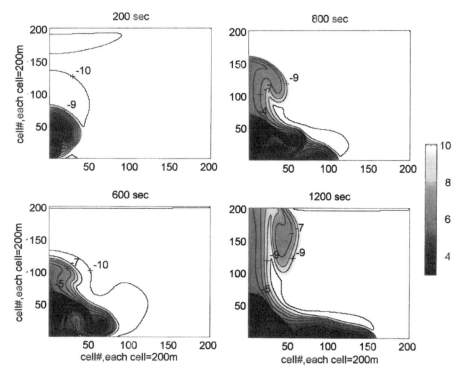

Figure 7.9. Plume simulations. Shown here in contours and gray-level shading is the log of the void fraction of particles, for each of four times. The snapshots correspond crudely to visual cross sections. See text for further explanation.

throughout the eruption domain. These effects are, unfortunately, extremely difficult to model numerically because of scaling problems.

The plume simulation described here is just one example; different initial conditions produce a wide variety of plumes. We chose this example (Figure 7.9) because it highlights a very interesting new result, that pyroclastic flows are possible in spite of Io's extremely tenuous global atmosphere. Although previous workers (e.g., Wilson and Head, 1983) have assumed that pyroclastic flows will not form on planetary bodies that lack appreciable atmospheres, this simulation shows that volatile-rich plumes may create their own local atmospheres and fountain collapse becomes possible. Io turns the premise of this book on its head: Instead of the environment effecting volcanic eruptions, the eruption creates its own environment.

7.4. SULFUR FLOWS AND LAKES?

One of the major debates about the interpretation of Voyager images centered on the composition of the materials being erupted as flows: sulfur or silicates (reviewed in Nash *et al.*, 1986). Recent observations have made it apparent that silicate volcanism is widespread on Io and that sulfur volcanism has, at most, a secondary role. The extent of secondary sulfur volcanism, however, has not yet been established.

The presence of sulfur compounds on Io's surface, including SO_2 frost and an ultraviolet-absorbing material such as elemental sulfur, is evident from the reflectance spectrum. There have also been observations of SO_2 gas in the plumes and atmosphere and of ionized sulfur in the plasma torus. The red deposits surrounding Pele and some other hot spots may be metastable short-chain S_3/S_4 molecules mixed with other sulfurous materials (Spencer et al., 1997a). Although various objections have been raised to the presence of pure elemental sulfur on Io, impure volcanogenic sulfur is far more likely and has spectral properties consistent with much of Io's surface (Kargel et al., 1999).

There may be sulfur flows at active volcanic centers not yet identified as hot spots, such as Ra Patera, the location studied by Pieri et al. (1984). Ra is the only plume observed by Galileo without a high-temperature hot spot, but the plume became inactive (or undetectable) after the first Galileo encounter. Hence, it is unclear if Ra is a different kind of eruption or if it is typical for the plume to persist for a few days to months after the high-temperature lava has ceased to erupt at detectable rates.

Secondary sulfur volcanism—that is, melting and remobilization of sulfur deposits by the injection of hot silicates from below—is a process that occurs on Earth. A particularly good example is the Mauna Loa sulfur flow studied by Skinner (1970) and by Greeley et al. (1984). This flow is thought to have been emplaced by the mobilization of fumarolic sulfur that had accumulated on the flank of a cone. Skinner (1970) attributed the mobilization of the sulfur to heat generated by the 1950 eruption of Mauna Loa, by means of proximity to molten lavas, a rise in the local thermal gradient by heat conducted through rocks from subsurface magma, and/or an increase in heat from hot gases rising through fissures above the subsurface magma. Some of the bright flowlike features along the margins of dark flows at Ra Patera (as seen by Galileo) and Amirani may be example of secondary sulfur flows. However, sulfur flows may be difficult to identify from distant monitoring by Galileo. The close flybys of Io planned for the end of 1999 should provide a better opportunity, from morphologies seen in high-resolution images, temperature measurements, and spectral reflectance data.

A plausible scenario for sulfur lakes on Io was proposed by McEwen et al. (1985), in which intrusions of hot silicates in a silicate and sulfur crust can remobilize the sulfur, causing flows and lakes within caldera, sometimes maintained in a liquid state by underlying hot silicates. This idea was modeled in detail by Lunine and Stevenson (1985) to explain the thermal fluxes from Loki in terms of a convective sulfur lake heated by an underlying magma chamber. But there is little evidence (such as from color changes) for condensation of sulfur vapor over large regions surrounding hot spots as predicted by the sulfur lake model. Instead, the discovery of an absorption band near 0.9 µm (Geissler et al., 1999) suggests that the dark materials on caldera floors are silicate lavas rather than some form of sulfur.

7.5. GLOBAL DISTRIBUTION OF VOLCANISM

To first order, the active and recently active volcanic centers are distributed uniformly (and partly randomly) over Io's surface (Carr et al., 1998), but there is a strong equatorial concentration of active plumes and persistent hot spots (Lopes-Gautier et al., 1999; see Figure 7.6 and Table 7.2). There are several other apparent deviations from uniformity. There may be concentrations of active vents near the sub- and anti-Jupiter regions (McEwen et al., 1998b). (Io is in a synchronous orbit around Jupiter as in the Moon around Earth, with one hemisphere always facing the primary.) A field of bright vents seen in eclipse occurs over the sub-Jupiter region, and a region of diffuse glow seen in eclipse at low resolution suggests that a similar field may exist over the anti-Jovian region. The sub- and anti-Jovian regions should experience the greatest tensile (and compressional) stress at the surface. The region of Io from longitude 240 to 360° has distinctive color and albedo patterns and the plumes in this region (Pillan, Pele, Loki, Ra, and Acala) display a greater range of characteristics than plumes at other longitudes. There appears to be a concentration of hot spots in a ring surrounding the bright equatorial region named Bosphorus Regio (Lopes-Gautier et al., 1997). The distribution of volcanism may suggest that the heat flow is uniform as a function of longitude over time periods of centuries or longer. However, in recent decades Loki alone has accounted for ~25% of Io's hot spot heat flow, so the global pattern is nonuniform at present. It is unclear how best to explain these patterns, but the tidal heating mechanisms may provide part of the answer.

Two basic models have been considered in detail for the internal heating of Io; both assume that convection is the main mode of heat transfer within Io, but differ on where the tidal heating is concentrated (Segatz et al., 1988). If heating occurs mainly in an asthenosphere about 100 km thick, convection should occur globally, with centers of upwelling and downwelling separated by a few hundred kilometers. In the alternative model, tidal heating occurs in the deep mantle and results in larger-scale convection, perhaps with a relatively small number of mantle plumes spaced widely apart. The deep-mantle model predicts greater heat dissipation (and higher concentration of hot spots) toward the poles. The distribution of active volcanic centers on Io, especially the persistent hot spots that are emitting the most heat, is inconsistent with this prediction for tidal dissipation in a deep solid mantle. The distribution of active volcanism is generally consistent with the asthenosphere model, including the latitudinal distribution of persistent hot spots and the average spacing of active and recently active volcanic centers.

The major current exception to a uniform pattern of heat flow with longitude is Loki Patera (Veeder et al., 1994). Tidal heating in the deep mantle could lead to a few major volcanic centers similar to Tharsis on Mars (see Chapter 4), Hawaii on Earth (see Chapter 2), and Atla and Beta Regiones on Venus (see Chapter 5). Perhaps the properties of Io lavas or thermal environment preclude the building of high volcanoes, but instead result in especially large calderas such as Loki. In any case, the large concentration of heat at Loki could plausibly originate from a mantel plume.

7.6. DISCUSSION

7.6.1. Hypotheses to Explain Io's High Heat Flow

The lowest possible value for Jupiter's dissipation factor (Q_J) averaged over 4.5×10^9 years results in an upper limit to Io's average energy dissipation rate of 3.3×10^{13} W, or $0.8 \, \text{W m}^{-2}$ global average heat flow (Peale, 1986; Greenberg, 1989). Io's estimated power output in recent decades ($1-2 \times 10^{14}$ W, or 2.5 to $5.0 \, \text{W m}^{-2}$ heat flow) clearly exceeds this value. We consider five hypotheses to resolve this discrepancy.

1. Volcanism is inherently episodic over many time scales as a result of processes in the magma chambers and conduits, so perhaps we are just lucky enough to observe Io when many volcanoes are especially active. Although earlier workers thought that Io's heat flow was dominated by just a few major hot spots (Johnson et al., 1984;

McEwen et al., 1985), recent results show that the heat is emanating from more than 50 volcanic centres (Figure 7.6). It is unlikely that a high percentage of these centers have been unusually active in recent decades because of the random fluctuations of individual volcanoes. Loki is a significant thermal anomaly, but this is not sufficient to resolve the discrepancy.

2. Chaotic mantle convection could produce spikes in the heat flow (Fischer and Spohn, 1990). However, the approximately uniform distribution of volcanism and heat flow with longitude argues against this hypothesis, unless there is some type of globally synchronized mantle overturning, perhaps similar to scenarios suggested for Venus (Schubert et al., 1997) (see Chapter 5).

3. Periodic tidal heating and heat flow may be invoked via evolution from deep orbital resonance (Greenberg, 1982). However, the detailed model of Ojakangas and Stevenson (1986) results in predictions that are inconsistent with the observations (see Section 7.6.2).

4. It is conceivable that Q_J has recently become smaller than its average over the age of the solar system. One mechanism for lowering Q_J is condensation of helium in the deep metallic hydrogen zone (Stevenson, 1983); recent Galileo probe results confirm a small depletion of He in Jupiter's atmosphere compared with the protosolar value (Von Zahn et al., 1998).

5. Io's current high heat flow could be explained if the Io–Europa resonance is not primordial (Yoder, 1979). For example, using the preferred Io interior model of Fischer and Spohn (1990), 1.5×10^{14} W would be consistent with a 1 Gyr age for the Io–Europa resonance. This hypothesis could probably be ruled out if it can be shown that Europa currently has or once had a subsurface layer of liquid water (Cassen et al., 1979) (see Chapter 8).

In summary, although we consider hypotheses 2, 4, or 5 more likely than 1 or 3, this remains an unsolved mystery.

7.6.2. Interior Structure of Io

Models for the interior of Io have evolved in parallel with our understanding of the volcanism on this remarkable body. In the original tidal heating concept of Peale et al. (1979), all of the tidal energy was dissipated uniformly within a thin (8 to 18 km) rigid shell, decoupled from an interior magma ocean resulting from "runaway" tidal melting. The tidal heat was envisioned to be conducted through the silicate shell, without silicate volcanism, although the energy would be available to drive volcanic recycling of more volatile species such as sulfur and SO_2. However, the presence of mountains 10–15 km tall is inconsistent with the idea of such a thin lithosphere, and it was shown that magma transport could allow a thick lithosphere in spite of a very high heat flow (O'Reilly and Davies, 1980). Several theoretical objections to the magma ocean concept were also raised. Schubert et al. (1981) showed that solid-state convection is capable of removing tidally generated heat, so runaway melting is unlikely to have occurred. Furthermore, even if runaway melting occurred, the heat flux from the interior should keep the lithosphere as thin as necessary to remove heat transferred by efficient liquid-state convection, so the magma ocean should freeze within a few hundred million years (Schubert et al., 1986).

Most workers have favored one of three Io interior models, all of which include a thick (30–100 km) lithosphere and a large core (700–1000 km): model A has a near-solidus mantle, model B includes a relatively thin (20–100 km) partially molten asthenosphere between the lithosphere and mantle, and model C has a mantle that is partially molten throughout. All three models are capable of producing heat flows consistent with Io's current output, depending on properties of Jupiter and Io's thermal–orbital history. Schematic illustrations of each of these models, along with compositional interpretations, are shown in Figure 7.10.

Model A is an entirely solid Io except for transient pockets of melt, favored by Ojakangas and Stevenson (1986) because (1) the migration of partial melts from the mantle is rapid (McKenzie, 1984) and (2) they assume that the magma is sufficiently buoyant to reach the surface, so Io never becomes appreciably molten. The dissipation rate in solid bodies is a very strong temperature-dependent quantity, so in Model A the thermal and orbital evolution should be coupled, leading to periodic variations in heat flow and orbital eccentricity. This model allows for significantly higher tidal heating rates (up to $\sim 6\,\mathrm{W\,m^{-2}}$) for relatively short periods of time (10–30 Myr) separated by more quiescent periods of ~ 60 to 200 Myr. However, several key Io observations are inconsistent with this model. (1) The magma migration rate from the near-soildus mantle is about two orders of magnitude less than Io's resurfacing rate by lava (Fischer and Spohn, 1990; Carr et al., 1998). (2) Most of Io's heat flow will be concentrated in the polar regions via model A (Segatz et al., 1988), which differs from the observed uniform distribution of volcanic centers and equatorial concentration of plumes and persistent hot spots (Table 7.2). (3) Model A implies that Io must be highly differentiated and have a low-density crust ~ 50 km thick (Keszthelyi and McEwen, 1997b), but the presence of this crust makes the assumption questionable that dense mantle melts will easily reach the surface. If mantle melts stall at the base of the crust, then a partially molten layer will begin to form, and tidal dissipation will be concentrated in this layer and cause runaway thickening (Segatz et al., 1988).

Model B contains a partially molten asthenosphere sandwiched between the lithosphere and near-solidus mantle (Schubert et al., 1986). Fluid in the viscous layer is forced to circulate by the tidal distortion of the decoupled lithosphere, and heat is generated in the fluid by viscous dissipation. In contrast to model A, model B is stable with respect to thermal and orbital evolution, because an increase in eccentricity and dissipation rate causes the asthenosphere to thicken, which reduces the dissipation rate (Segatz et al., 1988). Model B predicts an equatorial concentration of heat flow, in agreement with the equatorial concentrations of plumes and persistent hot spots.

Is model B consistent with the high-temperature hot spots and spectral reflectances suggesting abundant orthopyroxene? The lithosphere and asthenosphere have been thought to consist of crustal materials with a lower density and solidus temperature than the residual mantle, given the inevitable segregation of partial melts from a solid mantel, and because the change in solidus temperature acts to keep the asthenosphere partially molten (Ross et al., 1990). This model suggests that the eruption of ultramafic lava should be a rare occurrence, because the tidal heating is concentrated in a partially molten layer of crustal composition, and because the crust (lithosphere plus asthenosphere) forms a density barrier to dense melts that might rise from the deep mantle. If the lithosphere and asthenosphere are relatively thick (~ 100 km each) and are well mixed, then the asthenospheric melts may be sufficiently Mg-rich to explain the observations.

In model C the entire mantle consists of a dense suspension of crystals in magma (Ross and Schubert, 1985). It is unclear how Io might initially achieve this configuration, but once formed, the viscous dissipation would prevent Io's interior from solidifying. If Io has evolved from deeper resonance with higher tidal dissipation rates, then large-scale melting may have been easily achieved. Crystal settling is unlikely because, at depth, the first mineral on the

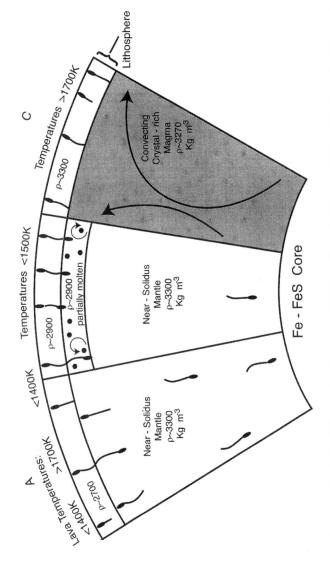

Figure 7.10. Three interior structure models for Io. All three models show a core of 980 km radius and a lithosphere 100 km thick. The main differences between the models are in the thermal state of the mantle and the composition of the crust. Model A has a near-solidus mantle, heated throughout by tidal dissipation, overlain by a low-density crust resulting from magmatic differentiation. Most of the lavas reaching the surface should be relatively low-temperature crustal melts, but an occasional high-temperature eruption may occur if the magma source is sufficiently deep. Model B contains a partially molten and convecting asthenosphere between the lithosphere and near-solidus mantle, and most of the tidal heating is concentrated in the partially molten zone. Both the crust and asthenosphere consist of lower-density rocks as a result of magmatic differentiation, but if the lithosphere and asthenosphere are sufficiently thick and well mixed the bulk composition may be more mafic than the crust in model A. Melts should originate largely in the asthenosphere and have intermediate temperatures. In model C the lithosphere overlies a convecting, crystal-rich magma "ocean," and the lithosphere has a mafic to ultramafic composition as a result of rapid resurfacing by high-temperature ultramafic lavas and foundering of lithospheric blocks.

liquidus is a magnesian pyroxene with a relatively low density of $3130\,kg\,m^{-3}$ (Keszthelyi et al., 1999). Thus, crystals separating from the mushy magma would rise and mix back into the magma ocean. The global distribution of heat flow is expected to be nearly uniform because of convection (Segatz et al., 1988). However, the greatest percentage of melt would be in an upper layer only about 200 km thick (Keszthelyi et al., 1999), similar to a thick asthenosphere model, so perhaps an equatorial concentration of heat flow can be explained. The main advantage of this model is that it could most easily explain abundant ultramafic volcanism, if that proves to be what is indeed occurring on Io.

7.6.3. Composition of Crust and Erupting Magmas

The most critical outstanding question for understanding the volcanism and interior structure of Io is: "What are the compositions of the crust and erupting magmas?" In the absence of large-scale mixing, the crust should be dominated by alkali and silica-rich compositions as a result of repeated "distillation" by the vigorous magmatic activity (Keszthelyi and McEwen, 1997b). Io's heat flow indicates that, on average, each part of Io has been subjected to at least 400 episodes of partial melting. This repeated fractionation is modeled to produce a low-density crust 30–50 km thick dominated by alkali-rich silicate minerals such as nepheline and feldspars. The mantle could also be differentiated into an upper residuum rich in Mg ("depleted mantle") and a lower "fertile mantle" rich in Fe and Ca.

However, the most recent data from the Galileo spacecraft suggest that many of the lavas on Io have a magnesium-rich, ultramafic composition (see Section 7.2.2). Two ideas were initially presented for the origin of these lavas: (1) that they consist of relatively rare mantle melts from a highly differentiated and compositionally layered satellite or (2) that the lavas result from large degrees of partial melting ($\sim 40\%$) of a "primitive" (broadly chondritic) mantle, perhaps similar to terrestrial komatiites erupted in the Archean (see Chapter 8). Komatiites are described as the most primitive lavas on the Earth. In this context, *primitive* means most geochemically similar to the parent rocks from which the melts were derived, and does not mean that they were derived from an undifferentiated Earth. In fact, by the time komatiite lavas were formed, the Earth had well-established subduction zones and the formation of continental crust was under way.

Still, the suggestion that Io could be largely undifferentiated remains intriguing. Io could retain an undifferentiated crust if the rapid resurfacing and subsidence of surface layers can lead to complete mixing of the volcanic products back into the source region (Carr et al., 1998). On the Earth, efficient mixing takes place when oceanic crust is subducted deep into the mantle. The formation of continental crust may be associated with the water escaping from the downgoing oceanic slab (see Chapter 2). If this is true, then the lack of water on Io might allow subduction to mix the erupted lavas back into the crust without producing a silica-rich crust. But there is no evidence for subduction zones (or plate tectonics in general) on Io. The centers of volcanic activity are evenly or randomly scatted, with no analogues to midocean ridge or arc volcanism. The "Ring of Fire" described by Lopes-Gautier et al. (1997) is an annulus of hot spots surrounding an SO_2-rich region, but is probably not analogous to the circum-Pacific "Ring of Fire" on the Earth (see Chapter 2). The mountains of Io appear as isolated massifs, and do not form linear chains such as those formed on the Earth by the collision of plates. Thus, if there is to be efficient recycling of Io's volcanic products, it must take place by some other mechanism.

Carr *et al.* (1998) suggest that such recycling can be achieved by 100% melting of the base of the crust as it subsides in response to rapid burial. For this to operate, the melting of the crust must be so rapid that magmas are not able to migrate away while the crust is only partially molten. Carr *et al.* (1998) calculate that the prodigious magma production rates on Io require >20% partial melting. While this is substantially greater than the ~10% partial melting common at terrestrial midocean ridges (e.g., Plank *et al.*, 1995; see Chapter 2), these calculations demonstrate the enormous difficulty of producing 100% melting of rock before the melt escapes. Komatiites are though to result from ~40% partial melting of the mantle, which appears to require very rapidly rising mantle plumes (see Chapter 8). Keszthelyi and McEwen (1997b) demonstrated that a highly differentiated Io is inevitable even if most lavas are produced by 25% partial melting, and qualitatively showed that even >50% partial melting will not allow Io to remain undifferentiated. A vigorously convecting magma ocean may be able to mechanically mix the soft, partially molten, base of the crust back into the mantle. It is difficult to imagine another process that can provide wholesale entrainment of the crust back into the mantle without either plate tectonics or 100% melting.

The differentiated Io model does not predict that ultramafic lavas should be common on Io's surface. Ascent of the very dense (~ 3100; kg m^{-3}) ultramafic lavas through a silica-rich crust (density ~ 2800 kg m^{-3}) is plausible if the magma rises from deep within the still denser solid mantle (~ 3300 kg m^{-3}). The minimum depth from which such dense ultramafic melts are predicted to be able to reach the surface is about 125 km. Such melts could occasionally erupt, but lower-temperature crustal melts should be much more common. Although we have evidence for lavas hotter than basalt from only 12 of the hot spots, with only one certain example (Pillan), we suspect that ultramafic lavas may be much more common for three reasons. First, we can only determine lower limits to the lava temperatures via remote sensing. Second, the 0.9 μm absorption band is seen in the dark materials at all volcanic centers, suggesting lavas of similar composition. Third, a predominance of low-viscosity ultramafic lava could explain the caldera morphologies on Io (see Section 7.1.3). If the majority of the erupting lavas are ultramafic, then the entire crust must be ultramafic, including the mountains (interpreted as tilted and foundering crustal blocks). The presence of a crystal-rich magma ocean below the lithosphere may be the best way to explain an ultramafic world with high mountains (Keszthelyi *et al.*, 1999), if this proves to be the case for Io.

7.7. CONCLUSION AND FUTURE EXPLORATION

Many important questions about Io remain unanswered, but there has been significant progress. We now know that the dominant volcanism on Io involves silicate magma rather than sulfur. There are more than 50 volcanic centers on Io that are at least as active as Kilauea, the most active volcano on Earth. Ultramafic volcanism probably occurs on Io, and may in fact be the most abundant lava erupted onto the surface. If Io is dominated by ultramafic volcanism, then we must reconsider the possibility of a magma ocean. There is good evidence for magma effusion rates as high as 10^5 m^3 s^{-1}, greatly exceeding the highest effusion rates that have been observed on Earth. The plumes are dramatic manifestations of the interaction between magma and volatiles (dominantly SO$_2$).

The Galileo spacecraft reached Jupiter and made a close flyby of Io in 1995, but a tape recorder anomaly precluded the acquisitions of high-resolution images or spectra of Io. With the extended mission now in progress, the plan is to return to Io in late 1999. There will be at least two close flybys of Io, one a near-equatorial pass providing a close look at volcanoes such as Pele and Pillan, and the second a pass under the south pole to determine if Io has a self-sustained magnetic field. Other key issues that will be addressed are (1) temperatures, emplacement mechanisms, and compositions of erupting lavas; (2) composition and structure of the crust; (3) improved heat flow estimates; (4) origins of the large mountains and other landforms; (5) dynamics of the plumes; and (6) polar volatile compositions.

We anticipate that a future small spacecraft mission will be dedicated to this dynamic moon. The severe radiation environment close to Jupiter is a significant impediment, but the uniqueness of this world is a strong incentive. Io is an incredible natural experiment that should teach us many lessons about volcanism on other planets.

7.8. REFERENCES

Anderson, J. D., W. L. Sjogren, and G. Schubert, Galileo gravity results and the internal structure of Io, *Science, 272*, 709–712, 1996.

Blaney, D. L., T. V. Johnson, D. L. Matson, and G. J. Veeder, Volcanic eruptions on Io: Heat flow, resurfacing, and lava composition, *Icarus, 113*, 220–225, 1995.

Carlson, R. W., W. D. Smythe, R. M. C. Lopes-Gautier, A. G. Davies, L. W. Kamp, J. A. Mosher, C. A. Soderblom, F. E. Leader, R. Mehlman, R. N. Clark, and F. P. Fanale, The distribution of sulfur dioxide and other infrared absorbers on the surface of Io, *Geophys. Res. Lett., 24*, 2479–2482, 1997.

Carr, M. H., Silicate volcanism on Io, *J. Geophys. Res., 91*, 3521–3532, 1986.

Carr, M. H., A. S. McEwen, K. A. Howard, F. C. Chuang, P. Thomas, P. Schuster, J. Oberst, G. Neukum, and G. Schubert, Mountains and calderas on Io: Possible implications for lithosphere structure and magma generation, *Icarus, 135*, 146–165, 1998.

Cassen, P., R. T. Reynolds, and S. J. Peale, Is there liquid water on Europa? *Geophys. Res. Lett., 6*, 731–734, 1979.

Davies, A. G., Io's volcanism: Thermophysical models of silicate lava compared with observations of thermal emission, *Icarus, 124*, 45–61, 1996.

Davies, A. G., L. Keszthelyi, R. Lopes-Gautier, A. McEwen, and W. Smythe, Eruption style and thermal signature of eruptions at Pele and Pillan Patera, Io, Fall AGU abstract P22B-05, 1998.

Davies, A. G., R. Lopes-Gautier, W. D. Smythe, and R. W. Carlson, Silicate cooling model fits to Galileo NIMS data of volcanism on Io, *Icarus*, 2000, in press.

Fanale, F. P., W. P. Barnerdt, L. S. Elson, T. V. Johnson, and R. W. Zurek, Io's surface: Its phase composition and influence on Io's atmosphere and Jupiter's magnetosphere, in *Satellites of Jupiter*, edited by D. Morrison, pp. 756–781, University of Arizona Press, Tucson, 1982.

Fischer, H.-J., and T. Spohn, Thermal-orbital histories of viscoelastic models of Io (J1), *Icarus, 83*, 39–65, 1990.

Geissler, P. E., A. S. McEwen, L. Keszthelyi, R. Lopes-Gautier, J. Granahan, and D. P. Simonelli, Global color variations on Io, *Icarus, 140*, 265–282, 1999.

Greeley, R., E. Theilig, and P. Christensen, The Mauna Loa sulfur flow as an analogy to secondary sulfur flows(?) on Io, *Icarus, 60*, 189–199, 1984.

Greenberg, R., Orbital evolution of the Galilean satellites, in *Satellites of Jupiter*, edited by D. Morrison, pp. 65–92, University of Arizona Press, Tucson, 1982.

Greenberg, R., Time-varying orbits and tidal heating of the Galilean satellites, in *Time-Variable Phenomena in the Jovian System*, edited by M. J. S. Belton, R. A. West, and J. Rahe, pp. 100–115, NASA SP-494, Washington, DC, 1989.

Hon, K., J. Kauahikaua, R. Denlinger, and K. Mackay, Emplacement and inflation of pahoehoe sheet flows: Observations and measurements of active lava flows on Kilauea Volcano, Hawaii, *Geol. Soc. Am. Bull., 106*, 351–370, 1994.

Howell, R. R., Thermal emission from lava flows on Io, *Icarus, 127*, 394–407, 1997.

Johnson, T. V., D. Morrison, D. L. Matson, G. J. Veeder, R. H. Brown, and R. M. Nelson, Volcanic hotspots on Io: Stability and longitudinal distribution, *Science, 226*, 134–137, 1984.

Johnson, T. V., G. J. Veeder, D. L. Matson, R. H. Brown, R. M. Nelson, and D. Morrison, Io: Evidence from silicate volcanism 1986, *Science, 242*, 1280–1283, 1988.

Johnson, T. V., D. L. Matson, D. L. Blaney, G. J. Veeder, and A. G. Davies, Stealth plumes on Io, *Geophys. Res. Lett.*, *22*, 3293–3296, 1995.

Kargel, J. S., P. Delmelle, and D. B. Nash, Volcanogenic sulfur on Earth and Io: Composition and spectroscopy, *Icarus*, *124*, 248–280, 1999.

Keszthelyi, L., and A. McEwen, Thermal models of basaltic volcanism on Io, *Geophys. Res. Lett.*, *24*, 2463–2466, 1997a.

Keszthelyi, L., and A. McEwen, Magmatic differentiation of Io, *Icarus*, *130*, 437–448, 1997b.

Keszthelyi, L., A. S. McEwen, and G. J. Taylor, Reviving the hypothesis of a mushy global magma ocean in Io, *Icarus*, *141*, 415–419, 1999.

Kieffer, S. W., Dynamics and thermodynamics of volcanic eruptions: Implications for the plumes of Io, in *Satellites of Jupiter*, edited by D. Morrison, pp. 647–723, University of Arizona Press, Tucson, 1982.

Lellouch, E., Io's atmosphere: Not yet understood, *Icarus*, *124*, 1–21, 1996.

Lopes-Gautier, R., A. G. Davies, R. Carlson, W. Smythe, L. Kamp. L, Soderblom, F. E. Leader, R. Mehlmen, and the Galileo NIMS team, Hot spots on Io: Initial results from Galileo's near-infrared mapping spectrometer, *Geophys. Res. Lett.*, *24*, 2439–2442, 1997.

Lopes-Gautier, R., A. S. McEwen, W. B. Smythe, P. E. Geissler, L. Kamp, A. G. Davies, J. R. Spencer, L. Keszthelyi, R. Carlson, F. E. Leader, R. Mehlman, L. Soderblom, and The Galileo NIMS and SSI Teams, Active volcanism on Io: Global distribution and variations in activity, *Icarus*, *140*, 243–264, 1999.

Lunine, J. I., and D. J. Stevenson, Physics and chemistry of sulfur lakes on Io, *Icarus*, *64*, 345–367, 1985.

Matson, D. L., G. A. Ransford, and T. V. Johnson, Heat flow from Io (J1), *J. Geophys. Res.*, *86*, 1664–1672, 1981.

McEwen, A. S., and L. A. Soderblom, Two classes of volcanic plumes on Io, *Icarus*, *55*, 191–217, 1983.

McEwen, A. S., D. L. Matson, T. V. Johnson, and L. A. Soderblom, Volcanic hot spots on Io: Correlation with low-albedo calderas, *J. Geophys. Res.*, *90*, 12345–12379, 1985.

McEwen, A. S., J. I. Lunine, and M. H. Carr, Dynamic geophysics of Io, in *Time-Variable Phenomena in the Jovian System*, edited by M. J. S. Belton, R. A. West, and J. Rahe, pp. 11–46, NASA SP-494, Washington, DC, 1989.

McEwen, A. S., L. Keszthelyi, J. R. Spencer, G. Schubert, D. L. Matsou, R. Lopes-Gautier, K. P. Klaasen, T. V. Johnson, J. W. Head, P. Geissler, S. Fagents, A. G. Davies, M. H. Carr, H. H. Breneman, and M. J. S. Belton, High-temperature silicate volcanism on Jupiter's moon Io, *Science*, *281*, 87–90, 1998a.

McEwen, A. S., L. Keszthelyi, P. Geissler, D. P. Simonelli, M. H. Carr, T. V. Johnson, K. P. Klaasen, H. H. Breneman, T. L. Jones, J. M. Kaufman, K. P. Magee, D. A. Senske, M. J. S. Belton, and G. Schubert, Active volcanism on Io as seen by Galileo SSI, *Icarus*, *135*, 181–219, 1998b.

McKenzie, D., The generation and compaction of partially molten rocks, *J. Petrol.*, *25*, 713–765, 1984.

Nash, D. B., and R. R. Howell, Hydrogen sulfide on Io: Evidence from telescopic and laboratory infrared spectra, *Science*, *244*, 454–457, 1989.

Nash, D. B., M. H. Carr, J. Gradie, D. M. Hunten, and C. F. Yoder, Io, in *Satellites*, edited by J. A. Burns and M. S. Matthews, pp. 629–688, University of Arizona Press, Tucson, 1986.

Ojakangas, G. W., and D. J. Stevenson, Episodic volcanism of tidally heated satellites with application to Io, *Icarus*, *66*, 341–358, 1986.

O'Reilly, T. C., and G. F. Davies, Magma transport of heat on Io: A mechanism allowing a thick lithosphere, *Geophys. Res. Lett.*, *8*, 313–316, 1980.

Peale, S. J., Orbital resonances, unusual configurations and exotic rotation states among planetary satellites, in *Satellites*, edited by J. A. Burns and M. S. Matthews, pp. 159–223, University of Arizona Press, Tucson, 1986.

Peale, S. J., P. Cassen, and R. T. Reynolds, Melting of Io by tidal dissipation, *Science*, *203*, 892–894, 1979.

Pearl, J. C., and W. M. Sinton, Hot spots of Io, in *Satellites of Jupiter*, edited by D. Morrison, pp. 724–755, University of Arizona Press, Tucson, 1982.

Pieri, D., S. M. Baloga, R. M. Nelson, and C. Sagan, Sulfur flows of Ra Patera, Io, *Icarus*, *60*, 685–700, 1984.

Plank, T., M. Spiegelman, C. H. Langmuir, and D. W. Forsyth, The meaning of "mean F": Clarifying the mean extent of melting at ocean ridges, *J. Geophys. Res.*, *100*, 15045–15052, 1995.

Ross, M. N., and G. Schubert, Tidally forced viscous heating in a partially molten Io, *Icarus*, *64*, 391–400, 1985.

Ross, M. N., G. Schubert, T. Spohn, and R. W. Gaskell, Internal structure of Io and the global distribution of its topography, *Icarus*, *85*, 309–325, 1990.

Schaber, G. G., The geology of Io, in *Satellites of Jupiter*, edited by D. Morrison, pp. 556–597, University of Arizona Press, Tucson, 1982.

Schenk, P. M., and M. H. Bulmer, Origin of mountains on Io by thrust faulting and large-scale mass movements, *Science*, *279*, 1514–1517, 1998.

Schenk, P., A. McEwen, T. Davenport, A. Davies, K. Jones and B. Fessler, Geology and topography of Ra Patera, Io, in the Voyager era: Prelude to eruption, *Geophys. Res. Lett., 24*, 2467–2470, 1997.

Schubert, G., D. J. Stevenson, and K. Ellsworth, Internal structures of the Galilean satellites, *Icarus, 47*, 46–59, 1981.

Schubert, G., T. Spohn, and R. T. Reynolds, Thermal histories, compositions, and internal structures of the moons of the solar system, in *Satellites*, edited by J. A. Burns and M. S. Mathews, pp. 224–292, University of Arizona Press, Tucson, 1986.

Schubert, G., V. S. Solomatov, P. J. Tackley, and D. L. Turcotte, Mantle convection and the thermal evolution of Venus, in *Venus*, edited by S. W. Bougher, D. M. Hunten, and R. J. Phillips, pp. 1245–1287, University of Arizona Press, Tucson, 1997.

Segatz, M., T. Spohn, M. N. Ross, and G. Schubert, Tidal dissipation, surface heat flow, and figure of viscoelastic models of Io, *Icarus, 75*, 187–206, 1988.

Skinner, B. J., A sulfur lava flow in Mauna Loa, *Pac. Mag., 24*, 144–145, 1970.

Smith, B. A., and the Voyager Imaging Team, The Jupiter system through the eyes of Voyager 1, *Science, 204*, 951–972, 1979a.

Smith, B. A., E. M. Shoemaker, S. W. Kieffer, and A. F. Cook, The role of SO_2 in volcanism on Io, *Nature, 280*, 738–743, 1979b.

Smythe, W. D., R. Lopes-Gautier, A. Davies, R. Carlson, L. Kamp, L. Soderblom, and the Galileo NIMS team, Galilean satellite observation plans for the near-infrared mapping spectrometer experiment on the Galileo spacecraft, *J. Geophys. Res., 100*, 18957–18972, 1995.

Spencer, J. R., and N. M. Schneider, Io on the eve of the Galileo mission, *Annu. Rev. Earth Planet Sci., 24*, 125–190, 1996.

Spencer, J. R., A. S. McEwen, M. A. McGrath, P. Sartoretti, D. B. Nash, K. S. Noll, and D. Gilmore, Volcanic resurfacing of Io: Post-repair HST imaging, *Icarus, 127*, 221–237, 1997a.

Spencer, J., J. Stansberry, C. Dumas, D. Vakil, R. Pregler, and M. Hicks, A history of high-temperature Io volcanism: February 1995 to May 1997, *Geophys. Res. Lett., 24*, 2451–2454, 1997b.

Spencer, J. R., P. Sartoretti, G. E. Ballester, A. S. McEwen, J. T. Clarke, and M. McGrath, The Pele plume (Io): Observations with the Hubble Space Telescope, *Geophys. Res. Lett., 24*, 2471–2474, 1997c.

Stansberry, J. A., J. R. Spencer, R. R. Howell, C. Dumas, and D. Vakil, Violent silicate volcanism on Io in 1996, *Geophys. Res. Lett., 24*, 2455–2458, 1997.

Stevenson, D. J., Anomalous bulk viscosity of two-phase fluids and implications for planetary interiors, *J. Geophys. Res., 88*, 2445–2455, 1983.

Stevenson, D. J., and S. C. McNamara, Background heatflow on hotspot planets: Io and Venus, *Geophys. Res. Lett., 15*, 1455–1458, 1988.

Strom, R. G., N. M. Schneider, R. J. Terrile, A. F. Cook, and C. Hansen, Volcanic eruptions on Io, *J. Geophys. Res., 86*, 8593–8620, 1981.

Thomas, P. C., M. E. Davies, T. R. Colvin, J. Oberst, P. Schuster, G. Neukum, M. H. Carr, A. McEwen, G. Schubert, M. J. S. Belton, and the Galileo Imaging Team, The shape of Io from Galileo limb measurements, *Icarus, 135*, 175–180, 1998.

Thordarson, T., and S. Self, The Laki (Skaftar Fires) and Grimsvotn eruptions in 1783–1785, *Bull. Volcanol., 55*, 233–263, 1993.

Veeder, G. J., D. L. Matson, T. V. Johnson, D. L. Blaney, and J. D. Goguen, Io's heat flow from infrared radiometry: 1983–1993, *J. Geophys. Res., 99*, 17095–17162, 1994.

Von Zahn, U., D. M. Hunten, and G. Lehmacher, Helium in Jupiter's atmosphere: Results from the Galileo probe Helium Interferometer Experiment, *J. Geophys. Res., 103*, 22815–22829, 1998.

Webb, E. K., and D. J. Stevenson, Subsidence of topography on Io, *Icarus, 70*, 348–353, 1987.

Wilson, L., and J. W. Head, A comparison of volcanic eruption processes on Earth, Moon, Mars, Io, and Venus, *Nature, 302*, 663–669, 1983.

Witteborn, F. C., J. C. Bregman, and J. B. Pollack, Io: An intense brightening near 5 micrometers, *Science, 203*, 643–646, 1979.

Wood, C. A., Calderas: A planetary perspective, *J. Geophys. Res., 89*, 8391–8406, 1984.

Woodward, R. C., F. L. Roesler, R. J. Oliversen, F. Scherb, and H. W. Moos, Imaging spectroscopy of Io in 1997 using STIS G140L: O I, S I and now H I, *Bull. Am. Astron. Soc., 30*, 1116, 1998.

Yoder, C. F., How tidal heating in Io drives the Galilean orbital resonance locks, *Nature, 279*, 767–770, 1979.

Zahnle, K., L. Dones, and H. F. Levison, Cratering rates on the Galilean satellites, *Icarus, 136*, 202–222, 1999.

8

Exotic Lava Flows

Harry Pinkerton, Sarah A. Fagents, Louise Prockter, Paul Schenk, and David A. Williams

8.1. INTRODUCTION

On Earth, most lavas have silica contents in the range from 30 to 78 wt%, and they are erupted at temperatures between 800 and 1170°C (see Chapter 2). However, some lavas have compositions and eruption temperatures outside this range. These are the exotic lavas that we will describe in this chapter. One type of exotic high-temperature, low-viscosity lavas called *komatiites* were important at an early stage in the development of the Earth's crust. Another group of exotic lavas are forming only on a volcano in northern Tanzania—Oldoinyo Lengai—at the present time. These fascinating lavas, termed *carbonatites*, are renowned for their very low eruption temperatures and viscosities. The first people to see these strange lavas understandably mistook them for mudflows, and the vents were considered to be boiling mud pools.

In addition to their importance in studies of terrestrial volcanic processes, carbonatites and komatiites may also be important elsewhere. For example, the discovery of over 200 channel and valley landform complexes in the 1992 Magellan data led to a rapid reappraisal of the way that the surface of Venus had formed (see Chapter 5). Channels identified included sinuous rilles similar to those described on the Moon (see Chapter 6) and long, wide sinuous canali up to 6800 km in length. Komatsu *et al.* (1992) used a radiative cooling model to show that basaltic magma would solidify before reaching the end of the longest of these Venusian canali. This modeling, together with the recognition of streamlined features similar to flood channels, suggested that a very low viscosity material was involved in the formation of these features. Candidates included water, ultramafic silicate melts, sulfur, and carbonate lavas (Baker *et al.*, 1992).

We also include in this chapter a type of volcanism not recognized until recently. The discovery of what look like volcanic features on the icy satellites of planets in the outer part of the solar system has resulted in studies of cryovolcanism, a new type of planetary volcanism. The strange features formed by this and other types of exotic volcanism will challenge the imagination of many planetary geologists in the new millennium.

8.2. COOL LAVAS: CARBONATITES

As we mentioned above, carbonatite lavas have been suggested as possible fluids that created channels on the surface of Venus and other features formed by the flow of low-viscosity fluids. While Venusian lavas are mostly tholeiitic (Weitz and Basilevsky, 1993), some are highly alkaline, particularly the sample from Venera 13, which resembles olivine nephilinite (Baker *et al.*, 1992). The close association between nephilinites and many terrestrial carbonatites, together with the presence of up to 19% calcite and anhydrite on the surface of Venus (Kargel *et al.*, 1994) have therefore been used to argue that carbonatites were the exotic lavas that formed some of the landforms that characterize the surface of Venus.

8.2.1. What Are Carbonatites?

Carbonatites are igneous rocks with more than 50% carbonate minerals (Streckeisen, 1980). These unusual rocks have a wide range of compositions, and understanding how they form has been a major challenge facing petrologists since they were first recognized almost 80 years ago in the Fen Ring Complex in southeastern Norway (Middlemost, 1985). Since their first discovery, intrusive carbonatites have been recognized in a number of other alkaline ring complexes. Half of the 350 known occurrences of carbonatites (Bailey, 1993) in different parts of the world are in Africa (mostly in eastern Uganda, northern Tanzania, and western Kenya), and many are closely linked with rift zones. The majority of carbonatites are intrusive, though recent work in many parts of the world has led to the recognition of many explosive and effusive carbonatites.

Many carbonatites are the last melts to be erupted from alkaline volcanic complexes (e.g., Oldoinyo Lengai and Shombole in northern Tanzania); others are associated with ultramafic rocks such as peridotite; others with carbonate-rich kimberlites; and it is now recognized that others have no apparent association with any silicate magmas. Petrologists recognize four different types of carbonatites: calciocarbonatites, magnesiocarbonatites, ferrocarbonatites, and natrocarbonatites. The dominant types are calciocarbonatites, followed in volumetric importance by magnesiocarbonatites and ferrocarbonatites. The dominant minerals present in these carbonatites are calcite, dolomite, and ankarite. Finally, natrocarbonatites are the main lavas that have been erupted on Oldoinyo Lengai in historic times. Natrocarbonatites are essentially composed of the two carbonate minerals nyererite and gregoryite, and they contain in excess of 30 wt% Na_2O.

Silicate–carbonate immiscibility has been demonstrated in nephilinitic lavas on Shombole Volcano in northern Tanzania (Kjarsgaard and Peterson, 1991). Evidence of coexisting carbonatite and silicate magmas has also been described for alkali carbonatite lavas and ashes erupted on Oldoinyo Lengai, Tanzania, in June 1993 (Dawson *et al.*, 1994). These lavas and ashes contain immiscible silicate spheroids, which in turn contain carbonatite segregations. Bell and Simonetti (1996) and Simonetti *et al.* (1997) used isotopic evidence to demonstrate that more recent viscous lavas from Oldoinyo Lengai formed either by liquid immiscibility or by the mixing of melts with similar isotopic compositions. Isotopic evidence has also been used to show that natrocarbonatite lavas have a mantle signature (Bell and Keller, 1995). Experimental evidence (e.g., Lee and Wylie, 1998) has confirmed that melts with compositions similar to natrocarbonatites and calciocarbonatite magmas can form by liquid immiscibility from silicate magmas. This work also concluded that magnesiocarbonatite magmas form either during fractionation of calciocarbonatites or as primary magmas.

Research into the genesis of carbonatites is important because of their association with economically important minerals, and because it will lead to an improved understanding of important geologic processes such as metasomatism, immiscibility, the role of carbon dioxide during volcanic eruptions, and the relationship between magma composition and tectonic environment. Since many carbonatites are mantle-derived, they also provide an insight into mantle compositions and processes during ascent from the upper mantle.

8.2.2. Carbonatite Locations

The best-known example, indeed the only example of an active carbonatite volcano, is Oldoinyo Lengai in northern Tanzania (Figure 8.1), a volcano whose importance was first recognized in the early 1960s by J. B. Dawson. The bulk of this 3-km-high stratovolcano is composed of extrusive and explosive nephelinite/phonolite volcanics, and it is currently erupting carbonatite lavas in the summit crater. However, Oldoinyo Lengai is chemically unique. It is the only locality where natrocarbonatites are found. Carbonatites from other parts of the Earth are composed mainly of Ca-Mg-Fe carbonates with negligible alkali contents. Oldoinyo Lengai is also unusual because lavas are the dominant eruptive products. Most of

Figure 8.1. The summit region on Oldoinyo Lengai, October 27, 1993. The dark material in the foreground is covered by blocky viscous flows and ash. A viscous lava flow can be seen emerging from a breached cone in the right (east) foreground. The lighter-colored material in the far side of the crater consists of fluid carbonatite lavas. The arachnoidlike feature dominates the left (south) foreground. Radial and circumferential fractures and circumferential fold structures on the "arachnoid" are clearly visible.

the other carbonatite localities on Earth are composed of intrusives or pyroclastics, though carbonatite lavas have been identified in Uganda and the Kola Peninsula (Middlemost, 1985).

Carbonatites have been found in association with carbonated kimberlites in the Igwisi Hills on the Tanzania Craton (Dawson, 1994). They have been found as centimeter-sized spheroidal carbonate lapilli in unconsolidated ash in a diatreme in the Eifel volcanic province of Germany where they are associated with melilite-bearing nephelinitic magma (Riley et al., 1996). Carbonatite extrusives, tuffs, and diatremes are present in alkaline complexes in the Precambrian Gardar rift, South Greenland (Anderson, 1997; Pearce et al., 1997). Carbonatite diatremes have also been found in association with a dome and radial dikes in southern Namibia (Kurszlaukis and Lorenz, 1997). A maar and tuff ring in Umbria, central Italy, are considered to be the surface expression of carbonate diatremes (Stoppa, 1996). Whereas Kurszlaukis and Lorenz (1997) found evidence supporting phreatomagmatic activity in diatremes in Namibia, the maar in Umbria described by Stoppa (1996) contains no evidence of phreatomagmatic activity, and the explosivity is considered to result from the violent exsolution of CO_2.

From this discussion, it is clear that carbonatite eruptions on Earth are not always effusive. Explosive events appear to be a characteristic of many carbonatite eruptions.

8.2.3. Physical Properties of Carbonatites and Other Volcanic Melts

When carbonatites were first seen on Oldoinyo Lengai, Tanzania, almost 100 years ago, they were described as soda mudflows (Dawson et al., 1995a). Indeed, when viewed from the crater rim of Oldoinyo Lengai, this is a mistake that could be made by even the most experienced of geologists. The flow behavior of natrocarbonatite lava is very similar to that of mud, and the low eruptive temperatures (500–700°C) mean that the flows are barely incandescent at night. Lavas on Oldoinyo Lengai were first recognized as carbonatites by Dawson (1962) who compared their flow behavior with very fluid basalts. He described both pahoehoe and aa (see Chapter 2) carbonatite lava, and he recognized that the jet black lava began to hydrate and change color shortly after eruption.

Wolff (1994) estimated the viscosities and densities of natrocarbonatite and calcium-rich carbonatite magmas from molten salt data. Natrocarbonatites have an estimated viscosity and density of 0.01 Pa s and 2.0–2.1×10^3 kg m^{-3} at 700–800°C, while calcium-rich carbonatites have viscosities and densities of 0.1 Pa s and 2.3–2.5×10^3 kg m^{-3}, respectively. Carbonatites will therefore be negatively buoyant in some silicate magmas. Dobson et al. (1996) carried out a systematic series of measurements of viscosity and density of carbonate melts at mantle pressures by measuring the terminal fall velocities of spheres in the melts. The measured viscosities of 1.5×10^{-2} to 5×10^{-3} Pa s are the lowest of any known magma. Measured melt densities are in agreement with the predictions of Wolff (1994).

Field measurements on Oldoinyo Lengai during June 1988 by Krafft and Keller (1989) and in November of the same year by Dawson et al. (1990) revealed that eruptive temperatures of the carbonatites in the summit crater of Oldoinyo Lengai ranged from 500–590°C, and that the lavas were very fluid. Eruptive temperatures and viscosities measured in the field were considerably lower than for any known terrestrial lava. A systematic series of measurements of the rheologic properties of these lavas has revealed how these properties are influenced by changes in composition, temperature, crystallinity, and vesicularity (Norton and Pinkerton, 1997). Aphyric natrocarbonatites are Newtonian with eruptive viscosities over an order of magnitude lower than the most fluid basaltic melt. For an eruption temperature of 590°C, apparent viscosities of the 1988 Oldoinyo Lengai natrocarbonatites measured in the laboratory using a rotational viscometer ranged from 0.15 Pa s for degassed, phenocryst-poor lava to 85 Pa s for a degassed lava with a high phenocryst content. These values are similar to the predicted viscosities of komatiite lavas (see Section 8.3.3). In addition to their low effusion temperatures and viscosities, the 1988 natrocarbonatite lavas had significantly lower thermal diffusivities and specific heat capacities than basaltic lavas (Pinkerton et al., 1995), and measured values are comparable with estimates by Treiman and Schedl (1983).

Volatiles are important in the generation, crystallization, and flow behavior of carbonatites. The importance of CO_2 in alkaline magmatism was confirmed experimentally (Bailey and Hampton, 1990) when volcanic glasses were fused and the released volatiles were analyzed. Dawson et al. (1995b) confirmed that volatiles, especially CO_2, are concentrated in the parental magmas on Oldoinyo Lengai. The role of fluorine is also important. An aphyric sample that had been filter-pressed from the front of a lava flow on Oldoinyo Lengai in 1988, and was enriched in halogens, had an apparent viscosity of 0.018 Pa s at 590°C (Norton and Pinkerton, 1997).

8.2.4. Morphology of an Active Carbonatite Volcano

Historic eruptions on Oldoinyo Lengai, the only active carbonatite volcano on Earth, are described in detail by Dawson et al. (1995a) and by Nyamweru (1997). Highly fluid carbonatite lava was erupted between 1958 and 1966, and this style of activity terminated in 1966 to be followed by explosive eruptions in 1966 and 1967 (Dawson et al., 1995a). Ash erupted during the most explosive events attained heights of 10.7 km (Dawson et al., 1995a), and ash falls during the initial part of this eruption were reported in Nairobi, 190 km northeast of the volcano (Dawson et al., 1968, 1995a). Activity resumed in 1983 and during the next decade it was similar to that of the 1960s. Low-volume lava flows were erupted from small, open vents for periods of a few hours to several years. Mild strombolian activity took place in many of these vents (Figure 8.1), and breadcrust bombs up to 800 mm long were erupted (Pinkerton et al., 1995).

Detailed examination of the small lava flows erupted on Oldoinyo Lengai in 1988 confirmed Dawson's (1962) observations that typical natrocarbonatite lavas on Oldoinyo Lengai are smaller, though morphologically similar to basaltic lava flows (Pinkerton et al., 1995). Maximum flow lengths of less than 200 m were observed for lavas during the 1988 activity on Oldoinyo Lengai, and thicknesses of many flows were in the range of a few millimeters to a few centimeters. Carbonatite lavas are similar in size and in flow behavior to the natural and industrial flows of molten sulfur described by Greeley et al. (1990) (see Chapter 7).

All of the processes and features that characterize basaltic lavas (see Chapter 2) were observed during the flow of the miniature flows on Oldoinyo Lengai. The formation of pahoeoe, a′a, and blocky lava (Figures 8.1–8.3); compound flow fields (e.g., Pinkerton and Sparks, 1976); inflated flows (e.g., Hon et al., 1994); channels (Figure 8.3); ephemeral vents and tubes (e.g., Peterson and Swanson, 1974; Peterson et al., 1994); Calvari and Pinkerton, 1998); and many other features and processes were observed at close quarters in these flows. In addition, some of the channel depths increased by combined mechanical and thermal erosion on a time scale of a few minutes (Dawson et al., 1990; Pinkerton et al., 1995). Figure 8.3 clearly shows that the channel below the lens cap has been incised by the more recent channel running from the top to the bottom of the photograph. Some of the resulting channels were remarkably similar to sinuous rilles (see Figure 3 in Pinkerton et al., 1995). Calculations

Figure 8.2. Aa carbonatite lavas formed on Oldoinyo Lengai in June 1993. See hammer for scale.

by Pinkerton *et al.* (1990) have shown that the measured and calculated erosion rates are in agreement, and these reveal the importance of thermal erosion for carbonatitic and basaltic lavas under laminar flow conditions.

Effusive activity on Oldoinyo Lengai came to an abrupt end in June 1993. Between June 14 and 25, small explosive eruptions occurred, and ash was deposited on the upper flanks of the volcano. Following partial collapse of major ash cones, three large lava flows were extruded in the southern (previously inactive) part of the crater (Figure 8.1). This period of unusual activity was followed by the renewed eruption of small, fluid carbonatite lava flows. Since June 1993, apart from two periods in 1995 when viscous lava flows were erupted, the style of activity has reverted to the style that characterized activity in the 1980s (Nyamweru, 1997).

The largest of the June 1993 flows (Figure 8.1) was 150 m long and 6 m thick with crevasse structures and slickensides. The flow dimensions and structures found on this flow are similar to those of small highly evolved silicate lava flows (see Pinkerton and Wilson, 1994). However, thin section examination of the rocks collected from this thick flow has shown that its unusual dimensions and morphology were not the result of a change from carbonatite to silicate chemistry; instead, the flow was a natrocarbonatite with up to 60% crystals and occasional silicate globules. These lavas, which play a key role in understanding the evolution of Oldoinyo Lengai, are described by Dawson *et al.* (1994–1996).

A structure that formed during June 1993 on Oldoinyo Lengai had an unusual fracture pattern that bears a striking resemblance to arachnoids on Venus (see Chapter 5), although it is much smaller [140-m diameter compared with diameters of ~90 km for Venusian arachnoids (Pinkerton *et al.*, 1994)]. Radial and circumferential fractures were well developed; the structure had a depression similar to the moats seen on Venus; and there were

Figure 8.3. Two proximal channels that fed pahoehoe carbonatite lava flows on Oldoinyo Lengai in June 1993 (camera lens cap for scale). The earlier channel (beneath the lens cap) is clearly cut by the later channel running from the top to the bottom of the photograph. This is clear evidence of thermal erosion, a process that was measured during an earlier phase of activity by Dawson *et al.* (1990) and reported in more detail by Pinkerton *et al.* (1990, 1995). The first channel has subsequently been occupied by a small pahoehoe flow, and the flow front of this flow is clearly visible in the upper half of the first channel.

circumferential compression ridges on the outer rim of the structure. Although this structure was observed in a carbonatite volcano, Pinkerton *et al.* (1994) emphasized that they are not inferring that this has any bearing on the compositions of rocks underlying the much larger arachnoids on Venus, though it may be instructive to reconsider arachnoid formation in the light of these observations on Oldoinyo Lengai.

While it is commonly accepted (e.g., Kargel *et al.*, 1994) that carbonatite volcanism is characterized by the eruption of low-viscosity lavas, we have demonstrated that carbonatite volcanism on Earth may form either low- or high-viscosity lavas, or highly explosive deposits, and sometimes these can both occur on the same volcano. For example, every one to three-decades there are explosive eruptions on Oldoinyo Lengai (Dawson *et al.*, 1995a). Indeed, there is a much closer association between carbonatites and explosive eruptions than between carbonatites and extensive eruptions of fluid lava.

8.2.5. Possible Planetary Occurrences

Komatsu *et al.* (1992) and Kargel *et al.* (1994) argue that many features on the volcanic plains of Venus resemble fluvial landforms (see Chapter 5). Some canali resemble meandering river channels and flood plains on Earth, outflow channels are reminiscent of the Channeled Scablands in Washington, lava delta are similar to the Mississippi delta, and Venusian valley systems show signs of sapping. This evidence, together with the very smooth surface of many parts of the plains, suggest that a low-viscosity fluid was responsible for the formation of some features on Venus. Kargel *et al.* (1994) argue that the fluid involved had a viscosity comparable to water, although ambient surface temperatures on Venus reveal that it could not have been water itself. Their suggestion that carbonatites are likely candidates is supported by the presence of 4–19% calcite and anhydrite on the surface of Venus, and by the need to raise the surface temperatures on Venus (660–760 K) by only a few tens to a few hundred degrees Kelvin to allow melting of the surface. Melts could be produced relatively easily, either by internal heating in the lower crust or mantle, or by heating during impact. Once erupted, these melts would remain mobile for considerably longer than a silicate melt. They proposed that a carbonatite "aquifer" may have existed at a depth of a few hundred meters to several kilometers beneath the surface of Venus.

In conclusion, there is evidence that carbonatites may have been erupted on the surface of Venus. Further research into the flow behavior and morphology of volcanics formed from these unusual rocks will undoubtedly help us in our quest to understand the evolution of Venus and possibly other planetary bodies.

As was mentioned in the introduction to this chapter, carbonatites are not the only low-viscosity lavas to have been erupted on Earth. Other lavas with low viscosities on eruption are komatiites. However, unlike carbonatites, these are erupted at temperatures well in excess of those encountered during current terrestrial volcanic activity. As we will see in the following section, these higher eruption temperatures are responsible for thermal erosion on a massive scale compared with similar features on carbonatites.

8.3. PRIMITIVE LAVA: KOMATIITES

Komatiites are another example of unusual and interesting types of volcanic rocks found on Earth. They have remarkably high magnesium contents (≥ 18 wt% MgO), they often possess more exotic textures (e.g., spinifex, cumulate) than other volcanic rocks, and they are almost completely restricted to Precambrian terrains (Arndt and Nisbet, 1982). Komatiites are the subject of ongoing geologic interest, primarily because they provide an insight into four key areas: (1) the early volcanic history of the Earth (e.g., Hill *et al.*, 1995), (2) the geochemical and thermal structure of the Archean mantle (e.g., Nisbet *et al.*, 1993), (3) the formation of magmatic Fe–Ni–Cu–(PGE) sulfide ore deposits (e.g., Lesher, 1989), and (4) the nature of some extraterrestrial lava flows (e.g., Warren, 1983; Longhi and Pan, 1989). The purpose of this section is to review the fundamentals of komatiite volcanology, including a discussion of where and when komatiites were emplaced, how komatiitic magmas were formed and erupted on the Earth's surface, and what role they may have had in the emplacement of extraterrestrial lava flows.

8.3.1. Komatiite Locations and Ages

Komatiites were first identified in the Barberton greenstone belt of South Africa in the late 1960s, and they were interpreted as the metamorphosed remnants of submarine, ultramafic lava flows (Viljoen and Viljoen, 1969a,b). At about this time, nickel sulfide ore deposits associated with ultramafic rocks were discovered in the Precambrian shield of

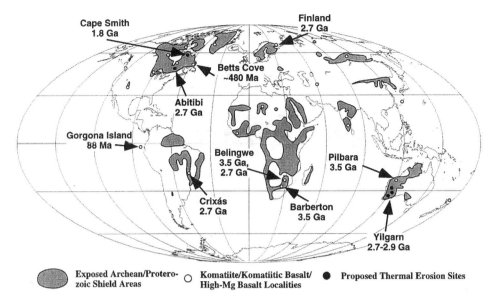

Figure 8.4. Distribution of komatiites and komatiitic basalts, with proposed sites of thermal erosion. Adapted from Arndt and Nisbet (1982).

Western Australia. These ultramafic rocks were later identified as komatiite lava flows, and continued mineral exploration in the 1970s and 1980s has identified komatiites (both with and without associated sulfides) in the Precambrian terrains of most continents (Figure 8.4). Komatiite lavas were erupted at multiple localities in two major pulses in the Archean (one ~3.0–3.5 Ga, the other ~2.7–2.9 Ga), and at one locality in one major pulse in the Proterozoic (~1.8 Ga). The reasons why komatiitic volcanism is basically restricted to short pulses at several points in the Precambrian is unknown, and the subject of much debate. Komatiites tend to be a volumetrically small component (<10%) of most Archean greenstone belts (e.g., de Wit and Ashwal, 1997), and are often associated with more voluminous mafic and felsic volcanic materials. One leading theory is that komatiites were associated with the hot tails of Archean mantle plumes (e.g., Campbell *et al.*, 1989), in which the larger, cooler plume heads produced the more voluminous basaltic flows found in association with komatiites in many greenstone belts. The only accepted occurrence of a komatiitic eruption since the Precambrian occurred at Gorgona Island, off the coast of Colombia, in which both explosive and effusive komatiites were emplaced during the Cretaceous period (~88–90 Ma), which may have been associated with a mantle plume source for the Caribbean–Colombian Cretaceous Large Igneous Province (Kerr *et al.*, 1997).

8.3.2. Melt Generation

Komatiitic liquids are ultramafic with MgO contents >18 wt% (Arndt and Nisbet, 1982). There has been some controversy regarding the maximum MgO contents of komatiitic liquids (Nisbet *et al.*, 1993; Parmen *et al.*, 1997), because of uncertainties in: (1) the degree of olivine accumulation during emplacement, (2) the degree of Mg metasomatism during seafloor alteration, and (3) the presence of magmatic water in komatiitic liquids. However, the high Mg contents of rare igneous olivines at several localities ($\leq Fo_{94}$–Fo_{95}) indicate that many lavas had maximum MgO contents of 28–30 wt% (Nisbet et al., 1993). Because of the nearly complete restriction of komatiites to the Precambrian, these high-MgO liquids are thought to have been produced when the mantle was several hundred degrees hotter than at present. They are considered to be the product of relatively high degrees (<30–80%) of partial melting of Archean mantle peridotite (e.g., Cawthorn and Strong, 1974; Arndt, 1977; Takahashi and Scarfe, 1985; Herzberg, 1992).

Petrologic studies have identified two primary type of komatiites: (i) Munro type (also called Group I or Al-undepleted), characterized by $Al_2O_3/TiO_2 = 20$, $CaO/Al_2O_3 = 1.0$, and chondritic ratios of heavy REEs; and (ii) Barberton type (also called Group II or Al-depleted), characterized by $Al_2O_3/TiO_2 = 10$, $CaO/Al_2O_3 > 1.0$, and depleted heavy REEs (e.g., Arndt et al., 1997). Munro-type komatiites are most abundant in late Archean (~2.7 Ga) rocks, whereas Barberton-type komatiites are most abundant in early Archean (>3.0 Ga) rocks, although both types are found in rocks of both age groups. The implication is that there were at least two distinct source compositions for these komatiites, perhaps indicative of a heterogeneous Archean mantle on a local and possibly a global scale.

Recent studies have used fluid dynamics modeling and equation of state calculations to better understand the role of komatiites in Archean mantle plumes that may have propagated upward from the lower mantle or similar depths. Thus, considerable insight on mantle dynamics has been gained from komatiites, and there is considerable work in progress in this area (e.g., Campbell and Griffiths, 1993; Xie et al., 1993; Arndt et al., 1997).

8.3.3. Lava Emplacement

A typical komatiite lava flow has a complex stratigraphy of spinifex and cumulate textural zones (Figure 8.5), which are thought to result from variations in cooling rate and olivine fractionation/accumulation during emplacement. Although the range of individual flows may not exhibit all of these textural zones, they are all characterized by olivine ± chromite ± pyroxene ± glass mineral assemblages. In addition to their higher MgO contents, komatiites are noted for their lower alkali and silica contents, and higher normative olivine contents than tholeiitic basalts. Because the chemical composition of komatiitic lavas (Table 8.1) implies higher liquidus temperatures (1360–1600°C) and lower dynamic viscosities (0.1–10 Pa s) relative to modern basaltic lavas (cf. 1050–1200°C, 50–100 Pa · s). Nisbet (1982) suggested that these lavas should have been emplaced as hot, turbulent flows. As a result, komatiite lavas are thought to have been capable of forming highly mobile sheet and channelized flows (Lesher, 1989; Hill et al., 1995) that flowed turbulently during emplacement (Huppert and Sparks, 1985), and that convected vigorously during crystallization (Turner et al., 1986).

A range of features suggest a volcanic origin for komatiites, including polyhedral jointing, chilled flow tops, pillows, and fragmental tuffs and breccias (Arndt and Nisbet, 1982). A wide variety of lava facies have been inferred, including massive, pillowed, and spinifex-textured lava lobes, sheet flows, channelized sheet flows, lava channels, and lava ponds (Arndt et al., 1979; Lesher et al., 1984; Hill et al., 1995), as well as associated dikes and sills. Volcaniclastic komatiites are rare in most areas but common in others (e.g., Saverikko, 1990; Schafer and Morton, 1991). Regional stratigraphic studies in greenstone belts (Lesher et al., 1984; Hill et al., 1995; Jackson et al., 1994; Perring et al., 1995) suggest that some komatiite sequences may be correlated over very great distances (tens to hundreds of kilometers). Consequently, Hill et al. (1995) suggested that some Western Australia komatiites may have formed very large flow fields perhaps akin to the lunar maria (see Chapter 6) or continental flood basalts (see Chapter 2). Several submarine emplacement

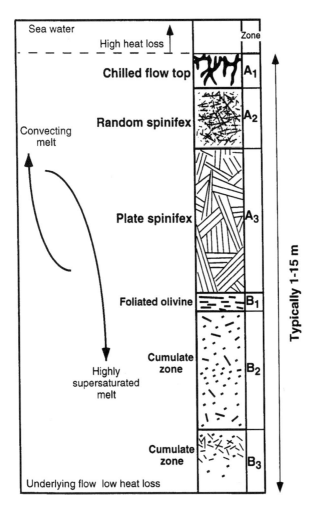

Figure 8.5. Schematic cross section through a typical komatiite flow. After Hill *et al.* (1990) based on formation models of Pyke *et al.* (1973), Arndt (1986), and Turner *et al.* (1986).

environments have been suggested for Precambrian komatiite flows, including shallow marine environments (e.g., Saverikko, 1990), and deep marine (≥ 1 km depth) environments such as intracratonic rift zones (e.g., Weaver and Tarney, 1979; Perring *et al.*, 1995), or the Archean equivalent of modern ocean plateau LIPs like Ontong Java (Arndt *et al.*, 1997) (see Chapter 2). Recently, Hill and Perring (1996) suggested that inflationary processes such as those observed in Hawaii (e.g., Hon *et al.*, 1994), and suggested for the Columbia River flood basalts (Self *et al.*, 1997), might have played a role in the emplacement of some of the large, komatiite flow fields in Western Australia, although no field evidence has been found to support this hypothesis.

Because there have been no historic komatiite flows, the nature of komatiite lava flow emplacement has been studied using mathematical modeling and analogue experiments. The physics of low-viscosity, high-temperature, turbulent komatiite lava flow emplacement was first modeled in the landmark work of Huppert *et al.* (1984) and Huppert and Sparks (1985), who predicted that turbulent emplacement should have caused high heat loss to the under-

Table 8.1. Compositions and Thermal/Rheologic Properties of Komatiitic Lavas[a]

Component	Komatiite	Komatiite	Komatiite	Komatiitic basalt
SiO_2	47.24	45.40	46.8	46.90
TiO_2	0.22	0.35	0.43	0.61
Al_2O_3	4.10	7.00	7.66	9.81
Fe_2O_3	—	1.20	—	—
FeO	10.79	10.80	10.50	14.40
MnO	—	0.17	0.17	0.30
MgO	30.30	27.90	25.20	18.90
CaO	5.84	6.50	7.50	8.60
Na_2O	0.56	0.29	0.01	0.30
K_2O	0.21	0.10	0.01	0.05
P_2O_5	—	—	0.04	0.15
T_{Liq} (°C)	1612	1578	1528	1419
T_{Sol} (°C)	1170	1170	1170	1150
ρ_l @T_{Liq} (kg m^{-3})	2756	2776	2758	2799
ρ_l @T_{Sol} (kg m^{-3})	2836	2857	2828	2860
c_l (J kg^{-1} K^{-1})	1764	1745	1720	1637
μ_l @T_{Liq} (Pa s)	0.12	0.14	0.30	0.81
μ_l @T_{Sol} (Pa s)	1.99	1.97	3.78	6.93
L_l @T_{Liq} (J kg^{-1})	6.84E+05	6.69E+05	6.46E+05	5.96E+05
L_l @T_{Sol} (J kg^{-1})	4.83E+05	4.83E+05	4.83E+05	4.74E+05
Composition location	?	Alexo, Ontario	Mickel, Ontario	Katinniq, New Québec
Composition reference[b]	1	2	3	4

[a] For all compositions, liquidus temperature was calculated using MELTS (Ghiorso and Sack, 1995), the solidus temperature for komatiite is from Arndt (1976) and is estimated for komatiitic basalt, liquid density was calculated using the method of Bottinga and Weill (1970), liquid viscosity was calculated using the method of Shaw (1972), specific heat was calculated from the heat capacity data of Lange and Navrotsky (1992), and heat of fusion for komatiite liquids is approximated using the expression for forsterite of Navrotsky (1995) and for komatiitic basalt liquids is approximated using the expression for diopside of Stebbins et al. (1983).
[b] References: 1, Huppert and Sparks (1985); 2, Arndt (1986); 3, Davies (1997); 4, Barnes et al. (1982).

lying substrate by forced convection of the lava, resulting in the formation of deep thermal erosion channels (Figure 8.6). Thermal erosion is here defined as the breakup and removal of substrate by hot flowing lava, and may include both *ablation* (i.e., melting) of substrate resulting from heating by the lava, and *physical degradation* (i.e., mechanical erosion) of unconsolidated or partly consolidated material and melt resulting from shearing or plucking by the moving lava, followed by partial or complete assimilation of melted substrate with the liquid lava. Thermal erosion has been inferred to play a major role in the formation of several terrestrial lava tubes (e.g., Cruikshank and Wood, 1972; Greeley and Hyde, 1972; Peterson et al., 1994; Kauahikaua et al., 1998; Greeley et al., 1998a), and in the formation of some lunar sinuous rilles (Hulme, 1973), some Martian lava channels (Carr, 1974; Cutts et al., 1978; Baird, 1984), and some Venusian canali (Head et al., 1991; Baker et al., 1992). Thermal erosion has been measured directly during the emplacement of carbonatite lavas at Oldoinyo Lengai, Tanzania (Dawson et al., 1990; Pinkerton et al., 1990) and in possible planetary lava analogues like industrial sulfur flows (Greeley et al., 1990). Other terrestrial evidence of thermal erosion includes field evidence of downcutting by tube-fed basaltic lavas in Hawaii and geophysical measurements of active tube-fed lavas that suggest erosion rates of basalt substrate by tube-fed basalt lava at rates of $\sim$10 cm day^{-1} (Kauahikaua et al., 1998). This suggests that, for sustained tube-fed eruptions of several weeks, lava could erode up to 3 m of substrate.

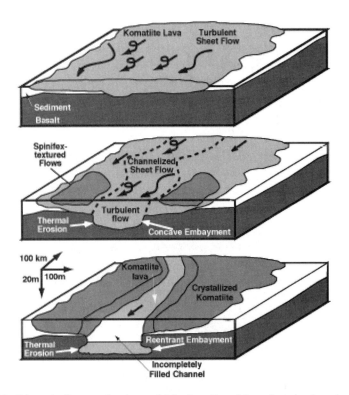

Figure 8.6. Schematic diagrams showing model for formation of thermal erosion lava channels by hot, turbulent, submarine komatiite lava flows. Adapted from Hill *et al.* (1990).

Thermal erosion by komatiites was suggested to explain a series of unusual sulfide ore-bearing embayments underlying komatiite flows at Kambalda, Western Australia (Huppert *et al.*, 1984; Lesher *et al.*, 1984; Hupper and Sparks, 1985). Since these initial studies, further research on komatiites led to the discovery of additional localities (Figure 8.4) with features that are indicative of thermal erosion channels underlying komatiitic lavas (e.g., Barnes *et al.*, 1988; Lesher, 1989), and further modeling of the erosional potential of komatiitic lavas is being carried out (Jarvis, 1995; Williams *et al.*, 1998). Of perhaps greater importance is the apparent relationship between thermally eroding komatiite lavas and the production of komatiite-hosted sulfide ore deposits. Komatiite-hosted Fe–Ni–Cu-(PGE) sulfide ore deposits are common in Archean greenstone belts (Lesher, 1989), and make up $\sim$25% of the world's total nickel (>0.8% wt% Ni) resources. These deposits are considered by most workers to have formed by thermal erosion of sulfur-rich substrate rocks and sediments by metal-rich komatiite lavas, followed by segregation of immiscible sulfides and miscible silicate components, scavenging of metals from komatiites, and accumulation of dense metal-sulfide melts at the base of the komatiite lavas, often in what are considered to be thermal erosion channels (Lesher, 1989). Understanding the process of thermal erosion by komatiites is required to understand the genesis of komatiite-hosted magmatic sulfides.

8.3.4. Possible Planetary Occurrences

Moon. Analysis of lunar basalt samples returned by the Apollo program suggests that lunar mare lavas and terrestrial komatiite lavas have similar chemical, petrologic, and thermal/rheologic properties. Several categories of lunar mare basalts have mineralogies similar to pyroxenitic komatiites (Warren et al., 1990). Studies of the viscosities of lunar lavas (Murase and McBirney, 1970) show that they have similar viscosities to komatiitic basalts, and Hulme (1982) suggested that lunar basalts might have been capable of turbulent flow and thermal erosion. Thermal erosion has been proposed as an explanation for some of the lunar sinuous rilles (e.g., Hulme, 1973, 1982). Although no pure lunar komatiite has been found, samples collected to date represent only 4–5% of the lunar surface area (Warren and Kallemeyn, 1990) (see Chapter 6).

Mars. There have been several lines of evidence suggesting that ultramafic komatiite lava flows may have been emplaced on Mars, including: (1) Viking Orbiter images of long lava flows, perhaps indicative of effusion of low-viscosity komatiitic lavas; (2) Viking XRF analyses of Martian soils that may indicate the presence of mafic to ultramafic protoliths (e.g., Baird and Clark, 1981); (3) experimental petrology on a synthetic Martian mantle composition that suggests picritic to komatiitic compositions as abundant products of volcanism throughout Martian history (Bertka and Holloway, 1994); (4) textural and chemical similarities between some SNC meteorites and komatiites, which would support parental magma compositions (e.g., Longhi and Pan, 1988; Treiman, 1988); and (5) Phobos 2 ISM measurements of Syrtis Major, perhaps indicative of a mineralogy consistent with komatiitic basalt (Reyes and Christensen, 1994). Furthermore, some analyses of Martian fines coupled with geochemical modeling suggest that sulfide ore deposits could be widespread on Mars (Burns and Fisher, 1990a,b). Thus, because the presence of komatiitic rocks is suggested from a wide variety of sources, a search for these rocks on Mars may be justified (see Chapters 3 and 4).

Venus. The Magellan mission (1989–1994) imaged almost 99% of the surface of Venus, and a large variety of volcanic features were observed, including long lava flows, long sinuous rilles, and canali, all indicative of low-viscosity volcanism. The largest canali-type lava channel on Venus is >6800 km long, and some have suggested that it may have been produced by thermal erosion (Head et al., 1991; Baker et al., 1992; Komatsu et al., 1993; Komatsu and Baker, 1994), possibly by komatiitic lavas (see Chapter 5).

Io. The Galileo spacecraft currently orbiting Jupiter has obtained images of Io, one of the four large Galilean satellites of Jupiter, and provided temperature measurements of active volcanic eruptions on its surface. These data, obtained by the Solid-State Imaging System (SSI), and Near-Infrared Mapping Spectrometer (NIMS), suggest that at least some of the eruptions on Io have temperatures ~1500–2000 K (~1230–1730°C) (McEwen et al., 1998). These temperatures are consistent with the eruption of ultramafic melts. On Io, tidal heating from Jupiter's gravitational influence could be a potential source for generating hot ultramafic melts, and some images show large eruptions of dark material consistent with silicate volcanism. Thus, studies of terrestrial komatiites may help in understanding volcanic activity on Io (Matson et al., 1998) (see Chapter 7).

The lavas described thus far, though relatively rare on Earth, have played an important role in the Earth's evolution. In the remainder of this chapter we consider a type of volcanism that, though unknown on Earth, has been very important in the evolution of other bodies in the solar system.

8.4. ICE AS LAVA: CRYOVOLCANISM

Although water is perhaps the most familiar liquid on Earth, in the outer solar system water may behave in a quite unfamiliar manner, constituting another type of exotic volcanism.

Most of the satellites of the outer planets are inferred, on the basis of their bulk densities and spectral properties, to consist of a large proportion of water ice (Clark et al., 1986; Schubert et al., 1986). The surfaces of these icy satellites contain a wide variety of features, some of which are morphologically similar to volcanic features on the terrestrial planets, such as lava domes, flows, calderas, cones, and sinuous rilles. Plausible compositions and internal heat sources on the icy satellites suggest that resurfacing can occur through eruptions of water, possible in combination with various other nonwater components in a solid, slushy, or liquid state, or as gas-driven particle sprays. Such activity is commonly called *cryovolcanism*. This term may, at first sight, appear to refer to two processes at opposite ends of the temperature spectrum. However, it is compatible with the American Geological Institute definition of volcanism, which refers to extrusion of molten rock without reference to composition. Also, as we will see below, there is evidence that, on the icy satellites, the extreme conditions environment (low temperatures, low gravity, zero external pressure) in the outer solar system mean that ice/water mixtures may behave in similar ways to silicate magmas.

Whereas volcanic activity on icy satellites has been directly observed only in Voyager images of plumes rising from the surface of Neptune's satellite Triton (Kirk et al., 1995), landforms and surfaces that are considered to be volcanic have been identified on satellites of Jupiter (Europa and Ganymede), Saturn (Enceladus, Dione, Tethys, and Iapetus), and Uranus (Miranda and Ariel) (see Table 8.2). In this section, we review geologic evidence for volcanism on the icy satellites, we discuss the properties of these strange magmas, and we propose possible eruption mechanisms that could produce the observed features.

8.4.1. Magma Generation and Eruption Mechanisms

The heat generated by accretion of a satellite is important in early differentiation of bodies having radii >1000 km, but probably plays little role thereafter. Heating arising from the decay of radioactive isotopes and dissipation of tidal energy will have different relative importance depending on the satellite's silicate mass fraction, degree of differentiation, and proximity to the parent planet (Schubert et al., 1986). These heat generation mechanisms are balanced by

Table 8.2. Comparison of Earth and Satellite Physical and Environmental Characteristics[a]

Body[b]	Radius (km)	Mass (10^{20} kg)	Density (g cm^{-3})	Gravity (m s^{-2})	Surface temperature (K)[c]	Note
Earth	6380	5974	5.52	9.81	288	$\sim 10^5$ Pa atmosphere
Ganymede[J]	2630	1480	1.94	1.42	110	Negligible atmosphere
Europa[J]	1560	480	3.04	1.31	93	Negligible atmosphere
Iapetus[S]	720	18.8	1.21	0.24	89	Negligible atmosphere
Dione[S]	560	10.5	1.44	0.22	72	Negligible atmosphere
Tethys[S]	520	7.6	1.26	0.19	~ 80	Negligible atmosphere
Enceladus[S]	250	0.8	1.24	0.09	~ 80	Negligible atmosphere
Ariel[U]	580	14.4	1.65	0.29	66	Negligible atmosphere
Miranda[U]	240	0.7	1.26	0.08	66	Negligible atmosphere
Triton[N]	1350	214	2.06	0.78	38	~ 1.6 Pa N_2 atmosphere

[a] Sources: Gaffney, E. S. and D. L. Matson (1980), Water ice polymorphs and their significance on planetary surfaces. *Icarus* 44, 511–519, Morrison, D. (ed.) 1982. Satellites of Jupiter. University of Arizona Press, Tucson, 972 pp, Stevenson (1982), Burns, J. A. and M. S. Matthews (eds.) 1986. Satellites. University of Arizona Press, Tucson, 1021 pp, Cruikshank, D. P. (ed.) 1995. Neptune and Triton. University of Arizona Press, Tucson, 1249 pp.
[b] Superscripts indicate planet: (J)upiter, (S)aturn, (U)ranus, (N)eptune.
[c] Satellite surface temperature values are global averages.

losses from conduction, subsolidus convection, and magma migration. Convection may be sufficient to preclude large-scale melting of the satellite, unless tidal heating is significant, as may be the case for Europa and Enceladus (Schubert et al., 1986). The presence of compounds that suppress melting temperatures (e.g., ammonia or salts) may lead more readily to the formation of liquid bodies (Kargel et al., 1991). Isolated liquid reservoirs could also form by the gradual freezing of more extensive layers, or from interaction with silicate magmas generated in the rocky interior of differentiated satellites (Wilson et al., 1997).

Pure water contained within or beneath a pure ice crust is negatively buoyant as a result of its greater density. These conditions imply that the buoyant-rise mechanism that commonly operates in silicate magmas on Earth may not be possible on icy satellites. However, a number of mechanisms can provide the necessary driving force to deliver melt to the surface:

"Contamination" of Water or Ice. The presence of possible nonwater compounds in the ice or liquid water phase may change their density and produce a positive buoyancy force. The density of the ice crust can be increased with the addition of metal salts [e.g., $MgSO_4$ and $NaSO_4$ (Kargel, 1991; Hogenboom et al, 1997)] or dense particulate matter derived from the silicate interior or implanted exogenically. Alternatively, gases such as CO_2, CO, CH_4 (methane), and SO_2 combine with ice at high pressures (i.e., at depth within the crust) to form clathrates, which have a greater density than pure ice (Lunine and Stevenson, 1985). Decomposition of clathrates may also provide a mechanism for explosive eruptions (Stevenson, 1982). Low-density (and hence buoyant) liquids may consist of ammonia–water mixtures, with the possible addition of CH_4 or CH_3O (methanol); nitrogen–methane mixtures might be a possibility for Triton (Kargel, 1991).

Volatiles Dissolved within Water. Similar to the way that gases such as H_2O and CO_2 exsolve from silicate magmas to drive eruptions on terrestrial planets, volatiles in water magmas may provide a crucial mechanism for cryovolcanic eruptions. Likely candidates, based on cosmic abundances and spectral evidence, include CO_2, CO, SO_2, N_2, CH_4, and NH_3 (ammonia). If the pressure exerted on a water body by ice overburden is released during the opening of a fissure in the ice, bubbles of gas will nucleate and grow rapidly by decompression and coalescence (Sparks, 1978), leading to disruption of the water phase into a collection of droplets entrained in a gas stream. With the exception of Triton, the icy satellites do not possess significant atmospheres. There will therefore be rapid expansion of released gas and consequent acceleration of particles. The lack of aerodynamic resistance to motion in the vacuum environment, in combination with the low acceleration resulting from gravity, will permit the droplets, ice crystals, and condensing vapor to follow long, largely ballistic paths, producing an umbrellalike plume, perhaps similar in morphology to the gas-rich plumes observed on Io. Deposition of the erupted products would produce a mantle of material surrounding the vent.

Pressurization of Melt. In the absence of buoyancy or volatiles to provide the necessary driving force, pressurization of isolated liquid bodies may drive watery melts to the surface. This could be achieved through tidal stresses in the crust of the satellite, or through volume changes associated with phase changes in a magma reservoir. For example, the volumetric increase on forming ice from pure water can produce significant excess pressures in the remaining liquid phase. During fracture propagation the pressure could drive water to the surface to produce low fountains and water/ice flows (Wilson et al., 1997).

8.4.2. Surface Flows

The surface of a body of water exposed to a vacuum environment will experience vigorous "boiling" as a result of the vapor pressure of H_2O exceeding the ambient (effectively zero) pressure. This will produce a fine spray of droplets and/or ice crystals from the water surface. The consequence extraction of latent heat from the water will promote ice formation, initially as crystals which will mix throughout the water as a result of motions induced by the vigorous vaporization, but eventually sufficient ice will form to produce a competent crust. When this ice is thick enough for the hydrostatic pressure at the base of the crust to exceed the H_2O vapor pressure, further vaporization is suppressed, and the water may continue to flow under the icy crust (Allison and Clifford, 1987). The front of this mobile icy lava flow is likely to be characterized by a chaotic jumble of ice blocks, vaporizing H_2O and water "breakouts." Once formed, the ice crust should significantly control flow propagation and resulting flow morphology.

In addition to influencing buoyancy and melting temperatures, the presence of nonwater components in a water magma may have significant implications for rheology. Brines, ammonia, methane, and methanol solutions have liquid viscosities ranging up to six orders of magnitude greater than liquid water (Kargel et al., 1991). As an alternative to fluid volcanism, effusions of warm ice or ammonia–water in the solid state, perhaps softened by the inclusion of interstitial volatiles such as methane, have been proposed to explain the significant relief of some features (Stevenson and Lunine, 1986; Jankowski and Squyres, 1988). However, Schenk (1991) proposed that the changing rheologic properties during cooling and partial crystallization of a fluid ammonia–water lava may be sufficient to produce the observed features.

8.4.3. Europa

Analysis of gravity field data indicates that Europa, Jupiter's smallest Galilean satellite, has a predominantly H_2O ice shell ~100 km thick, surrounding a rocky mantle and perhaps a metallic core (Anderson et al., 1998). Tidal dissipation might produce sufficient heat to ensure that there is liquid water beneath Europa's icy surface, perhaps as a global ocean (Ojakangas and Stevenson, 1989). The existence, past or present, of liquid water is of key significance in providing a potential niche for the evolution of life, and also has important implications for resurfacing processes. Resolving the timing, extent, and volume of liquid water is a primary goal for the Galileo Europa Mission and future Europa missions.

Images returned by the two Voyager spacecraft in 1979 revealed two main units on the surface of Europa: bright, uniform plains and dark, mottled terrain, with a superposed global pattern of lineaments, broad bands, and regions of dark wedge-shaped features (Lucchitta and Soderblom, 1982; Malin and Pieri, 1986). The relative scarcity of impact craters indicates a youthful surface [at least compared with Ganymede and Callisto (Smith et al., 1979)] and a possible role for cryovolcanism as a resurfacing mechanism. Volcanic activity in the form of muddy slurries or dark ice erupted onto the surface or confined as intrusions in the ice crust was proposed to explain mottled terrain and dark spots (Lucchitta and Soderblom, 1982). The formation of triple bands, prominent lineaments consisting of a bright central stripe flanked by broader dark margins, was attributed to block faulting and flooding by water or slush (Buratti and Golombek, 1988), intrusion and subsequent dehydration of hydrated minerals (Finnerty et al., 1981), or extrusion of material onto the surface. Crawford and Stevenson (1988) proposed that resurfacing of Europa could occur as a consequence of gas-rich eruptions driven by exsolution of volatiles in cracks propagating upward from the interface between a liquid ocean and the overlying ice, but did not relate their model to specific features in the Voyager data. However, Thrace Macula, a dark, ~160-km-long lobate feature, was proposed as a volcanic flow on the basis of its morphologic characteristics, stratigraphic relationships, and theoretical modeling of eruption processes (Wilson et al., 1997).

The primary phase of the Galileo mission revealed a staggering diversity and density of surface features on Europa, and has done much to refine our interpretations of geologic features and models of volcanic processes (Greeley et al., 1998b). At image resolutions superior to those of Voyager, the bright plains appear far from smooth and uniform. The systems of lineaments, ridges, and fractures are visible down to the limit of resolution of Galileo data ($\sim 10\,\text{m}\,\text{pixel}^{-1}$). Clearly, extensive tectonism had modified the surface. However, other features suggest dynamic internal processes. Mottled terrains consist of lenticulae (domes, pits, depressions, and dark spots), disrupted surfaces, and chaotic jumbles of translated and tilted blocks, possibly an indication of a liquid or ductile interior overlain by a thin ice crust (Carr et al., 1998). Figure 8.7 illustrates some models for Europa's internal structure, together with possible eruption mechanisms. The following are features with possible volcanic or magmatic origins, along with tentative interpretations of their mechanisms of formation.

Triple Bands. Data from Galileo's first orbit revealed that the dark margins of the triple bands have diffuse edges (Figure 8.8a). This is inconsistent with models of formation that would produce sharp margins (Finnerty et al., 1981; Buratti and Golombek, 1988), and mantling by an explosively vented "cryoclastic" deposit containing a dark nonice component has been suggested (Greeley et al., 1998b; Figure 8.7b,e). Rhadamanthys Linea (Figure 8.8b) consists of a series of dark patches aligned along a central lineament, and has been proposed as an immature triple band (Belton et al., 1996). The strong resemblance of Rhadamanthys to halos of scoria centered along an eruptive fissure hints at an eruptive origin for triple bands. Alternatively, conductive heating by a subsurface heat source (e.g., a diapir or intrusion) might promote changes in the characteristics of the surface, such as annealing of ice grains or sublimation leading to formation of a dark lag deposit of nonice material (Fagents et al., 1999).

Ridges. Ridges dominate Europa's surface. At Galileo resolutions their morphologies range from single ridges, to closely spaced parallel double ridges (Figure 8.8c), to complex multiple ridges. They have heights on the order of 100 m, and extend for hundreds to thousands of kilometers (Greeley et al., 1998b). The central bright "stripes" of triple bands are typically ridge pairs or multiple ridges. Their origin is controversial; current models include tidal opening/closing of fractures and extrusion of ice or slush (Greenberg et al., 1998; Figure 8.7c), upwarping of the surface by ascent of linear ice diapirs (Head et al., 1999), or explosive eruption and local deposition of water droplets and/or fragmental ice from an open fissure (Kadel et al., 1998). All of the models make some assumptions about the interior of Europa that have yet to be validated; further data are required to resolve the problem.

Smooth, Dark Surfaces. A number of localized deposits occupy topographically low areas, and embay the surrounding terrain. Perhaps the best example is shown in Figure 8.9a; a 3-km-diameter "puddle" clearly younger than the ridged terrain. The lack of relief and apparent shallow depth of these features suggest that the surface was locally flooded by a small volume of low-viscosity liquid (Carr et al., 1998; Figure 8.7d).

Low-Albedo Spots. Dark halos associated with circular to elliptical structures (e.g., Figure 8.9b), domes, and surface cracks may have an origin similar to those proposed for triple band margins, i.e., mantling by explosive deposits or thermal modification of surface materials. In some cases the preexisting terrain appears subdued or embayed, lending support for the existence of a volcanic mantling deposit or fluid (Fagents et al., in press).

Lenticulae. Lenticulae are circular to elliptical in planform but display a variety of textures, albedos, and relief, and have been attributed to the surface manifestation of diapirism

Exotic Lava Flows

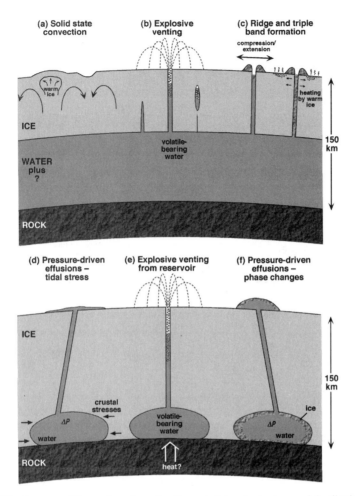

Figure 8.7. Some possible configurations for Europa's internal structure and implications for magmatic and volcanic activity. (a) Diapirisim associated with compositional instabilities or thermally induced solid-state convection might produce domes and depressions. (b, e) Explosive exsolution of volatiles caused by depressurization of the water layer by the opening of a fracture or at the tip of an upward-propagating crack (Crawford and Stevenson, 1988) might produce ballistic plumes. (c) Water exposed and frozen during tidal opening and closing of fractures might be extruded onto the surface to form ridges, or promote sublimation of surface ice to form a lag deposit of nonice material. Pressurization of a water reservoir by crustal stresses (d) or volume changes (f) associated with ice crystallization, could promote surface effusions. Diagram not to scale.

associated with solid-state convection in an ice shell overlying a liquid ocean (Pappalardo *et al.*, 1998a). Some of the many domical features appear to be simple upwarping of the crust: Preexisting features can be traced continuously across the dome. Others, however, appear to have broken through the surface and spread laterally, and are associated with fractures that might provide a pathway to the surface for an ascending magma (Figure 8.9c,d). Hence, these features might represent icy analogues of silicic lava domes (see Chapter 2). Dome morphology and dimensions (~3–15 km in diameter and several tens of meter high) are consistent with a material of significant viscosity and/or strength, suggesting that liquid water does not dominate their emplacement. Instead, possible models for their emplacement include

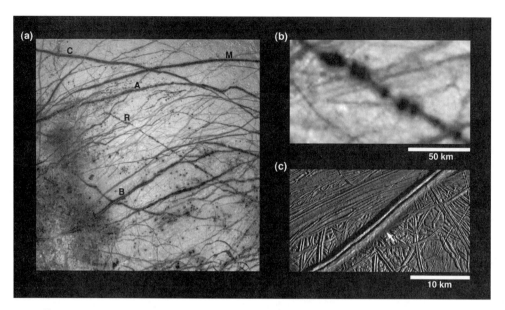

Figure 8.8. (a) Major lineaments seen in an area of Europa's surface of ~1900 by 1500 km: Cadmus (C), Minos (M), Asterius (A), Belus (B), and Rhadamanthys (R) Lineae. Image resolution is 1.6 km pixel^{-1}. (b) Detail of a section of Rhadamanthys Linea showing several of the dark patches. (c) High-resolution (22 m pixel^{-1}) image showing a doublet ridge. The adjacent smooth dark surface (arrow) has been interpreted as flooding of locally depressed crust resulting from the ridge load, mantling by volcaniclastic material, or mass wasting of ridge materials. North is to top in all images.

injection of water under an ice shell, analogous to endogenous silicate dome growth on Earth (Iverson, 1990; Fink, 1993; Figure 8.7f), or viscous effusion of contaminant- or ice-rich slush.

Flowlike Features. Early Galileo data revealed lobate features that apparently emanated from lineaments, and are suggestive of effusive flows (Figure 8.9e). While these appear to be some of the best examples of flow morphology yet seen on the icy satellites, some of the geologic relations are unclear (some ridges appear to be breached or overflowed by the "flows"), and high-resolution data (e.g., Figure 8.9f) have yet to resolve whether fluid effusion from fissures is responsible for the observed morphology, or whether some other mechanism might operate, such as solid-state upwelling associated with irregularly shaped diapiric bodies. If the former is the case, then, in common with the domes described above, the dimensions and morphology imply that the fluid possesses significant viscosity and/or yield strength.

A range of volcanic styles and magma properties may therefore be represented on Europa. More definitive determination of the origins of the features in question is precluded by lack of constraints on the presence and composition of volatile species and other nonwater components, and the existence of liquid water. While a volcanic origin is still equivocal, some tentative inferences may be drawn for Europa's interior. For example, if buoyancy mechanisms cannot be invoked to explain liquid effusions, then the requirement for pressure-driven eruption is met by having discrete water reservoirs, perhaps the last vestiges of a freezing ocean, or produced in response to heat from the silicate mantle (Figure 8.7). It would be difficult or impossible to achieve local pressurization of a global water layer. Explosive volcanism simply requires that a volatile-bearing liquid be depressurized by the opening of a fracture, which could conceivably occur in the presence of an ocean or a pressurized liquid

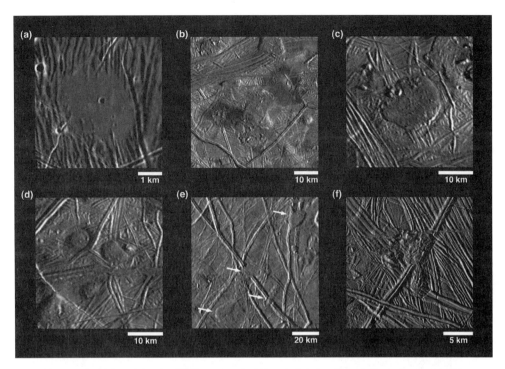

Figure 8.9. (a) Low-lying "puddle" seen here in 25 km pixel^{-1} data from Galileo's fourth orbit. (b) Patches of low-albedo material associated with central structures lying along fractures. (c) Pancakelike dome with lobate margins associated with ridge system. Surface texture is distinct from surrounding terrain. (d) Cluster of three domes seen at moderate resolution. Images in b to d have a resolution of ~ 200 m pixel^{-1}. (e) Lobate features apparently emanating from and disrupting the doublet ridges. Resolution is ~ 550 m pixel^{-1}. (f) High-resolution (~ 30) m pixel^{-1}) image showing a ridge disrupted by a positive relief feature. Note that preexisting lineaments can be traced up onto this feature, casting doubt on its origin as a volcanic flow. North is to the top in all images.

reservoir. Although the question of Europa's ocean remains unresolved on the basis of the potentially cryovolcanic features discussed here, they likely indicate the presence of some amount of liquid water in Europa's recent past, if not currently.

8.4.4. Ganymede

Ganymede is divided almost equally into bright terrain, composed of relatively clean water ice, and more ancient dark terrain, containing a greater percentage of low-albedo contaminant (Clark, 1980; Spencer, 1987; Figure 8.10). Cryovolcanism has been suggested as important in the histories of both of these terrain types. Other potential volcanic structures on Ganymede's surface include features related to impact cratering.

Bright Terrain. Common within bright terrain are sets of subparallel ridges and troughs (giving rise to the term *grooved terrain*), organized into structural cells. Morphologic evidence from Voyager and Galileo imaging suggests an extensional tectonic origin for these features, as evidenced by the sharp margins between bright terrain and the surrounding units (e.g., Squyres and Croft, 1986). The global distribution of bright terrain, along with the scale of the major groove lanes, suggest a global driving mechanism. One possibility is deep

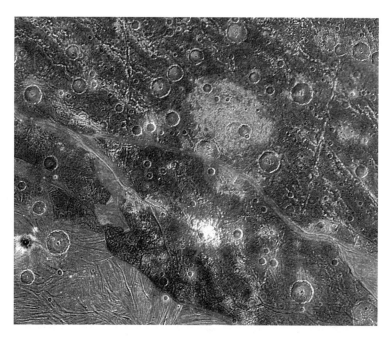

Figure 8.10. Voyager image (~800 km in width) showing the boundary between dark terrain of southwestern Galileo Regio, and the bright terrain of Uruk Sulcus. Note the sharpness of the contact between the two terrain types. The bright ~180-km subcircular patch in the center of the image is a palimpsest. Several of the larger craters have broad, higher-albedo deposits within their floors (e.g., 40-km crater at the top of the image). In the southwest of the image, the bright terrain is crisscrossed by ridges and troughs organized into discrete polygons. Running W–SE across the image are two grooves associated with bright terrain; the westernmost of these has a diffuse margin on either side of the groove, which may be a cryovolcanic deposit. The bright, linear features trending NW–SE across the dark terrain are furrows. North is toward the top; illumination is from the west.

mantel convection, with restricted convection in a shallow layer underlying the lithosphere, producing the observed patterns of surface structural cells (Shoemaker et al., 1982). Some workers have suggested that bright terrain formation accompanied global expansion of the satellite associated with homogeneous differentiation, which would result in a surface area increase of up to 7% (e.g., McKinnon and Parmentier, 1986; Mueller and McKinnon, 1988).

Several models for the origin of bright terrain have been proposed. One model suggests that silicate-poor liquid water, warm ice, or an ice/water mixture is extruded onto the surface through narrow fissures or tension fractures (Parmentier et al., 1982). This model may explain some of the smoother bright terrain regions, but does not explain the sharp boundaries seen at the margins of some grooved swaths, or the faulting that forms ridges and troughs. The model best supported by the geologic evidence is one in which the grooved terrain is created through the formation of normal fault-bounded rift zones, into which silicate-poor liquid water or warm ice has been extruded (e.g., Allison and Clifford, 1987). Evidence supporting this theory comes from observations of dark halo craters on Ganymede's bright terrain (Schenk and McKinnon, 1985), interpreted as having excavated through the bright material into underlying dark terrain, producing low-albedo ejecta deposits. The majority of the bright terrain imaged at high resolution ($<200 \text{ m pixel}^{-1}$) by Galileo shows little evidence of volcanic source vents, flow fronts, embayment relationships, or associated structures, and the groove lanes appear to be largely tectonic in nature (Head et al., 1997). However, some lanes of very smooth plains are observed, and these are the best candidates for cryovolcanism on Ganymede. Some of these smooth swaths are associated with caldera-like features. These depressions (Figure 8.11) have a single inward-facing scarp that forms a broad unclosed arc and is either truncated or breached by a lane of brighter material (Lucchitta, 1980; Schenk and Moore, 1995; Head *et al.*, 1998a).

Dark Terrain. Voyager imaging of Ganymede has led several workers to interpret dark terrain as an older, heavily cratered surface buried by multiple blankets of cryovolcanic material (e.g., Croft and Goudreau, 1987; Murchie *et al.*, 1989; Figure 8.10). This interpretation was supported by depleted densities of small craters, apparent embayment of large craters, and complex relationships between the ages of Ganymede's dark materials and probably impact-related furrows. Some workers suggested that some materials within dark terrain may have extruded from furrows (Murchie *et al.*, 1990; Lucchitta *et al.*, 1992). Increased resolution from the Galileo spacecraft shows that some contacts between different units are lobate in form, and some craters imaged at high resolution show evidence of modification by flowlike features (Prockter *et al.*, 1998). However, there is a lack of recognizable volcanic edifices or source vents, and little or no ponding is seen, as might be expected from the extrusion of a low-viscosity cryovolcanic material. Furthermore, no obvious cryovolcanic relationships are observed between the furrows and the dark smooth material. Although some plains units may be volcanic in origin, Prockter *et al.* (1998) instead conclude that Ganymede's dark material is composed of a low-albedo lag deposit, concentrated onto the surface by processes such as sublimation, mass wasting, and impact cratering,

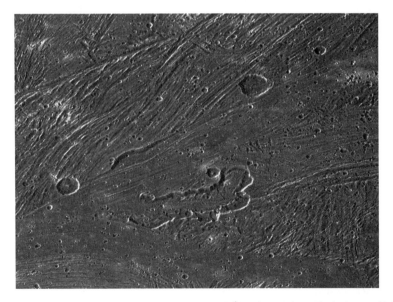

Figure 8.11. A caldera-like feature imaged at 172 m pixel^{-1} in Sippur Sulcus. The background bright terrain is organized into separate swaths or packets, each having a smooth texture, or grooves oriented in similar direction. The caldera-like feature has scalloped walls and a lobate flowlike deposit occupying the floor of the depression. The central deposit appears to flow eastward out into the adjacent smoother terrain, where it is crosscut by an ENE-trending groove lane. The image is approximately 200 m across; south is toward the top; illumination is from the left. Photo courtesy of NASA/JPL Galileo project.

while smoothing of topography and removal of small craters has resulted from emplacement of impact ejecta, rather than cryovolcanic flows.

Impact-Related Features. Large craters (diameter >50 km) on the dark terrain known as *palimpsests* have persistent high-albedo deposits (Figure 8.10), thought by some workers to result from post-impact flooding by clean water or ice (e.g., Thomas and Squyres, 1990). However, high-resolution imaging by Galileo has revealed that the high-albedo deposits probably correspond to the extent of the continuous ejecta blanket from the palimpsest-forming impact (Jones et al., 1997), rather than being of cryovolcanic origin. Some Ganymede craters are seen to have distinct high-albedo steep-sided topographic domes within their central pits, suggested to result from the ascent and extrusion of warm ice diapirs (Moore and Malin, 1988). However, other workers (Schenk, 1993) suggest that these domes form during the impact process and are the result of uplift of a compositionally distinct layer at depth. This conclusion is supported by Galileo observations (Head et al., 1998b). Schenk and Moore (1995) suggested that broad, flat-topped domes that occupy many of the craters within the older dark terrain (Figure 8.10), may imply high-viscosity magmas, although these features could also result from post-impact isostatic relaxation.

Ganymede's surface exhibits many enigmatic features (bright grooved terrain, dark lobate smooth areas, crater domes, and palimpsests) that have been proposed as cryovolcanic features, but which may be better explained by other processes. The most compelling evidence currently identified for cryovolcanism on Ganymede is the presence of calderalike features associated with smooth bright plains.

8.4.5. Volcanism on Small Icy Satellites of the Outer Planets

As Voyager surveyed the outer planets beyond Jupiter, a continuing surprise was the extent of endogenic volcanism and resurfacing evidence on the small and middle-sized icy satellites of these planets (e.g., Smith et al., 1982, 1986). With the exception of smog-shrouded Titan, the satellites of Saturn, Uranus, and Neptune range from ~400 to 2700 km in diameter, and are all smaller than the four Galilean satellites and significantly smaller than Earth's Moon. Because of their small sizes, it was thought that these satellites would lose their internal heat within several hundred million years of formation and would be heavily cratered. Low bulk densities, high albedos, and water-ice-rich spectra all indicate that the crusts and mantles of these bodies are ice-rich and that volcanism is dominated by various icy materials. Understanding not only the composition of resurfacing materials, but also the sources of heat necessary to melt them and bring them to the surface remains one of the central problems in our understanding of outer solar system dynamics. Although Voyager made critical discoveries, a thorough understanding of the origins of these features awaits future infrared spectra and images of volcanic deposits that exceed a resolution of 100 m.

Landforms. A wide variety of volcanic landforms have been recognized on the smaller icy satellites (see review of Schenk and Moore, 1998). Smooth plains of presumed volcanic origin occur on Dione, Tethys, Enceladus, and Triton. On Dione and Tethys (Plescia, 1983; Moore and Ahern, 1983), these plains are moderately cratered (Figure 8.12). On Enceladus and Triton, these plains are very sparsely cratered, indicating relatively youthful ages of possibly <1 Ga. Although it is not possible to date the formation of these plains, the relative lack of craters compared with older terrains on these satellites indicates they probably formed after the period of heaviest impact bombardment. No definitive volcanic landforms such as lava flow margins are observable in these plains, except on Triton. These plains may be

Exotic Lava Flows 231

Figure 8.12. Voyager view of smooth plains on Dione. Moderately cratered but lacking prominent relief, this material is typical of smooth plains observed on other small icy satellites. Older heavily cratered terrain is visible to left. North is to the top. NASA.

morphologically (but not compositionally) analogous to basaltic lava plains on the Moon (see Chapter 6) and might be related to low-viscosity volcanism.

In addition to low-relief plains, a number of distinctive high-relief volcanic features are evident on several smaller icy satellites. These include unusual crater-filling domes on Enceladus (Schenk and Moore, 1995), and ridges and valley floor fill on Miranda and Ariel (Jankowski and Squyres, 1988; Schenk, 1991). The steep-sided domical peaks in craters on Enceladus can be up to 1 km high (Figure 8.13). Ridges on Miranda are several kilometers wide and several hundred meters high; those on Ariel are up to 2 km high. Ridges on Miranda occur within three large ovoid-shaped coronae, which were most likely formed by rising material in the interior that triggered deformation and volcanism in the overlying crust (Pappalardo et al., 1997). These volcanic features and especially the ridges are interpreted as fissure extrusions of apparently viscous material. These features may be icy satellite analogues of steep-sided viscous volcanic domes such as those at Mount St. Helens (Lipman and Mullineaux, 1981) (see Chapter 2).

Evidence for explosive volcanism is controversial at best. The most likely candidates are elliptical dark-haloed craters or vents on Triton (Croft et al., 1995). Caldera-like features are observed on Ganymede (Schenk and Moore, 1995; see Section 8.4.4.) and on Triton (Croft et al., 1995; Figure 8.14). At ~80 km across, the Triton caldera is the largest in the solar system as a percentage of planet surface area. Calderas need not form, however, as a result of

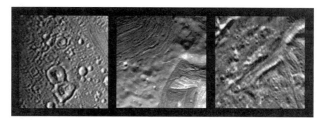

Figure 8.13. Examples of anomalous peaks on Enceladus (left) and volcanic ridges on Ariel (center) and Miranda (right). These ridges may have formed by the extrusion of relatively viscous ammonia–water lavas through linear fractures in the crust. Ridges can reach up to 2 km high on Ariel. Note the medial ridge visible along the crest of the Ariel ridge. On Miranda, the ridges form concentric bands within Elsinore Corona (top left) and Inverness Corona (bottom right). Voyager images, NASA.

explosive volcanic activity, and the caldera on Triton has evidently been resurfaced by renewed volcanism. Crater chains and oblong pits are also seen within the smooth plains surrounding this caldera. Together these features resemble those observed in basaltic lava plains such as the Snake River Plains (Greeley and King, 1977) (see Chapter 2).

Composition. Once volcanic landforms were identified on satellites such as Ariel and Miranda, one of the first issues to be raised was whether these flows were erupted in the solid state (unmelted ice mobilized by warm temperatures) or as liquid or liquid/solid mixtures (Jankowski and Squyres, 1988; Schenk, 1991). The issue is significant because mobilization of solid water ice would require different temperature conditions from those required to melt water or other ices, many of which have melting temperatures of <200 K. It is worth noting

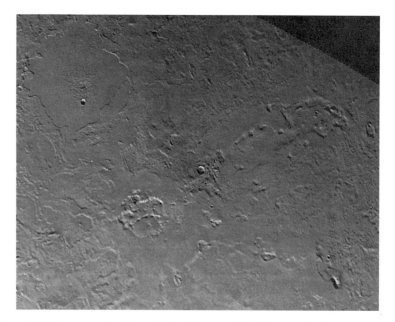

Figure 8.14. These smooth plains on Triton form one of the largest volcanic terrains on the icy satellites. They are centered around an 80-km-wide caldera, which itself has been partially filled with volcanic deposits. Several volcanic pit chains are also present, as well as several oblong craters, which may be volcanic vents. Voyager 2 mosaic, NASA.

that the Mount St. Helens dome might be considered a "solid-state" extrusion but is in reality a highly viscous dacite flow laden with flow-inhibiting crystals (see Chapter 2). Thus, the distinction between solid-state and liquid extrusion may not be distinct, depending on the degree of crystallization in a melt or of partial melting within rising ice.

There is no observational evidence regarding the composition of flows on the outer icy satellites. Only water ice has been detected on the satellites of Saturn and Uranus, although carbon dioxide and the highly volatile ices of methane, nitrogen, have also been detected on Triton. Liquid water is not generally considered a likely volcanic fluid on these satellites (except possibly in the smooth plains) because of the high melting temperature required (which would presumably soften or melt large parts of the crust), and the rheology of liquid water. Liquid water has a negligible viscosity and does not polymerize like silicate melts. This would make it difficult for liquid water to form large volcanic edifices hundreds of meters high.

The prime candidate for the composition of many of these flows has been ammonia/water mixtures (see review by Kargel, 1998), which have melting temperatures of 174 K. The justification for this composition has been the apparent need to extrude fluids with a lower melting temperature than water ice (the presumed bulk crustal constituent), the presence of ammonia in significant quantities in some comets, and models of solar nebula composition in the early 1970s suggesting that ammonia could constitute up to a few tens of percent of the interiors of many icy satellites. Partial melting of this ammonia might allow the formation of significant bodies of magma, potentially enough to explain the small volumes of erupted material observed on these satellites.

The steep-sided volcanic landforms observed on some satellites, might be explained by the extrusion of ammonia/water liquids. The basic rheologic properties of ammonia/water melt (Kargel *et al.*, 1991; Kargel, 1998) mimic those of basalt. This might explain the crude similarity of volcanic landforms on Triton and basaltic deposits on the Earth and Moon (see Chapters 2 and 6). The steep-sided landforms on Enceladus, Ariel, and Miranda may require the admixture of crystals or supercooling of the ammonia/water lavas (Kargel *et al.*, 1991) to increase lava viscosity. The absence of ammonia/water in telescopic spectra of these satellites has been explained as a possible result of decomposition into residual dark materials. Spacecraft, such as Cassini, with high-resolution near-infrared detectors, may finally resolve the composition of individual volcanic features, possibly in the ejecta of recently formed impact craters. Until such identification occurs, the composition of lava flows on the smaller icy satellites remains in the realm of informed speculation.

Internal Heat Sources. The heat source driving melting and volcanism within the small icy satellites remains a mystery. The pre-Voyager paradigm suggested that size (or mass) and composition determined how geologically active a planetary body would be. Larger and more silicate-rich bodies would be expected to be more active because of greater volume and large amounts of heat-producing radionuclides. In smaller icy satellites, accretional and radiogenic heating are probably not significant heat sources after the first several hundred million years or so (Hillier and Squyres, 1991). This might be sufficiently long enough to explain some of the older smooth deposits, depending on their true age, but this mechanism probably cannot explain resurfacing on bodies such as Ariel or Enceladus, where deposits are considerably younger.

Voyager demonstrated the importance of position in determining a planetary body's heat budget (e.g., Smith *et al.*, 1979). Tidal interactions between a satellite and its neighbors, including the parent planet, can distort the shape of a satellite, resulting in varying degrees of internal heating. Some satellites, specifically the Galilean satellites, are currently in orbital resonance. Other such as the Uranian satellites Ariel and Miranda are not in resonance now but may have been in temporary resonance in the past, leading to intermittent or periodic heating (e.g., Tittemore and Wisdom, 1989; Malhotra and Dermott, 1990).

In the absence of viable alternatives, tidal heating in temporary orbital resonances remains the most likely candidate for triggering volcanism on the small icy satellites. It is difficult, however, to model these resonances precisely or estimate the amount of heat these resonances might have generated. In the case of Triton, the heat source was probably captured into Neptune orbit and the subsequent circularization of Triton's orbit (McKinnon *et al.*, 1995). Heat generated during these events probably melted most of Triton's interior, and the volcanic features we see today might represent the later stages of cooling (McKinnon *et al.*, 1995).

8.5. DISCUSSION AND CONCLUSION

As has been shown throughout this book, many volcanologic features on planetary surfaces can form by processes that can be observed at present on basaltic volcanoes such as Kilauea, or more silicic volcanoes such as Soufrière Hills volcano, Montserrat. However, as we have seen in this chapter, some volcanic landforms can be formed only by the flow of very low viscosity melts such as carbonatites and komatiites. The interpretation of surface features on many satellites of the outer planets has presented planetary geologists with a major intellectual challenge. Many of the features observed on these bodies cannot be explained using examples from Earth, and detailed mapping and observations have resulted in cryovolcanism, a new class of volcanism.

Additional progress in understanding icy satellite volcanism will come from two fronts. The first is continued research into the rheologic and chemical properties of likely solar system ice phases (e.g., Kargel, 1998), including complex mixtures of water and other volatile phases. The second lies in the continued exploration of icy satellites, especially those whose surfaces have not yet been observed by spacecraft instruments (Titan and Pluto). Differences in volcanic style or composition, or the presence of active or very recent volcanism on Titan, similar in size to Ganymede, may provide clues to the origin of Titan's unique atmosphere. The detection of volcanic activity on Pluto would also provide important clues into internal processes on volatile-rich bodies. Although similar in size and density to Triton, Pluto has undergone a much different dynamic history (McKinnon *et al.*, 1997). At the time of writing, Cassini is en route to the Saturn system, the Europa Orbiter Mission is being planned, and there is hope for a mission to Pluto in the next decade or so. These and other future missions will enable us to examine lava flows in detail on these bodies, and conduct high-resolution near-infrared spectroscopic observations, hopefully identifying the compositions of these volcanic materials.

8.6. REFERENCES

Allison, M. L., and S. M. Clifford, Ice-covered water volcanism on Ganymede, *J. Geophys. Res., 92.*, 7865–7876, 1987.

Anderson, T., Age and petrogenesis of the Qassiarsuk carbonatite–alkaline silicate volcanic complex in the Gardar Rift, South Greenland, *Mineral Mag., 61*, 499–513, 1997.

Anderson, J. D., G. Schubert, R. A. Jacobsen, E. L. Lau, W. B. Moore, and W. L. Sjogren, Europa's differentiated internal structure: Inferences from four Galileo encounters, *Science, 281*, 2019–2022, 1998.
Arndt, N. T., Melting relations of ultramafic lavas (komatiites) at one atmosphere and high pressure, *Yearbook, Carnegie Inst. Washington, 75*, 555–562, 1976.
Arndt, N. T., Ultrabasic magmas and high-degree melting of the mangle, *Contrib. Mineral. Petrol., 64*, 205–221, 1977.
Arndt, N. T., Differentiation of komatiite flows, *J. Petrol., 27*, 279–301, 1986.
Arndt, N. T., and E. G. Nisbet, *Komatiites*, 526 pp., Allen & Unwin, London, 1982.
Arndt, N. T., D. M. Francis, and A. J. Hynes, The field characteristics and petrology of Archean and Proterozoic komatiites, *Can. Mineral., 17*, 147–163, 1979.
Arndt, N. T., F. Albarède, and E. G. Nisbet, Mafic and ultramafic magmatism, in *Greenstone Belts, Oxford Monographs on Geology and Geophysics, 35*, edited by M. de Wit and L. D. Ashwal, 809 pp., Oxford University Press, London, 1997.
Bailey, D. K., Carbonate magmas, *J. Geol. Soc., 150.*, 637–651, 1993.
Bailey, D. K., and C. M. Hampton, Volatiles in alkaline magmatism, *Lithos, 26*, 157–165, 1990.
Baird, A. K., Did komatiitic lavas erode channels on Mars? *Nature, 311*, 18, 1984.
Baird, A. K., and B. C. Clark, On the original igneous source of Martian fines, *Icarus, 45*, 113–123, 1981.
Baker, V. R., G. Komatsu, T. J. Parker, V. C. Gulick, J. S. Kargel, and J. S. Lewis, Channels and valleys on Venus: Preliminary analysis of Magellan data, *J. Geophys. Res., 97.*, 13421–13444, 1992.
Barnes, S. J., C. J. A. Coats, and A. J. Naldrett, Petrogenesis of a Proterozoic nickel sulfide–komatiite association: The Katiniq Sill, Ungava, Quebec, *Econ. Geol., 77*, 413–429, 1982.
Barnes, S. J., R. E. T. Hill, and M. J. Cole, The Perseverance Ultramafic Complex, Western Australia: The product of a komatiite lava river, *J. Petrol., 29*, 302–331, 1988.
Bell, K., and J. Keller (Eds.), *Carbonatite Volcanism of Oldoinyo Lengai—Petrogenesis of Natrocarbonatite*, Springer-Verlag, Berlin, 1995.
Bell, K., and A. Simonetti, Carbonatite magmatism and plume activity: Implications from the Nd, Pb and Sr isotope systematics of Oldoinyo Lengai, *J. Petrol., 37*, 1321–1339, 1996.
Belton, M. J. S., J. W. Head, A. P. Ingersoll, R. Greeley, A. S. McEwen, K. P. Klaasen, D. Senske, R. Pappalardo, G. Collins, A. R. Vasavada, R. Sullivan, D. Simonelli, P. Geissler, M. H. Carr, M. E. Davies, J. Veverka, P. J. Geirasch, D. Banfield, M. Bell, C. R. Chapman, A. Anger, R. Greenberg, G. Neukum, C. B. Pilcher, R. F. Beebe, J. A. Burns, F. Fanale, W. Ip, T. V. Johnson, D. Morrison, J. Moore, G. S. Orton, P. Thomas, and R. A. West, Galileo's first image of Jupiter and the Galilean satellites, *Science, 274*, 377–385, 1996.
Bertka, C. M., and J. R. Holloway, Anhydrous partial melting of an iron-rich mantle II: Primary melt compositions at 15 kbar, *Contrib. Mineral. Petrol., 115*, 323–338, 1994.
Bottinga, Y., and D. F. Weill, Densities of liquid silicate systems calculated from impartial molar volumes of oxide components, *Am. J. Sci., 269*, 169–182, 1970.
Buratti, B., and M. Golombek, Geologic implications of spectrophotometric measurements of Europa, *Icarus, 75*, 437–449, 1988.
Burns, R. G., and D. S. Fisher, Evolution of sulfide mineralization on Mars, *J. Geophys. Res., 95.*, 14169–14173, 1990a.
Burns, R. G., and D. S. Fisher, Iron–sulfur mineralogy of Mars: Magmatic evolution and chemical weathering products, *J. Geophys. Res., 95.*, 14415–14421, 1990b.
Calvari, S., and H. Pinkerton, Formation of lava tubes and extensive flow field during the 1991–93 eruption of Mount Etna, *J. Geophys. Res., 103.*, 27291–27301, 1998.
Campbell, I. H., and R. W. Griffiths, The evolution of the mantle's chemical structure, *Lithos, 30*, 389–399, 1993.
Campbell, I. H., R. W. Griffiths, and R. J. Hill, Melting in an Archean mantle plume: Heads it's basalts, tails it's komatiites, *Nature, 339*, 697–699, 1989.
Carr, M. H., The role of lava erosion in the formation of lunar rilles and Martian channels, *Icarus, 22*, 1–22, 1974.
Carr, M. H., M. J. S. Belton, C. R. Chapman, M. E. Davies, P. E. Geissler, R. Greenberg, A. S. McEwen, B. R. Tufts, R. Greeley, R. Sullivan, J. W. Head, R. T. Pappalardo, K. P. Klaasen, T. V. Johnson, J. Kaufman, D. Senske, J. M. Moore, G. Neukum, G. Schubert, J. A. Burns, P. Thomas, and J. Veverka, Evidence for a subsurface ocean on Europa, *Nature, 391*, 363–365, 1998.
Cawthorn, R. G., and D. F. Strong, The petrogenesis of komatiites and related rocks as evidence for a layered upper mantle, *Earth Planet. Sci. Lett., 23*, 369–375, 1974.
Clark, R. N., Ganymede, Europa, Callisto and Saturn's rings: Compositional analysis from reflectance spectroscopy, *Icarus, 44*, 388–409, 1980.

Clark, R. N., F. P. Finale, and M. J. Gaffney, Surface composition of natural satellites, in *Satellites*, edited by J. A. Burns and M. S. Matthews, pp. 437–491, University of Arizona Press, Tucson, 1986.

Crawford, G. D., and D. J. Stevenson, Gas-driven water volcanism and the resurfacing of Europa, *Icarus, 73*, 66–79, 1988.

Croft, S. K., and B. N. Goudreau, Tectonism and volcanism in Ganymede's dark terrain (abstract), *Lunar Planet. Sci., XVIII*, 209–210, 1987.

Croft, S. K., J. S. Kargel, R. L. Kirk, J. M. Moore, P. M. Schenk, and R. G. Strom. The geology of Triton, in *Neptune and Triton*, edited by D. Cruikshank, pp. 879–948, University of Arizona Press, Tucson, 1995.

Cruikshank, D. P., and C. A. Wood, Lunar rilles and Hawaiian volcanic features: Possible analogs, *Moon, 3*, 412–447, 1972.

Cutts, J. A., W. J. Roberts, and K. R. Blasius, Martian channels formed by lava erosion, *Lunar Planet. Sci., IX*, 209, 1978.

Davis, P. C., *Volcanic stratigraphy of the Late Archean Kidd-Munro assemblage in Dundonald and Munro Townships and genesis of associated nickel and copper-zinc volcanogenic massive sulfide deposits, Abitibi Greenstone Belt, Ontario, Canada*, M. S. Thesis, 165 pp., University of Alabama, Tuscaloosa, 1997.

Dawson, J. B., The geology of Oldoinyo Lengai, *Bull. Volcanol., 24*, 349–387, 1962.

Dawson, J. B., Quaternary kimberlitic volcanism on the Tanzania Craton, *Contrib. Mineral. Petrol., 116*, 473–485, 1994.

Dawson, J. B., P. Bowden, and G. C. Clark, Activity of the carbonatite volcano Oldoinyo Lengai, *Geol. Rundsch., 57*, 865–879, 1968.

Dawson, J. B., H. Pinkerton, G. E. Norton, and D. M. Pyle, Physicochemical properties of alkali carbonatite lavas: Data from the 1988 eruption of Oldoinyo Lengai, Tanzania, *Geology, 18*, 260–263, 1990.

Dawson, J. B., H. Pinkerton, D. M. Pyle, and C. Nyamweru, June 1993 eruption of Oldoinyo Lengai, Tanzania: Exceptionally viscous and large carbonatite lava flows and evidence for co-existing silicate and carbonate magmas, *Geology, 22*, 799–802, 1994.

Dawson, J. B., J. Keller, and C. Nyamweru, Historic and recent eruptive activity of Oldoinyo Lengai, in *IAVCEI Proceedings in Volcanoloy 4. Carbonatite Volcanism of Oldoinyo Lengai—Petrogenesis of Natrocarbonatite*, edited by K. Bell and J. Keller, pp. 4–22, Springer-Verlag, Berlin, 1995a.

Dawson, J. B., H. Pinkerton, G. E. Norton, D. M. Pyle, P. Browning, D. Jackson, and A. E. Fallik, Petrology and geochemistry of Oldoinyo Lengai lavas extruded November 1988: Magma source, ascent and crystallisation, in *IAVCEI Proceedings in Volcanology 4. Carbonatite Volcanism of Oldoinyo Lengai—Petrogenesis of Natrocarbonatite*, edited by K. Bell and J. Keller, pp. 47–69, Springer-Verlag, Berlin, 1995b.

Dawson, J. B., D. M. Pyle, and H. Pinkerton, Evolution of natrocarbonatite from a wollastonite nephelinite parent: Evidence from the June 1993 eruption of Oldoinyo Lengai, Tanzania, *J. Geol., 104*, 41–54, 1996.

de Wit, M. J., and L. D. Ashwal (Eds.), *Greenstone Belts, Oxford Monographs on Geology and Geophysics, 35*, 809 pp., Oxford University Press, London, 1997.

Dobson, D. P., A. P. Jones, R. Rabe, T. Sekine, K. Kurita, T. Taniguchi, T. Kondo, T. Kato, O. Shimomura, and S. Urakawa, In-situ measurement of viscosity and density of carbonate melts at high-pressure, *Earth Planet. Sci. Lett., 143*, 207–215, 1996.

Fagents, S. A., R. Greeley, R. J. Sullivan, R. T. Pappalardo, and L. M. Prockter, Cryomagmatic mechanisms for the formation of Rhadamanthys Linea, triple band margins, and other low albedo features on Europa, *Icarus*, in press.

Fink, J. H., The emplacement of silicic lava flows and associated hazards, in *Active Lavas*, edited by C. R. J. Kilburn and G. Luongo, pp. 5–24, UCL Press, London, 1993.

Finnerty, A. A., G. A. Ransford, D. C. Pieri, and K. D. Collerson, Is Europa surface cracking due to thermal evolution? *Nature, 289*, 24–27, 1981.

Ghiorso, M. S., and R. O., Sack, Chemical mass transfer in magmatic processes IV. A revised and internally consistent thermodynamic model for the interpolation and extrapolation of liquid–solid equilibria in magmatic systems at elevated temperatures and pressures, *Contrib. Mineral. Petrol., 119*, 197–212, 1995.

Greeley, R., and J. H. Hyde, Lava tubes of the Cave Basalt, Mount St. Helens, Washington, *Geol. Soc. Am. Bull., 83*, 2397–2418, 1972.

Greeley, R., and J. King (Eds.), *Volcanism of the Eastern Snake River Plains*, (NASA CR-154621, 308 pp., 1977.

Greeley, R., S. W. Lee, D. A. Crown, and N. Lancaster, Observations of industrial sulfur flows: Implications for Io, *Icarus, 84*, 374–402, 1990.

Greeley, R., S. A. Fagents, R. S. Harris, S. D. Kadel, D. A. Williams, and J. E. Guest, Erosion by flowing lava: Field evidence, *J. Geophys. Res., 103.*, 27325–27345, 1998a.

Greeley, R., R. Sullivan, J. Klemaszewski, K. Homan, J. W. Head, R. T. Pappalardo, J. Veverka, B. E. Clark, T. V. Johnson, K. P. Klaasen, M. Belton, J. M. Moore, E. Asphaug, M. H. Carr, G. Neukum, T. Denk, C. R. Chapman, C. B. Pilcher, P. E. Geissler, R. Greenberg, and B. R. Tufts, Europa: Initial Galileo geological observations, *Icarus, 135*, 4–24, 1998b.

Greenberg, R., P. Geissler, G. Hoppa, B. R. Tufts, D. D. Durda, R. T. Pappalardo, J. W. Head, R. Greeley, R. Sullivan, and M. H. Carr, Tectonic processes on Europa: Tidal stresses, mechanical response, and visible features, *Icarus, 135*, 64–78, 1998.

Head, J. W., D. B. Campbell, C. Elachi, J. E. Guest, D. P. McKenzie, R. S. Saunders, G. G. Schaber, and G. Schubert, Venus volcanism: Initial analysis from Magellan data, *Science, 252*, 276–288, 1991.

Head, J. W., R. Pappalardo, R. Greeley, R. Greenberg, C. Chapman, G. Neukum, M. J. S. Belton, M. Carr, C. Pilcher, G. Collins, L. Prockter, K. Jones, J. Moore, D. Senske, K. Klaasen, K. Magee, and H. Breneman, Ganymede: Synthesis of solid state imaging results from the Galileo mission, *Eos. Trans. AGU, 78*, F417, 1997.

Head, J. W., R. Pappalardo, J. Kay, G. Collins, L. Prockter, R. Greeley, C. Chapman, M. Carr, M. J. S. Belton, and the Galileo Imaging Team, Cryovolcanism on Ganymede: Evidence in bright terrain from Galileo solid state imaging data, *Lunar Planet. Sci., XXIX*, #1666 (CD-ROM), 1998a.

Head, J. W., R. T. Papparlardo, L. M. Prockter, G. Collins, M. J. S. Belton, M. Carr, C. Chapman, R. Greeley, R. Greenberg, A. McEwen, G. Neukum, C. Pilcher, J. Veverka, T. Johnson, K. Klaasen, D. Senske, K. Magee, H. Breneman, J. Kaufman, T. Jones, P. Helfenstein, J. Oberst, B. Giese, T. Denk, D. Morrison, J. Moore, and the Galileo Solid State Imaging Team, Ganymede: Overview of solid state imaging (SSI) findings from the nominal mission, *Lunar Planet. Sci., XXIX*, #1774 (CD-ROM), 1998b.

Head, J. W., R. T. Pappalardo, R. Sullivan, and the Galileo SSI Team, Europa: Morphological characteristics of ridges and triple bands from Galileo data (E4 and E6) and assessment of a linear diapirism model, *J. Geophys. Res., 104*, 24223–24236, 1999.

Herzberg, C., Depth and degree of melting of komatiites, *J. Geophys. Res., 97*, 4521–4540, 1992.

Hill, R. E. T., and C. S. Perring, The evolution of Archean komatiite flow fields—Are they inflationary sheet flows?, in *Chapman Conference on Long Lava Flows: Conference Abstracts, Econ. Geol. Res. Unit Contrib. 56*, edited by P. W. Whitehead, pp. 18–21, James Cook University of N. Queensland, and Townsville, Australia, 1996.

Hill, R. E. T., S. J. Barnes, M. J. Cole, and S. E. Dowling, Physical volcanology of komatiites, *Excursion Guidebook #1*, 100 pp., Geological Society of Australia, Western Australia Division, Perth, 1990.

Hill, R. E. T., S. J. Barnes, M. J. Cole, and S. E. Dowling, The volcanology of komatiites as deduced from field relationships in the Norseman-Wiluna greenstone belt, Western Australia, *Lithos, 34*, 159–188, 1995.

Hillier, J., and S. Squyres, Thermal tectonic stresses on the satellites of Saturn and Uranus, *J. Geophys. Res., 96.*, 15665–15674, 1991.

Hogenbloom, D. L., J. S. Kargel, G. J. Consolmagno, T. C. Holden, L. Lee, and M. Buyyounouski, The ammonia–water system and the chemical differentiation of icy satellites, *Icarus, 128*, 171–180, 1997.

Hon, K., J. Kauahikaua, R. Denlinger, and K. Mackay, Emplacement and inflation of pahoehoe sheet flows: Observations and measurements of active lava flows on Kilauea Volcano, Hawaii, *Geol. Soc. Am. Bull., 106*, 351–370, 1994.

Hulme, G., Turbulent lava flows and the formation of lunar sinuous rilles, *Mod. Geol., 4*, 107–117, 1973.

Hulme, G., A review of lava flow processes related to the formation of lunar sinuous rilles, *Geophys. Surv., 5*, 245–279, 1982.

Huppert, H. E., and R. S. J. Sparks, Komatiites I: Eruption and flow, *J. Petrol., 26*, 694–725, 1985.

Huppert, H. E., R. S. J. Sparks, J. S. Turner, and N. T. Arndt, Emplacement and cooling of komatiite lavas, *Nature, 309*, 19–22, 1984.

Iverson, R. M., Lava domes modeled as brittle shells that enclose pressurized magma, with application to Mt. St. Helens, in *Lava Flows and Domes*, edited by J. H. Fink, pp. 47–69, Springer-Verlag, Berlin, 1990.

Jackson, S. L., J. A. Fyon, and F. Corfu, Review of Archean supracrustal assemblages of the southern Abitibi greenstone belt in Ontario, Canada: Products of microplate interaction within a large-scale plate-tectonic setting, *Precambrian Res., 65*, 183–205, 1994.

Jankowski, D. J., and S. W. Squyres, Solid-state ice volcanism on the satellites of Uranus, *Science, 141*, 1322–1325, 1988.

Jarvis, R. A., On the cross-sectional geometry of thermal erosion channels formed by turbulent lava flows, *J. Geophys. Res., 100*, 10127–10140, 1995.

Jones, K. B., J. W. Head, C. R. Chapman, R. Greeley, J. M. Moore, G. Neukum, R. T. Pappalardo, and the Galileo SSI Team, Morphology of palimpsests on Ganymede from Galileo observations, *Lunar Planet. Sci., XXVIII*, 679–680, 1997.

Kadel, S. D., S. A. Fagents, R. Greeley, and the Galileo SSI Team, Trough-bounding ridge pairs on Europa—Considerations for an endogenic model of formation (abstract), *Lunar Planet. Sci., XXIX*, #1078 (CD-ROM), 1998.

Kargel, J. S., Brine volcanism and the interior structures of asteroids and icy satellites, *Icarus, 94*, 368–390, 1991.

Kargle, J. S., Physical chemistry of ices in the outer solar system, in *Solar System Ices*, edited by B. Schmitt, C. deBergh, and M. Festou, pp. 3–32, Kluwer, Dordrecht, 1998.

Kargel, J. S., S. K. Croft, J. I. Lunine, and J. S. Lewis, Rheological properties of ammonia–water liquids and crystal–liquid slurries: Planetological implications, *Icarus, 89*, 93–112, 1991.

Kargel, J. S., R. L. Kirk, B. Fegley, and A. H. Treiman, Carbonate–sulfate volcanism on Venus, *Icarus, 112*, 219–252, 1994.

Kauahikaua, J., K. V. Cashman, T. N. Mattox, C. C. Heliker, K. A. Hon, M. T. Mangan, and C. R. Thornber, Observations on basaltic lava streams in tubes from Kilauea Volcano, Hawai'i, *J. Geophys. Res., 103*, 27303–27323, 1998.

Kerr, A. C., J. Tarney, G. F. Marriner, A. Nivia, and A. D. Saunders, Caribbean–Colombian Cretaceous igneous province: The internal anatomy of an oceanic plateau, in *Large Igneous Provinces: Continental, Oceanic, and Planetary Flood Volcanism*, edited by J. J. Mahoney and M. F. Coffin, pp. 123–144, *AGU Geophysical Monograph, 100*, 1997.

Kirk, R. L., L. A. Soderblom, R. H. Brown, S. W. Kieffer, and J. S. Kargel, Triton's plumes: Discovery, characteristics, and models, in *Neptune and Triton*, edited by D. P. Cruikshank, pp. 949–989, University of Arizona Press, Tucson, 1995.

Kjarsgaard, B., and T. Peterson, Nephelinite–carbonatite liquid immisicibility at Shombole Volcano, East-Africa—Petrographic and experimental evidence, *Mineral. Petrol., 43*, 293–314, 1991.

Komatsu, G., and V. R. Baker, Meander properties of Venusian channels, *Geology, 22*, 67–70, 1994.

Komatsu, G., J. S. Kargel, and V. R. Baker, Canali-type channels on Venus—Some genetic constraints, *Geophys. Res. Lett., 19*, 1415–1418, 1992.

Komatsu, G., V. R. Baker, V. C. Gulick, and T. J. Parker, Venusian channels and valleys: Distribution and volcanological implications, *Icarus, 102*, 1–25, 1993.

Krafft, M., and J. Keller, Temperature-measurements in carbonatite lava lakes and flows from Oldoinyo-Lengai, Tanzania, *Science, 245*, 168–170, 1989.

Kurszlaukis, S., and V. Lorenz, Volcanological features of a low-viscosity melt: The carbonatitic Gross Brukkaros Volcanic Field, Namibia, *Bull. Volcanol., 58*, 421–431, 1997.

Lange, R. A., and A. Navrotsky, Heat capacities of Fe_2O_3-bearing silicate liquids, *Contrib. Mineral. Petrol., 110*, 311–320, 1992.

Lee, W. J., and P. J. Wyllie, Petrogenesis of carbonatite magmas from mantle to crust, constrained by the system CaO-$(MgO + FeO^*)(Na_2O + K_2O)$-$(SiO_2 + Al_2O_3 + TiO_2)$-$CO_2$, *J. Petrol., 39*, 495–517, 1998.

Lesher, C. M., Komatiite-associated nickel sulfide deposits, in *Ore Deposits Associated with Magmas*, edited by J. A. Whitney and A. J. Naldrett, *Rev. Econ. Geol., 4*, 45–102, 1989.

Lesher, C. M., N. T. Arndt, and D. I. Groves, Genesis of komatiite-associated nickel sulphide deposits at Kambalda, Western Australia: A distal volcanic model, in *Sulphide Deposits in Mafic and Ultramafic Rocks*, edited by D. L. Buchanan and M. J. Jones, pp. 70–80, Institute of Mineralogy and Metallurgy, London, 1984.

Lipman, P. W., and D. Mullineaux (Eds.), The 1980 eruption of Mount St. Helens, Washington, *U.S. Geol. Surv. Prof. Pap., 1250*, 1981.

Longhi, J., and V. Pan, What SNC meteorites tell us about Martian magmatism, *LPI Tech. Publ. 88-05*, pp. 76–78, Lunar and Planetary Institute, Houston, 1988.

Longhi, J., and V. Pan, The parent magmas of the SNC meteorites, *Proc. Lunar Planet. Sci. Conf., XIX*, 451–464, 1989.

Lucchitta, B. K., Grooved terrain on Ganymede, *Icarus, 44*, 481–501, 1980.

Lucchitta, B. K., and L. A. Soderblom, The geology of Europa, in *Satellites of Jupiter*, edited by D. Morrison, pp. 521–555, University of Arizona Press, Tucson, 1982.

Lucchitta, B. K., C. W. Barnes, and M. F. Glotfelty, Geological map of the Memphis Facula quadrangle (Jg-7) of Ganymede, *U.S. Geol. Surv. Map.* I-2289, 1992.

Lunine, J. I., and D. J. Stevenson, Thermodynamics of clathrate hydrate at low and high pressures with application to the outer solar system, *Astrophys. J. Suppl., 58*, 493–531, 1985.

Malhotra, R., and S. Dermott, The role of secondary resonances in the orbital history of Miranda, *Icarus, 85*, 444–480, 1990.

Malin, M. C., and D. C. Pieri, Europa, in *Satellites*, edited by J. A. Burns and M. S. Matthews, pp. 689–716, University of Arizona Press, Tucson, 1986.

Matson, D. L., D. L. Blaney, T. V. Johnson, G. J. Veeder, and A. G. Davies, Io and the early Earth, *Lunar Planet. Sci., XXIX*, 1650–1651, 1998.
McEwen, A. S., L. Keszthelyi, J. R. Spencer, G. Schubert, D. L. Matson, R. Lopes-Gautier, K. P. Klaasen, T. V. Johnson, J. W. Head, P. E. Geissler, S. Fagents, A. G. Davies, M. H. Carr, H. H. Breneman, and M. J. S. Belton, High-temperature silicate volcanism on Jupiter's moon Io, *Science, 281*, 87–90, 1998.
McKinnon, W. B., and E. M. Parmentier, Ganymede and Callisto, in *Satellites*, edited by J. A. Burns and M. S. Matthews, pp. 718–763, University of Arizona Press, Tucson, 1986.
McKinnon, W., J. Lunine, and D. Banfield, Origin and evolution of Triton, in *Neptune and Triton*, edited by D. Cruikshank, pp. 807–878, University of Arizona Press, Tucson, 1995.
McKinnon, W., D. Simonelli, and G. Schubert, Composition, internal structure, and thermal evolution of Pluto and Charon, in *Pluto and Charon*, edited by S. Stern and D. Tholen, pp. 295–343, University of Arizona Press, Tucson, 1997.
Middlemost, E. A. K., *Magmas and Magmatic Rocks*, 266 pp., Longman, New York, 1985.
Moore, J., and J. Ahern, The geology of Tethys, *Icarus, 59*, 205–220, 1983.
Moore, J. M., and M. C. Malin, Dome craters on Ganymede, *Geophys. Res. Lett., 15*, 225–228, 1988.
Mueller, S., and W. B. McKinnon, Three-layered models of Ganymede and Callisto: Compositions, structures and aspects of evolution, *Icarus, 76*, 437–464, 1988.
Murase, T., and A. R. McBirney, Viscosity of lunar lavas, *Science, 167*, 1491–1493, 1970.
Murchie, S. L., J. W. Head, and J. B. Plescia, Crater densities and crater ages of different terrain types on Ganymede, *Icarus, 81*, 271–297, 1989.
Murchie, S. L., J. W. Head, and J. B. Plescia, Tectonic and volcanic evolution of dark terrain and its implications for the internal structure of Ganymede, *J. Geophys. Res., 95*, 10743–10768, 1990.
Navrotsky, A., Energetics of silicate melts, *Reviews in Mineralogy, 32*, 121–142, 1995.
Nisbet, E. G., The tectonic setting and petrogenesis of komatiites, in *Komatiites*, edited by N. T. Arndt and E. G. Nisbet, 526 pp., Allen & Unwin, London, 1982.
Nisbet, E. G., M. J. Cheadle, N. T. Arndt, and M. J. Bickle, Constraining the potential temperature of the Archaen mantle: A review of the evidence from komatiites, *Lithos, 30*, 291–307, 1993.
Norton, G. E., and H. Pinkerton, Rheological properties of natrocarbonatites from Oldoinyo Lengai, Tanzania, *Eur. J. Mineral., 9*, 351–364, 1997.
Nyamweru, C., Changes in the crater of Oldoinyo Lengai: June 1993–February 1997, *J. Afr. Earth Sci., 25*, 43–53, 1997.
Ojakangas, G. W., and D. J. Stevenson, Thermal state of an ice shell on Europa, *Icarus, 81*, 220–241, 1989.
Papparlardo, R., S. Reynolds, and R. Greeley, Extensional tilt blocks on Miranda: Evidence for an upwelling origin of Arden Corona, *J. Geophys. Res., 102*, 13369–13379, 1997.
Pappalardo, R. T., J. W. Head, R. Greeley, R. Sullivan, C. Pilcher, G. Schubert, W. B. Moore, M. H. Carr, J. M. Moore, M. J. S. Belton, and D. L. Goldsby, Geological evidence for solid-state convection in Europa's ice shell, *Nature, 391*, 365–367, 1998.
Parmen, S. W., J. C. Dann, T. L. Grove, and M. J. de Wit, Emplacement conditions of komatiite magmas from the 3.49 Ga Komati Formation, Barberton Greenstone Belt, South Africa, *Earth Planet. Sci. Lett., 150*, 303–323, 1997.
Parmentier, E. M., S. W. Squyres, J. W. Head, and M. L. Allison, The tectonics of Ganymede, *Nature, 295*, 290–293, 1982.
Pearce, N. J. G., M. J. Leng, C. H. Emeleus, and C. M. Bedford, The origins of carbonatites and related rocks from the Gronnedal-Ika Nepheline Syenite Complex, South Greenland: C-O-Sr isotope evidence, *Mineral. Mag., 61*, 515–529, 1997.
Perring, C. S., S. J. Barnes, and R. E. T. Hill, The physical volcanology of Archean komatiite sequences from Forrestania, Southern Cross Province, Western Australia, *Lithos, 34*, 189–208, 1995.
Peterson, D. W., and D. A. Swanson, Observed formation of lava tube during 1970–71 at Kilauea Volcano, Hawaii, *Stud. Speleol., 2*, 209–222, 1974.
Peterson, D. W., R. T. Holcomb, R. I. Tilling, and R. L. Christiansen, Development of lava tubes in the light of observations at Mauna Ulu, Kilauea Volcano, Hawaii, *Bull. Volcanol., 56*, 343–360, 1994.
Pinkerton, H., and R. S. J. Sparks, The 1975 sub-terminal lavas, Mount Etna: A case history of the formation of a compound lava field, *J. Volcanol. Geotherm. Res., 1*, 167–182, 1976.
Pinkerton, H., and L. Wilson, Factors controlling the lengths of channel-fed lava flows, *Bull. Volcanol., 56*, 108–120, 1994.
Pinkerton, H., L. Wilson, and G. E. Norton, Thermal erosion—Observations on terrestrial lava flows and implications for planetary volcanism, *Lunar Planet. Sci., XXI*, 964–965, 1990.

Pinkerton, H., J. B. Dawson, and D. M. Pyle, Arachnoid-like feature on Oldoinyo Lengai, an active carbonatite volcano in northern Tanzania, *Lunar Planet. Sci., XXV,* 1087–1088, 1994.

Pinkerton, H., G. E. Norton, J. B. Dawson, and D. M. Pyle, Field observations and measurements of the physical properties of Oldoinyo Lengai alkali carbonatite lavas, November 1988, in *IAVCEI Proceedings of Volcanology 4. Carbonatite Volcanism of Oldoinyo Lengai—Petrogenesis of Natrocarbonatite,* edited by K. Bell and J. Keller, pp. 23–36, Springer-Verlag, Berlin, 1995.

Plescia, J., The geology of Dione, *Icarus, 56,* 255–277, 1983.

Prockter, L. M., J. W. Head, R. T. Pappalardo, D. A. Senske, G. Neukum, R. Wagner, U. Wolf, J. Oberst, B. Giese, J. M. Moore, C. R. Chapman, P. Helfenstein, R. Greeley, H. H. Breneman, and M. J. S. Belton, Dark terrain on Ganymede: Geological mapping and interpretation of Galileo Regio at high resolution, *Icarus, 135,* 317–344, 1998.

Pyke, D. R., A. J. Naldrett, and O. R. Eckstrand, Archean ultramafic flows in Munro Township, Ontario, *Geol. Soc. Am. Bull., 84,* 955–978, 1973.

Reyes, D. P., and R. Christensen, Evidence for komatiite-type lavas on Mars from Phobos ISM data and other evidence, *Geophys. Res. Lett., 21,* 887–890, 1994.

Riley, T. R., D. K. Bailey, and F. E. Lloyd, Extrusive carbonatite from the Quaternary Rockeskyll Complex, West Eifel, Germany, *Can. Mineral., 34,* 389–401, 1996.

Saverikko, M., Komatiitic explosive volcanism and its tectonic setting in Finland, the Fennoscandian (Baltic) Shield, *Bull. Geol. Soc. Finl., 62,* 3–38, 1990.

Schafer, S. J., and P. Morton, Two komatiitic pyroclastic units, Superior Province, northwestern Ontario: Their geology, petrography, and correlation, *Can. J. Earth Sci., 28,* 1455–1470, 1991.

Schenk, P. M., Fluid volcanism on Miranda and Ariel: Flow morphology and composition, *J. Geophys. Res., 96,* 1887–1906, 1991.

Schenk, P., Central pit and dome craters: Exposing the interiors of Ganymede and Callisto, *J. Geophys. Res., 98,* 7475–7498, 1993.

Schenk, P., and W. B. McKinnon, Dark halo craters and the thickness of grooved terrain on Ganymede, *Proc. Lunar Planet. Sci. Conf., XVI, J. Geophys. Res. Suppl., 90,* C775–C783, 1985.

Schenk, P. M., and J. M. Moore, Volcanic constructs on Ganymede and Enceladus: Topographic evidence from stereo images and photoclinometry, *J. Geophys. Res., 100,* 19009–19022, 1995.

Schenk, P., and J. Moore, Geologic landforms and processes on icy satellites, in *Solar System Ices,* edited by B. Schmitt, C. deBergh, and M. Festou, pp. 551–578, Kluwer, Dordrecht, 1998.

Schubert, G., T. Spohn, and R. T. Reynolds, Thermal histories, compositions and internal structures of the moons of the solar systems, in *Satellites,* edited by J. A. Burns and M. S. Matthews, pp. 224–292, University of Arizona Press, Tucson, 1986.

Self, S., T. Thordarson, and L. Keszthelyi, Emplacement of continental flood basalt lava flows, in *Large Igneous Provinces: Continental, Oceanic and Planetary Flood Volcanism,* edited by J. J. Mahoney and M. Coffin, pp. 381–410, *AGU Geophysical Monograph, 100,* 1997.

Shaw, H. R., Viscosities of magmatic silicate liquids: An empirical method of prediction, *Am. J. Sci., 272,* 870–893, 1972.

Shoemaker, E. M., B. K. Lucchitta, J. B. Plescia, S. W. Squyres, and D. E. Wilhelms, The geology of Ganymede, in *Satellites of Jupiter,* edited by D. Morrison, pp. 435–520, University of Arizona Press, Tucson, 1982.

Simonetti, A., K. Bell, and C. Shrady, Trace- and rare-earth-element geochemistry of the June 1993 natrocarbonatite lavas, Oldoinyo Lengai (Tanzania): Implications for the origin of carbonatite magmas, *J. Volcanol. Geotherm. Res., 75,* 89–106, 1997.

Smith, B. A., L. A. Soderblom, R. F. Beebe, J. M. Boyce, G. A. Briggs, M. H. Carr, S. A. Collins, A. F. Cook, G. E. Danielson, M. E. Davies, G. E. Hunt, A. P. Ingersoll, T. V. Johnson, H. Masursky, J. F. McCauley, D. Morrison, T. Owen, C. Sagan, E. M. Shoemaker, R. G. Strom, V. E. Suomi, and J. Veverka, The Galilean satellites and Jupiter: Voyager 2 imaging science results, *Science, 206,* 927–950, 1979.

Smith, B. A., L. A. Soderblom, R. Batson, P. Bridges, J. Inge, H. Masursky, E. M. Shoemaker, R. F. Beebe, J. M Boyce, G. A. Briggs, A. Bunker, S. A, Collins, C. J. Hansen, T. V. Johnson, J. L. Mitchell, R. J. Terrile, A. F. Cook, J. Cuzzi, J. B. Pollack, G. E. Danielson, A. P. Ingersoll, M. E. Davies, G. E. Hunt, D. Morrison, T. Owen, C. Sagan, J. Veverka, R. G. Strom, and V. E. Suomi, A new look at the Saturn system: The Voyager 2 images, *Science, 215,* 504–536, 1982.

Smith, B. A., L. A. Soderblom, R. F. Beebe, D. Bliss, J. M. Boyce, A. Brahic, G. A. Briggs, R. H. Brown, S. A. Collins, A. F. Cook, S. K. Croft, J. Cuzzi, G. E. Danielson, M. E. Davies, T. E. Dowling, D. Godfrey, C. J. Hansen, C. Harris, G. Hunt, A. P. Ingersoll, T. V. Johnson, R. J. Krauss, H. Masursky, D. Morrison, T. Owen, J. B. Plescia, J. B. Pollack, C. C. Porco, K. Rages, C. Sagan, E. M. Shoemaker, L. A. Stromovsky, C. Stoker, R. G.

Strom, V. E. Suomi, S. P. Synnott, R. J. Terrile, P. Thomas, W. R. Thompson, and J. Veverka, Voyager 2 in the Uranian system, *Science, 233*, 43–64, 1986.

Sparks, R. S. J., The dynamics of bubble generation and growth in magmas: A review and analysis, *J. Volcanol. Geotherm. Res., 3*, 1–13, 1978.

Spencer, J. R., Icy Galilean satellite reflectance spectra: Less ice on Ganymede and Callisto? *Icarus, 70*, 99–110, 1987.

Squyres, S. W., and S. K. Croft, The tectonics of icy satellites, in *Satelites*, edited by J. A. Burns and M. S. Matthews, pp. 292–341, University of Arizona Press, Tucson, 1986.

Stebbins, J. F., I. S. E. Carmichael, and D. E. Weill, The high temperature liquid and glass heat contents and the heats of fusion of diopside, albite, sanidine and nepheline, *Am. Mineral., 68*, 717–730, 1983.

Stevenson, D. A., and J. I. Lunine, Mobilization of cryogenic ice in the outer solar system, *Nature, 323*, 46–48, 1986.

Stevenson, D. J., Volcanism and igneous processes in small icy satellites, *Nature, 298*, 142–144, 1982.

Stoppa, F., The San-Venanzo Maar and tuff ring, Umbria, Italy—Eruptive behavior of a carbonatite–melilitite volcano, *Bull. Volcanol., 57*, 563–577, 1996.

Streckeisen, A. L., Classification and nomenclature of volcanic rocks, lamprophyres, carbonatites, and melilitic rocks, *Geol. Rundsch., 69*, 194–207, 1980.

Takahashi, E., and C. M. Scarfe, Melting of periodotite to 14 GPa and the genesis of komatiite, *Nature, 315*, 566–568, 1985.

Thomas, P. J., and S. W. Squyres, Formation of crater palimpsests on Ganymede, *J. Geophys. Res., 95*, 19161–19174, 1990.

Tittemore, W., and J. Wisdom, Tidal evolution of the Uranian satellites: II, *Icarus, 78*, 63–89, 1989.

Treiman, A. H., Nahklites are nifty cool, *LPI Tech. Publ. 88-05*, pp. 127–128, Lunar and Planetary Institute, Houston, 1988.

Treiman, A. H., and A. Schedl, Properties of carbonatite magma and processes in carbonatite magma chambers, *J. Geol., 91*, 437–447, 1983.

Turner, J. S., H. E. Huppert, and R. S. J. Sparks, Komatiites II: Experimental and theoretical investigations of post-emplacement cooling and crystallization, *J. Petrol., 27*, 397–437, 1986.

Viljoen, M. J., and R. P. Viljoen, The geology and geochemistry of the lower ultramafic unit of the Onverwacht Group and a proposed new class of igneous rock, *Spec. Publ. Geol. Soc. S. Afr., 2*, 55–85, 1969a.

Viljoen, M. J., and R. P. Viljoen, Evidence for the existence of a mobile extrusive peridotitic magma from the Komati Formation of the Onverwacht Group, *Spec. Publ. Geol. Soc. S. Afr., 2*, 87–112, 1969b.

Warren, P. H., Lunar Mg-rich rocks as analogs of terrestrial komatiites: Implications of early outgassing of Earth's volatile elements, *Meteoritics, 18*, 417, 1983.

Warren, P. H., and G. W. Kallemeyn, Lunar mare meteorites, in *Abstracts from Mare Volcanism and Basalt Petrogenesis: Astounding Fundamental Concepts Developed Over the Last Fifteen Years*, pp. 55–56, Lunar and Planetary Institute, Houston, 1990.

Warren, P. H., E. A. Jerde, and G. W. Kallemeyn, A spinifex textured mare basalt: Comparison with komatiites, in *Abstracts from Mare Volcanism and Basalt Petrogenesis: Astounding Fundamental Concepts Developed Over the Last Fifteen Years*, pp. 57–58, Lunar and Planetary Institute, Houston, 1990.

Weaver, B. L., and J. Tarney, Thermal aspects of komatiite generation and greenstone belt models, *Nature, 279*, 689–692, 1979.

Weitz, C. M., and A. T. Basilevsky, Magellan observations of the Venera and Vega landing site regions, *J. Geophys. Res., 98*, 17069–17097, 1993.

Williams, D. A., R. C. Kerr, and C. M. Lesher, Emplacement and erosion by Archean komatiite lava flows at Kambalda: Revisited. *J. Geophys. Res., 103*, 27533–27550, 1998.

Wilson, L., J. W. Head, and R. T. Pappalardo, Eruption of lava flows in Europa: Theory and application to Thrace Macula, *J. Geophys. Res., 102*, 9263–9272, 1997.

Wolff, J. A., Physical-properties of carbonatite magmas inferred from molten-salt data, and application to extraction patterns from carbonatite silicate magma chambers, *Geol. Mag., 131*, 145–153, 1994.

Xie, Q., R. Kerrich, and J. Fan, HFSE/REE fractionations recorded in three komatiite–basalt sequences, Archean Abitibi greenstone belt: Implication for multiple plume sources and depths, *Geochim. Cosmochim. Acta, 57*, 4111–4118, 1993.

9

Volcanic Vestiges

Pulling it Together

Tracy K. P. Gregg and James R. Zimbelman

9.1. INTRODUCTION

Commonly, students of volcanology study the effect large volcanic eruptions have on the local and global environment—as well they should, because it is within this arena that society and science meet to determine risk assessment and volcanic hazard mitigation. However, investigating the reverse is equally valuable and intriguing. By constraining how the environment effects the final morphology of volcanic deposits, we can quantitatively use volcanic geomorphology to reveal something about the environment in which the deposits were emplaced. In this way, volcanoes provide information about climate change on Earth (indicating the past location of a lacustrine shoreline, for example, or the previous extent of an alpine glacier; see Chapters 3 and 5) as well as the past and present conditions on other planets.

If a person's eyes can be thought of as windows into the soul, then a volcano can be seen as a window into a planet's deep interior. Until we are technologically capable of drilling to the center of the Earth or any other planet, volcanoes remain the single most important clue to the thermal, physical, and chemical behavior of the interior of a planet. For the Earth and the Moon, we have hand-sample analyses to give us direct information of lava compositions on those planetary bodies (see Chapters 2 and 6); and analyses of meteorites from Mars that have landed on Earth provide limited information of lava compositions there (see Chapter 4), although we cannot pinpoint the spot on Mars from whence these meteorites originated. Therefore, for the vast majority of solid bodies in the solar system, we must infer the lava and magma composition, as well as eruption and emplacement parameters (e.g., rheology, effusion rate, eruption duration), from the resulting volcanic morphologies. To interpret these morphologies accurately, however, "planetary volcanologists" must be able to decipher and remove the environmental effects, much the way a remote sensor analyzing the surface composition of a planet must remove the effects of the atmosphere on the spectra.

Because active eruptions have only been observed on Earth and Io, analytical and numerical models are used to predict the behavior and interpret the morphologies of extraterrestrial and deep-sea volcanic deposits, as well as for those terrestrial, land-based volcanoes that were not observed while active. In this chapter, we provide the reader with some basic numerical models obtained from the literature, as well as information on the thermophysical properties of various lavas and different environments likely to be encountered within the solar system.

9.2. INFORMATION

Most extant numerical models require that the user input various parameters (e.g., surface pressure, gravity, eruption temperature) to solve for the desired unknown (such as effusion rate, viscosity, or yield strength). In this section we include basic information on the thermophysical properties of a range of lava types, as well as the ambient conditions encountered in the solar system.

Table 9.1 lists typical parameters for lava types that have been proposed to exist on the terrestrial planets. Table 9.2 lists the thermophysical properties of the various bodies in the solar system that have been discussed in this book.

9.3. MODELS

Several numerical and analytical models designed to predict or interpret volcanic morphologies have been presented in the literature, and a thorough review would fill a book by itself! Here, we have selected a few of these models for the reader to experiment with. There is no room to document fully each model, so we strongly encourage interested students to go to the original references and investigate the assumptions that are an intimate part of any model. Natural volcanic systems are inherently complex, and currently no models can accurately predict the behavior of every aspect of a volcanic eruption, so almost by

Table 9.1. Thermophysical Properties of Common Lava Types

Eruption temperature (K)	Glass transition temperature (K)	Thermal diffusivity ($m^2\ s^{-1}$)	Eruption viscosity (Pa s)	Unvesiculated density[a] ($kg\ m^{-3}$)	Heat capacity ($J\ kg^{-1}\ K^{-1}$)	Associated composition
1425	1000	5.0×10^{-7}	10^2–10^5	2900	1200	Basalt[b]
1300	900	3.0×10^{-7}	10^5–10^7	2600	1125	Andesite[c]
1200	850	2.0×10^{-7}	10^7–10^9	2500	1050	Dacite[d]
1100	800	1.4×10^{-6}	$\geq 10^9$	2400	1000	Rhyolite[e]

[a] The actual erupted density may be a factor of 2 or 3 lower than listed here. Densities listed here assume no vesiculation. Typical densities for Hawaiian basalts are 1200–1500 $kg\ m^{-3}$ (Keszthelyi and Self, 1998). After Gregg and Fink (1996).
[b] Midocean ridge basalt (Griffiths and Fink, 1992a).
[c] Mount Hood andesite (Murase and McBirney, 1973).
[d] From Anderson and Fink (1992).
[e] Newberry rhyolite (Murase and McBirney, 1973).

Table 9.2. Thermophysical Properties of Environments Encountered in the Solar System[a]

Environment	Temperature (K)	Thermal expansion (K^{-1})	Thermal diffusivity ($m^2\,s^{-1}$)	Kinematic viscosity ($m^2\,s^{-1}$)	Density ($kg\,m^{-3}$)
Moon	4	—	—	—	—
Mercury	440	—	—	—	—
Io	120	—	—	—	—
Mars	200	5.0×10^{-3}	7.2×10^{-4}	6.3×10^{-5}	0.2
Venus	730	1.3×10^{-3}	7.3×10^{-7}	4.5×10^{-7}	62.5
Earth (subaerial)	300	3.4×10^{-3}	2.3×10^{-5}	1.6×10^{-5}	1.2
Earth (submarine)	275	1.5×10^{-4}	1.0×10^{-7}	1.0×10^{-6}	1000

[a]After Griffiths and Fink (1992b) and Gregg and Fink (1996).

definition, a numerical model seeks to determine which parameters of the volcanic system are dominant. In what first appears to be a case of severe circular reasoning, however, each model must begin by making reasonable assumptions about which parameters can be neglected to simplify the system and make it tenable.

In this somewhat historical review, we begin by presenting models for lava flow emplacement, followed by the generation of explosive eruptions, Plinian eruption columns, and pyroclastic flows. In each case, the interested reader is strongly encouraged to find the original papers.

9.3.1. Lava Flow Emplacement

Modeling the emplacement of lava flows has advanced considerably in the past 25 years or so, but because lavas are complex mixtures of liquids, solids, and gases, it may well be another 25 years before a "universal" model of lava behavior can be adequately derived. Typically, the terrestrial volcanologist, investigating the emplacement of active flows, uses these models to determine properties that are the most difficult to measure in the field—commonly, lava viscosity. Similarly, the extraterrestrial volcanologist has only the solidified lava flow, and, most commonly, wishes to be able to determine the lava composition based on easily measured parameters (such as flow length, width, and thickness). No extant model yields lava composition, but many give yield strength and/or lava viscosity, and much research has been devoted to relating these properties to composition.

Nichols (1939) introduced Jeffrey's equation, given by

$$\eta = \frac{g\rho \sin\theta d^2}{nu} \quad (1)$$

where η is lava viscosity, g is gravity, ρ is lava density, θ is underlying slope, d is flow thickness, u is lava flow velocity, and n is an empirical constant that equals 3 for a broad flow and 4 for narrow flows. This model assumes a Newtonian rheology, and, therefore, is unlikely to be appropriate for lavas with a high yield strength, such as dacites and rhyolites (see Williams and McBirney, 1979, and Cas and Wright, 1987, for a more general discussion of lava rheology). This model has long been used to determine lava viscosities for extraterrestrial flows, assuming a "reasonable" flow velocity and density. At the simplest level, the problem with this application is twofold, however. First, reasonable flow velocities usually reflect observations made for Hawaiian lavas; it is important to note that Hawaiian volcanism is but one of many styles on Earth and may not be "typical" when compared with volcanic activity on other planets. Second, it is known that lava flow velocity varies with time (reflecting changes at the vent, or local changes in flow geometry, for example) and along the length and width of a flow. Although Jeffrey's equation may do an adequate job of estimating viscosity for active flows (where flow velocity, underling slope, thickness, and density can be accurately measured), but it can only produce estimates as accurate as are the assumptions for lava properties in the extraterrestrial context.

Hulme (1974) assumed a Bingham rheology to explain the formation of leveed lava flows on Earth and the Moon. He used the following relations to constrain the yield strength of leveed flows:

$$\Phi = \tfrac{2}{15} W^{2.5} - \tfrac{1}{4} W^2 + \tfrac{1}{6} W - \tfrac{1}{20} \quad (2a)$$

where Φ is a dimensionless quantity relating flow rate to liquid properties and external forces, and W is given by

$$W = \frac{w}{2 w_b} \quad (2b)$$

where w is the half-width of a channelized flow and w_b is the width of the stationary levee. The quantity Φ can be related to yield strength through the relation

$$\Phi = Q \eta (g \rho)^3 \left(\frac{\theta}{\tau_y} \right)^4 \quad (2c)$$

where Q is volumetric flow rate, τ_y is lava yield strength, and other variables are as previously defined. Thus, by measuring the channel width and levee widths, and making "reasonble assumptions" about effusion rate, density, and lava viscosity, a yield strength could be obtained. Probably the single most difficult assumption to prove here is that the observed levees are simple levees, and were not formed through repeated lava overflows, or through accretion of rubble along the flow margin (cf. Sparks *et al.*, 1976).

Zimbelman (1985) used this model, and others also assuming a Bingham rheology, to constrain the yield strength of lavas on Ascraeus Mons, Mars. The relations he used are (Moore *et al.*, 1978)

$$\tau_y = \rho g d \sin \theta \quad (3a)$$

$$\tau_y = \frac{\rho g d^2}{2w} \quad (3b)$$

$$\tau_y = \rho g (2w - w_b) \sin^2 \theta \quad (3c)$$

where d is flow thickness. Although it is clear that these relations provide the user with "a number," it is less clear precisely what this number reflects. Recent work by Peitersen and Crown (1998) shows that flow width ($2w$) varies dramatically along the length of a single basalt flow. It may be that these relations give some sort of "average" value for the yield strength of the entire flow. In other words, by using the above relations to compare different lava flows, you may be able to say something about the relative behavior of those two flows— but it is not obvious exactly what that is. Additionally, the reader should note the incompatibility between these three equations (leading to absurd equalities for some parameters). This is a direct consequence of the assumptions involved in the derivation of each equation, which vary drastically. Again, the utility of the results are only as good as the assumptions of each derivation.

It is important to note that these relations implicitly rely on a dimensionless parameter called the Graetz number, which is given by

$$\text{Gz} = \frac{Qd}{Dxw} \tag{4}$$

where d is flow depth, D is lava thermal diffusivity, and x is flow length. Observations of Hawaiian flows indicate that most basalt lavas there stop advancing when $\text{Gz} = 300$, which is roughly the time during which the solidified surface crust of a flow has a thickness of approximately $1/3d$ (e.g., Pinkerton and Wilson, 1994). However, it is unlikely that this relation can be blindly applied to lavas of other compositions in other environments, because this limiting number ($\text{Gz} = 300$) is an empirical relation derived from Hawaiian lava flows.

Cooling rate clearly plays an important role in the final morphology of lava flows: The presence of lava pillows, generated only in subaqueous conditions, attests to that (see Chapter 5 and Gregg and Fink, 1995). Many workers have attempted to model the cooling rate of lava flows to better predict and interpret downstream changes in lava rheology and flow length (e.g., Crisp and Baloga, 1990; Pinkerton and Wilson, 1994; Fink and Griffiths, 1990). Other researchers have been concentrating on constraining the cooling rate of lava flowing within a lava tube (e.g., Sakimoto and Zuber, 1998; Keszthelyi and Self, 1998). Intuitively, the faster a lava flow cools, the shorter and thicker the resulting flow should be. However, a rapidly cooled lava flow may actually travel farther than a slowly cooled flow because the presence of a solid surface crust insulates the molten core of a flow from additional heat loss. Gregg and Fornari (1998) show that, for identical lava flows, basalts emplaced on the seafloor will travel approximately 30% farther than those on Earth's surface because of the enhanced cooling on the seafloor.

The rate of cooling of a lava flow is controlled by the ambient conditions, the lava eruption temperature, viscosity, and velocity (a proxy for the rate of heat advection within a flow) (e.g., Crisp and Baloga, 1990; Fink and Griffiths, 1990). For subaerial terrestrial flows, Martian flows, as well as lunar, Mercurian, and Ionian lavas, the primary cooling mechanism is radiative cooling. The equation for heat flux from a radiating lava flow is given by

$$F_r = \varepsilon\sigma\{(1-f)(T^4 - T_a^4) + f(T_{sc}^4 - T_a^4)\} \tag{5}$$

where F_r is radiative heat flux from the flow; ε is lava emissivity; σ is the Stephan–Boltzmann radiative constant; f is the fraction of the flow covered with a solidified crust; T is temperature, and subscripts "a" and "sc" indicate "ambient" and "surface crust". Without even plugging in the numbers, it is evident that for basaltic lavas on Earth, the ambient temperature can be safely neglected, but that on Venus, it might be worthy of consideration.

For lavas emplaced in sufficiently thick atmospheres (such as the deep seafloor or the surface of Venus), convective cooling plays an important role. The heat flux from a connectively cooled body is given by (Fink and Griffiths, 1990; Gregg and Greeley, 1993)

$$F_c = \rho_a c_a \gamma \left(\frac{\alpha_a g D_a^2}{v_a}\right)\{(1-f)(T - T_a^{4/3})^{1/3} + f(T_{sc} - T_a)^{4/3}\} \tag{6}$$

where ρ is lava density, c is heat capacity, γ is an empirical constant equal to 0.1 (Turner, 1973), α is the coefficient of thermal expansion, g is gravitational acceleration, D is thermal diffusivity, v is kinematic viscosity, and other variables are as previously defined.

Griffiths and Fink (1992a) show that for lava flows on the seafloor, convective cooling dominates throughout lava flow emplacement. On land, radiative cooling dominates until the surface temperature of the lava flow $\leq 250°C$ (significantly lower than the solidification temperature of silicate lavas; Table 9.1). Convective cooling is the dominant process on Venus until the lava surface temperature falls below ~750°C, when radiative cooling begins to take over.

Once a surface crust has formed over a lava flow, the molten interior of the flow cools by thermal diffusion through the overlying crust (Crisp and Baloga, 1990). This is also the primary mechanism for cooling lava within lava tubes (e.g., Sakimoto and Zuber, 1998; Kesztheyli and Self, 1998). Therefore, although the formation of lava tubes may be preferentially enhanced on Venus and the seafloor (see Chapter 5) in comparison with other volcanic environments, once a tube forms, the behavior of the lava varies little in different ambient conditions. Gravity helps to control the velocity at which lava is capable of flowing within the tube, and also controls the width of an unsupported lava tube roof once the lava had drained away (Oberbeck et al., 1969). Thermally, however, lava tubes should behave similarly under various ambient conditions.

9.3.2. Explosive Eruptions

Pyroclasts can be generated by the violent depressurization of magmatic volatiles, or by the interaction of hot magma with near-surface ground water, ground ice, or shallow lakes, ponds, or seas. On Earth, H_2O and CO_2 are the most common magmatic volatiles (e.g., Cas and Wright, 1987; Sparks et al., 1997); although CO may dominate on the Moon (see Chapter 6) and CO_2 may prevail on Venus and Mars (see Chapters 4 and 5).

A combination of cooling rate, effusion rate, and lava viscosity may exert the strongest controls on the final morphology of effusive lava flows (Fink and Griffiths, 1990), although only one of these—cooling rate—is determined by the ambient conditions. The production and distribution of pyroclastic deposits, however, is closely tied to the atmospheric pressure and temperature, as well as the intrinsic magmatic properties such as volatile content, viscosity, and temperature. The surface pressures of the different environments examined in this book vary widely—from the vacuum of space to the intense pressure (>250 MPa) on the deep seafloor.

Models for the generation and emplacement of pyroclastic deposits have focused on understanding the generation and subsequent collapse of Plinian eruption columns (e.g., Sparks, 1978; Sparks et al., 1997) because these types of eruptions are historically most closely associated with loss of life and property damage (e.g., Tilling, 1989). Wilson et al. (1978) showed that the maximum height of a Plinian eruption column can be predicted from

$$H_p = 8.2 R^{1/4} \qquad (7a)$$

where R is the steady-state energy release in watts and is given by

$$R = \rho v \pi r^2 c (T - T_a) E \qquad (7b)$$

where ρ is the bulk density of the erupting fluid (a mixture of solids, liquid, and vapor), v is velocity, r is the vent radius, c is specific heat, T is temperature, and E is an "efficiency factor" that measures how efficiently heat is converted to potential or kinetic energy. Results from recent modeling show that the type of gas has a strong influence on the eruption column height. For example, Campbell et al. (1998) demonstrated that it would be very difficult to sustain a CO_2-dominated eruption column on Venus—the high density of the column would cause it to readily collapse (see Chapter 5). In contrast, SO and SO_2 have relatively low densities, contributing to the large eruption plumes observed on Io.

Additionally, the lower the surface pressure, the greater the amount of gas is able to exsolve from the magma by the time the magma reaches the surface—which should result in a greater range for the pyroclasts (cf. Wilson and Head, 1981). A thin or nonexistent atmosphere will also exert less drag on any ejected magma fragments, resulting in a greater dispersal for given eruption conditions. The range, X, of pyroclasts can be obtained as (Wilson and Head, 1981)

$$X = r - \frac{v}{g}\left(\frac{dr}{dh}\right)\left[v + \sqrt{v^2 - 2gh}\right] \tag{8}$$

where X is the range (distance from the vent), r is vent radius, v is rise velocity, and (dr/dh) is the slope of the vent wall at the fragmentation level at depth h in the conduit. Clearly, for a given set of eruption conditions, planets with gravity lower than Earth's will have more widely dispersed pyroclastic deposits (see Chapters 4, 6, and 7).

9.4. MAGMA INTRUSION

The means by which magma reaches the surface is a vital part of the volcanic system. For example, near-surface dikes have created spectacular landforms on Venus (see Chapter 5), and apparently created some linear rilles on the Moon (see Chapter 6). The formation of dikes and of magma storage systems within or beneath the crust (see Chapter 5) is apparently more closely controlled by density differences between the host rock and the magma and by the pressurization within the magma storage system, than by any "atmospheric" or superficial parameters (e.g., Wilson and Head, 1981; Ryan, 1994, and references therein). Although gravity weakly enters into most models designed to predict the ascent rate of magma within a dike of given dimensions, it is overshadowed by density differences and regional stresses (e.g., Wilson and Head, 1981; Ryan, 1994). And while a basaltic dike rising through hydrothermally cooled oceanic crust may cool more rapidly than an identical dike ascending through the hot Venusian crust, again, controls other than temperature (e.g., lava viscosity, velocity, and relative density) are more important.

9.5. SUMMARY AND CONCLUSION

Historic eruptions have clearly revealed how volcanism can strongly impact local and global climate, and much research has been devoted to predicting and understanding this phenomenon. For those interested in deciphering the volcanic clues left enticingly on the surfaces of other planets, however, investigating the converse relation—how the local and global environment effects eruption dynamics—is essential. Volcanoes have been identified on solid bodies throughout the solar system, but Earth remains the only place where actively flowing lava has been clearly observed, and Io is the only other place in the solar system where active volcanism has been witnessed. Thus, it is imperative that we learn to interpret extraterrestrial and submarine volcanic morphologies by first quantifying and removing the specific environmental effects.

Within this book, we have presented volcanism in the cold vacuum of space (e.g., Earth's Moon and Io; see Chapters 6 and 7), the wispy-thin atmosphere of Mars (Chapter 4), beneath tens to hundreds of meters of ice (Chapter 3), at the dark, cold bottom of Earth's oceans and on the scorching surface of Venus (Chapter 5). These chapters introduce an astonishing array of volcanic morphologies and eruptive styles, and when one considers the manifestations of "exotic" and "ice" lavas (Chapter 8), the morphologic possibilities are virtually endless.

We have tried to demonstrate that the kind of environment a volcano erupts in is just as important as many of the more commonly studied variables, such as lava composition and rheology. The models presented in this chapter are included so that the interested readers can insert the appropriate values to see for themselves the role ambient conditions play in the emplacement of volcanic deposits. Although this book contains a wealth of information on the environmental effects on volcanic eruptions, there is still a vast amount of research to be done. Until we, as volcanologists, are able to do thorough fieldwork on the surfaces of other planets, we will never know for sure if our models and interpretations are correct. And until then, we will continue to observe, model, and test—only to be required to observe everything again from a new perspective generated by ongoing analyses and missions.

9.6. REFERENCES

Anderson, S. W., and J. H. Fink, Crease structures: Indicators of emplacement rates and surface stress regimes of lava flows, *Geol. Soc. Am. Bull., 104*, 615–625, 1992.

Campbell, B. A., L. Glaze, and P. G. Rogers, Pyroclastic deposits on Venus: Remote-sensing evidence and modes of formation, *Lunar Planet. Sci. Conf., XXIX*, #1810, 1998.

Cas, R. A. F., and J. V. Wright, *Volcanic Successions: Modern and Ancient*, 528 pp., Allen & Unwin, London, 1987.

Crisp, J., and S. M. Baloga, A model for lava flows with two thermal components, *J. Geophys. Res., 95*, 1255–1270, 1990.

Fink, J. H., and R. W. Griffiths, Radial spreading of viscous-gravity currents with a solidifying crust, *J. Fluid Mech., 221*, 485–509, 1990.

Gregg, T. K. P., and J. H. Fink, Quantification of submarine lava-flow morphology through analog experiments, *Geology, 23*, 73–76, 1995.

Gregg, T. K. P., and J. H. Fink, Quantification of extraterrestrial lava flow effusion rates through laboratory simulations, *J. Geophys. Res., 101*, 16891–16900, 1996.

Gregg, T. K. P., and D. J. Fornari, Long submarine lava flows: Observations and results from numerical modeling, *J. Geophys. Res., 103*, 27517–27532, 1998.

Gregg, T. K. P., and R. Greeley, Formation of Venusian canali: Considerations of lava type and their thermal behaviors, *J. Geophys. Res., 98*, 10873–10882, 1993.

Griffiths, R. W., and J. H. Fink, Solidification and morphology of submarine lavas: A dependence on extrusion rate, *J. Geophys. Res., 97*, 19729–19737, 1992a.

Griffiths, R. W., and J. H. Fink, The morphology of lava flows in planetary environments: Predictions from analog experiments, *J. Geophys. Res., 97*, 19739–19748, 1992b.

Hulme, G., The interpretation of lava flow morphology, *Geophys. J. R. Astron. Soc., 39*, 361–383, 1974.

Keszthelyi, L. P., and S. Self, Some physical requirements for the emplacement of long basaltic lava flows, *J. Geophys. Res., 103*, 27447–27464, 1998.

Moore, H. J., W. G. Arther, and G. G. Schaber, Yield strengths of flows on the Earth, Mars, and Moon, *Proc. Lunar Planet. Sci. Conf., 9th*, 3351–3378, 1978.

Murase, T., and A. R. McBirney, Properties of some common igneous rocks and their melts and high temperatures, *Geol. Soc. Am. Bull., 84*, 3563–3592, 1973.

Nichols, B. L., Viscosity of lavas, *J. Geol., 47*, 290–302, 1939.

Oberbeck, V. R., W. L. Quade, and R. Greeley, On the origin of lunar sinuous rilles, *Mod. Geol., 1*, 75–80, 1969.

Peitersen, M. N., and D. A. Crown, Correlations between topography and intraflow width behavior in Martian and terrestrial lava flows, *Lunar Planet. Sci. Conf., XXIX*, #1382, 1998.

Pinkerton, H., and L. Wilson, Factors controlling the lengths of channel-fed lava flows, *Bull. Volcanol., 56*, 108–120, 1994.

Ryan, M. P., Neutral-buoyancy controlled magma transport and storage in mid-ocean ridge magma reservoirs and their sheeted-dike complex: A summary of basic relationships, in *Magmatic Systems*, edited by M. P. Ryan, pp. 97–138, Academic Press, San Diego, 1994.

Sakimoto, S. E. H., and M. T. Zuber, Flow convection cooling in lava tubes, *J. Geophys. Res., 103*, 27465–27488, 1998.

Sparsk, R. S. J., H. Pinkerton, and G. Hulme, Classification and formation of lava levees on Mount Etna, Sicily, *Geology, 4*, 269–271, 1976.

Sparks, R. S. J., The dynamics of bubble formation and growth in magmas: A review and analysis, *J. Volcanol. Geotherm. Res., 3*, 1–37, 1978.

Sparks, R. S. J., M. I. Bursik, S. N. Carey, J. S. Gilbert, L. S. Glaze, H. Sigurdsson, and A. W. Woods, *Volcanic Plumes*, 574 pp., Wiley, New York, 1997.

Tilling, R. I., Introduction and overview, in *Volcanic Hazards*, edited by R. I. Tilling, pp. 1–8, American Geophysical Union, Washington, DC, 1989.

Turner, J. S., *Buoyancy Effects in Fluids*, 368 pp., Cambridge University Press, London, 1973.

Williams, H., and A. R. McBirney, *Volcanology*, 397 pp., Freeman and Cooper, San Francisco, 1979.

Wilson, L., and J. W. Head, Ascent and eruption of basaltic magma on the Earth and Moon, *J. Geophys. Res., 86*, 2971–3001, 1981.

Wilson, L., R. S. J. Sparks, T. C. Huang, and N. D. Watkins, The control of eruption column heights by eruption energetics and dynamics, *J. Geophys. Res., 83*, 1829–1836, 1978.

Zimbelman, J. R., Estimates of rheologic properties for flows on the Martian volcano Ascraeus Mons, *Proc. Lunar Planet. Sci. Conf. 16th, Part 1, J. Geophys. Res., 90*, D157–D162, 1985.

Index

A'a, 18–19, 20, 31, 103, 134, 210–212
Abitibi, Canada, 215
Ablation, 218
Acala, Io, 86, 194, 197
Acidalia Planitia, Mars, 59–61, 102
Advection, 3
Aeolis, Mars, 102
Africa, 208
Aidne, Io, 188
Alaska, USA, 50, 55
Alba Patera, Mars, 64, 88, 90, 93–95, 102, 104
Albedo, 39, 145, 146, 148, 151, 154–156, 167–169, 171, 197
Alkalic, alkaline, 116, 208–211, 216
Alphonsus crater, Moon, 148, 149, 150, 151, 163
Altimetry, 114–115, 130
Altjirra, Io, 188
Amazonis Planitia, Mars, 63
Amirani, Io, 188, 196
Ammonia, 5, 222–223, 232–233
Amphitrites Patera, Mars, 97–98
Andes Mountains, Earth, 41
Andesite, 14, 18, 22, 25, 41, 50, 56, 79, 82, 83, 244
Anhydrite, 208, 214
Ankarite, 208
Anorthosite, 161, 167, 170
Antarctica, 39, 50–55
Apennine Bench, Moon, 171
Apollinaris Patera, Mars, 42, 64, 88, 95
Apollo, 9, 144, 146, 151, 152, 154, 161, 165
  spacecraft, 219
Apophyse, 46
Arachnoid, 209, 212–213
Archean, 201, 214–219
Ares Vallis, Mars, 60, 64
Ariel, 221, 231–234
Arisa Mons, Mars, 87–91, 102, 106, 246
Aristarchus Plateau, Moon, 147, 149, 150, 151, 152
Ascraeus Mons, Mars, 88, 91, 92
Armero, Columbia, 41
Ash, 10, 24–28, 31, 90, 96, 125, 182, 208–212
  fall, 90
Askja caldera, Iceland, 14

Asterius Linea, Europa, 226
Asthenosphere
  Earth, 11
  Io, 197–201
Atla Regio, Venus, 116
Atmosphere, dense (Venus), 4, 113–115, 133, 248
Augustine, USA, 55
Aureole, Olympus Mons, Mars, 42, 67–68, 96
Australia, 215–217
Avalanche, 54–56
Axarfjörður, Iceland, 57

Backscatter, 122, 134
Bacteria, 47
Balloons (Venus), 116
Barberton, South Africa, 214–215
Barberton-type komatiite, 216
Bárdarbunga, Iceland, 57
Basalt, basalts, basaltic, 3–4, 9–10, 12, 14, 16, 18, 20, 22, 24, 28, 29, 31, 43, 46, 50, 59, 76, 79, 82–83, 86–88, 95, 113–116, 124–125, 127, 134, 136, 144, 146, 155–156, 159–153, 165, 167, 168, 173, 182, 188–191, 207, 210–220, 231–234, 244, 247, 249
Basaltic andesite, 14, 18, 20, 42, 88
Base surge, 46
Bathymetry, 122, 130
Belingwe, South Africa, 215
Belus Linea, Europa, 226
Beta–Atla–Themis region, Venus, 116, 122, 129, 197
Betts Cove, Canada, 215
Biblis Patera, Mars, 76
Bingham: *see* Rheology, Bingham
Bláhnúkur, Iceland, 47
Blocky flow: *see* Flow, blocky
Boulder, 55, 57
Bosphorus Regio, Io, 197
Bomb, 25, 26
  breadcrust, 211
Breccia, 49, 54, 61
Brightness temperature, 185
Brine, 223
British Columbia, Canada, 50

253

Brittle–ductile transition, 85, 189
Brown Bluff, Antarctica, 50–55
Bulk density, 27, 83, 86, 145

Cadmus Linea, Europa, 226
Calciocarbonatite, 208–209
Calcite, 208, 21
Calcium (Ca), 181, 201, 209–210
Caldera, calderas, 20, 89, 90–99, 101–102, 106–107, 118, 126, 129, 131, 150, 156, 164, 182–184, 187, 191, 196–197, 202, 221, 229–232
Callisto, 2, 223
Caloris basin, Mercury, 166, 168
Canali: see Channels
Canary Islands, 131
Cape Smith, Canada, 215
Carbon dioxide (CO_2), 11, 85, 115, 124, 128, 136, 181, 192, 210–211, 222, 233
Carbonatite, 14, 207–214
Carbon monoxide, 151, 162, 164
Carbowax, 6
Cassini spacecraft, 233–234
Cauldron, subsidence, 45, 51, 57
Cayley Formation, Moon, 154
Ceraunius Tholus, Mars, 64, 76., 89, 90, 95
Channeled Scabland, USA, 57, 214
Channel, channels, 118–119, 134–135, 188, 207, 211–213, 216–220
Chondritic, 201–216
Chromite, 216
Chryse Planitia, Mars, 59
Cinder cone: see Cone, cinder
Cirque, 51–52
Clathrate, 222
Clay, 54
Clementine spacecraft, 151, 152, 161
Climate, 7, 32, 48, 75, 95
Clinopyroxene, 10
Cobbles, 55
Color units, Mercury, 166–167
Columbia River Basalts, USA, 15, 17–18, 28–30, 148, 156, 217
Composite volcano, 20–21, 40, 79, 209
Cone, 118, 130–131, 147, 154, 163, 196, 209, 221
 ash, 212
 cinder, 20, 22, 59, 86, 102
 littoral, 58
 spatter, 20, 59, 86
 tuff, 24, 51–53
Conglomerate, 51–54
Convection, 3
Corona, coronae, 114, 118, 232
Corrie: see Cirque
Crater Peak, USA: see Mount Spurr, USA
Creep rate, 116
Cretaceous, 216

Crisium basin, Moon: see Mare, Moon
Crixas, Brazil, 215
Creidne Patera, Io, 183
Crust
 brittle, 117
 Earth, 126–127, 133, 143
 flow, 134–135, 189, 223
 Io, 179, 183–184, 192–194, 199–203
 Moon, 143–146, 161–162, 165, 172
 oceanic, 113, 122–123, 126–127, 201
 small icy satellites, 230–231
 Venus, 128, 131, 137, 214
Cryovolcanism, 207, 220–234
Cryptodomes, 23
Cryptomaria, 147, 154–156, 163
Crystalized beads, 151–152
Culann, Io, 180, 188
Cumulate, 214, 216–217
Cupola lakes, 57–58

Dacite, 14, 18, 20, 22, 26, 84, 233, 244, 245
Dalsheidi, Iceland, 46
Dao Vallis, Mars, 64
Dark halo craters, Ganymede, 228
Dark-halo craters, Moon, 151, 155, 163
Dark mantle deposits, Moon, 147, 150, 153, 163
Debris flow, 42, 58
Deccan Traps, India, 28–29
Deflation, 118
Degas crater, Mercury, 167
Density, 2, 11, 18, 54, 75, 78, 85, 85, 87, 105, 113, 115, 126–128, 144–146, 156, 164, 165, 171, 173, 179–181, 189, 201, 210, 221–222, 230, 244–249
Descartes region, Moon, 154
Deuteronilus Mensae, Mars, 59
Diabase, 116
Diapir, diapirism, 11, 15, 85, 155, 158, 224–226, 230
Diatreme, 210
Dike, 11, 15, 16, 30, 85, 87, 122, 125, 156, 158–159, 161–165, 171–172, 210, 216, 249
Dike swarms, radiating, 15, 118, 122, 126, 128–129
Dione, 221, 230
Dissipation factor, 181–182, 197–198
Dolomite, 208
Dome, domes, 18, 20, 22–24, 27, 79, 118, 125, 129–132, 150, 154, 155, 164, 188, 210, 221, 224–227, 230–233
Ductile, 117

Earth, 2–5, 9–74, 75, 79, 85, 86, 91, 95, 113–137, 145, 147, 150, 161, 165, 179, 181, 186, 188, 190–192, 196, 201–202, 214–221, 226, 233–234
Earthquake, 41
East Pacific Rise (EPR), 13, 123, 125–127, 131–132
Eclipse, 186, 194, 196–197

Index

Effusion rate, 4, 6, 13, 20, 22, 30, 85–87, 102, 132–134, 148–149, 150, 161, 163, 172, 243, 246, 258,
Eifel, Germany, 210
Elsinore Corona, Miranda, 232
Elysium Fossae, Mars, 61
Elysium
  Mons, Mars, 41, 60–61, 63–64, 76, 88–91, 103
  Planitia, Mars, 76, 93, 102, 104
Emissivity, 114–115
Enceladus, 2, 221–222, 230–234
Endogenous, 118
Englacial, 47, 50–51
Entropy, 192–194
Euboea Montes, Io, 183
Europa, 2, 181, 197, 221–227
Europa Orbiter Mission, 234
Exsolution, 116, 151, 163, 249

Fen Ring Complex, Norway, 208
Ferrocarbonatite, 208
Finland, 215
Fire fountain, fire fountains, fire-fountaining, 11, 12, 27, 86, 102, 152
Flamsteed region, Moon, 159
Flood lavas, 15, 102, 133, 156, 158, 165, 170, 216–217
Floor-fractured craters, 150
Flow
  blocky, 18–19, 211
  festooned, 79
  field, 18, 22, 135–136, 211
  front, 5, 147–148, 171
  margin, 103, 230
Fluorine, 27, 211
Fluvioglacial, 47
Foreset breccia, 46
Fractional crystallization, 11, 14, 16, 87, 127, 144
Frigoris: *see* Mare, Moon
Fumarole, 42, 183, 196

Gabbro, gabbros, gabbroic, 168
Gaesafjöll, Iceland, 61, 63
Galapagos Islands, Ecuador, 20, 58
Galilean satellites: *see* Io, Europa, Ganymede, Callisto
Galileo Regio, Ganymede, 228
Galileo spacecraft, 116, 179–188, 152, 191, 196–198, 201–202, 223–230
Gamma-ray spectrometer, 115
Ganymede, 2, 181, 221, 223, 227–231, 234
Gardar rift, Greenland, 210
Gassendi crater, Moon, 148
Geothermal gradient: *see* Thermal gradient
Geyser, 192–194
Gish Bar, Io, 188
Gjálp, Iceland, 42–46, 50, 57
Glacier, 39–40, 44–46, 50, 55–57

Glass, 3, 44, 46, 124, 153, 211, 216
Glass beads, 151–152
Glass, 3, 153
GLORIA, 122
Gorgona Island, Columbia, 215
Goya Formation, Mercury, 168
Graben, 122
Graëtz number, 247
Grainflow, 54
Gravel, 52, 57
Gravitational sliding, 46
Gravity, 2, 3, 59, 84, 85, 87, 102, 143, 146, 151, 153, 165, 171–173, 179–181, 221–222, 244, 247, 248
Gravity anomaly, 122
Greenhouse, atmospheric, 7, 65
Greenland, 39
Greenstone belt, 215–216, 219
Gregoryite, 208
Grímsvötn, Iceland, 43–45, 57–58
Gruithuisen domes, Moon, 154–155

Hadriaca Patera, Mars, 64, 76, 97–98, 100–102
Hawaii, USA, 6, 18–21, 30, 50, 58, 96, 128, 131, 134, 161, 163, 197, 217–218, 247
Hawaiian
  fountains, 11, 152
  Emporer-Seamount chain, 16
  style eruptions, 26, 152, 163, 165
  thoeliite, 10
Heat capacity, 189, 211, 244, 247
Heat flow, 66, 181
Hecates Tholus, Mars, 64, 76, 89–91, 95–96, 105
Helium (He), 198
Herdubreidartögl, Iceland, 61–62
Hesperia Planum, Mars, 103
Hi'iaka, Io, 188
Highlands
  Moon, 144–145, 151, 154–156, 164, 167, 171
  Venus, 114, 124
Hlidarfjall, Iceland, 47
Holocene, 57
Homer
Hot spot, 10, 12, 15–17, 122, 182, 185–188, 191, 196–199
Höttur, Iceland, 47
Hubble Space Telescope spacecraft, 192
Humboldtianum: *see* Mare
Humorum: *see* Mare
Hyaloclastic, hyaloclastite, 40–55, 60–63, 68, 125
Hydrothermal, 41–44, 46, 54, 64–68, 125–126
Hypsometry, 114

Iapetus, 221
Ice, 1, 7, 39–68, 86–87, 102, 104, 220–234
Iceland, 13, 14, 39–74, 150

Icelandite (Ch 4), 14, 82
Igmimbrite, 26–28, 31, 137
Igwisi Hill, Tanzania, 210
Ilmenite, 153, 167
Imbrium basin, Moon, 155, 158, 160
Immiscibile, immiscibility, 208–209, 219
Impact crater, 133, 182, 228–233
Incidence angle, 114
Inferior conjunction, 114
Inflated, inflation, 6, 29–30, 103, 118, 134, 211
Infrared Interferometer Spectrometer (IRIS), Voyager spacecraft, 185–186, 189
Intrusiive, intrusion, 66, 113, 118, 125, 128, 208–210, 224
Io, 1–7, 179–203, 220, 222
Iopolis Planum, Io, 183
Iron (Fe), 78–82, 143, 167, 168, 171, 201, 209
  ferrous (II)(FeO), 79, 82, 168, 169, 170, 171
  ferric (III) (Fe_2O_3), 79
  maturity parameter, 168, 170
  sulfide (FeS), 78
Iridium crater, Moon, 155
Ishtar Terra, Venus, 135
Isum, Io, 188

Jan Mayen, Norwegian Sea, 45
Janus, Io, 188
Jeffrey's equation, 245, 246
Jökulhlaup, 40, 43–44, 49, 55–58, 60–64, 68
Jökulsá á Fjöllum, Iceland, 57
Jovis Tholus, 76
Juan de Fuca Ridge (JFR), 13, 123, 125
Jupiter, 179–182, 185–187, 191, 196–197, 220–221, 223, 230

Kalfstindar, Iceland, 47, 49, 61
Kambalda, Australia, 219
Kanehekili, Io, 188
Katla, Iceland, 57–58
Kenya, 208
Kilauea, USA, 18–21, 161, 164, 181, 202, 234
Kimberlite, 208, 210
Kirkjufell, Iceland, 47
Kola Peninsula, 210
Komatiite, komatiites, komatiitic, 3, 30, 83, 103, 171, 201–202, 207, 211, 214–220
Krakatau, 31
KREEP, Moon, 154, 164
Kuiper–Muraski crater complex, Mercury, 169
Kverkfjöll, Iceland, 57

Lacuastrine, 47
Lahar, 28, 40–42, 55–57, 60–63, 68, 104
Lake Myvatn, Iceland, 24, 59
Laki, Iceland, 6, 14, 22, 30–31, 150, 191

Lander spacecraft: see individual names
Landslide, 56
Langjökull, Iceland, 48
Lapilli, 25, 210
Large Igneous Province (LIP), 10, 28–30, 215, 217
Laugarvatn, Iceland, 52
Lava, 10, 32
  composition, 4–5, 82–84, 180, 189–190, 195–196, 201–202, 245
  flow, 13, 17–18, 20, 27–28, 54–55, 79, 83, 88, 91, 93, 95, 97, 100, 102–104, 106, 113, 118, 132–135, 147, 150, 160, 172, 183, 186–189, 245–248
  fountain, 11, 189–90; see also Fire-fountaining; Hawaiian fountains
  inflated flow: see Inflated, inflation
  lake, 188–192
  pillow, 7, 18, 24, 40, 43, 47–49, 51–53, 55, 247
  plains, 89, 95, 100, 102–104, 118
  ponds, 156–159
  tube, 7, 22, 87–88, 95, 103, 134–135, 148–149, 211, 218, 247
Lenticula, 224–226
Lermontov, Mercury, 168, 170
Lherzolite, 10, 79
Lichtenberg crater, Moon, 147
Life, evolution of, 223
Life on Mars, 41, 65
Light plains deposits, Moon, 154–155
Lithofacies, 52–54
Lithosphere
  Earth, 11, 16
  Ganymede, 228
  Io, 181–184, 198–202
  Mars, 84–85
  Mercury, 145
  Moon, 144
  Venus, 130
Littoral cone: see Cone, littoral
Loki Patera, Io, 186, 188, 191, 197–198
Luna spacecraft, 144, 146, 152, 161

Maar, maars, 24, 58–59, 64, 86, 102, 210
Maasaw Patera, Io, 183
Magellan spacecraft, 114–122, 127, 129, 133–136, 207, 220
Magma ocean, 198–202
Magma reservoir (or chamber), 11, 64, 85, 87, 90, 91, 95, 107, 118, 126–129, 137, 150, 156, 158, 161, 163–165, 170–171, 172, 249,
Magnesiocarbonatite, 208
Magnesium (Mg), 201, 209, 214–216
Magnetic field
  Io, 203
  Moon, 145
Malik, Io, 188

Index

Mantle
   Earth, 15–18, 79, 201, 208–210, 214–216
   Europa, 223–225
   Ganymede, 228–229
   Io, 179, 181–182, 185, 190, 197–202
   Mars, 78–79, 83–84, 220
   Mercury, 168, 170
   Moon, 144–145, 160–161, 163
   small icy satellites, 230
Marduk, Io, 188
Mare, maria, 102, 144–146, 151, 154–156, 158–161, 163–167, 171–173, 216, 219
   Australe, 159–160
   Crisium, 146, 159
   far side, 146
   Fecunditatus, 146
   Frigoris, 159
   Humboldtainum, 160, 171
   Humorum, 146, 159
   Imbrium, 146–148, 160, 163
   Muscoviense, 146
   Nectaris, 146
   Orientale, 152, 154, 156
   Procellarum, 159–160
   Serenitatis, 146, 148, 160
   Tranquillitatus, 146, 159–160
   Vaporum, 152
Mariner spacecraft, 116
   Mariner 9, 76, 97
   Mariner 10, 145, 154, 165–167, 169–171, 173
Marius Hills, Moon, 149–150, 154
Marquesas fracture zone, 131–132
Mars, 1–7, 9, 24, 39–44, 58–68, 75–107, 143, 150, 165, 181, 218, 220, 245–246, 249
Mars Global Surveyor spacecraft, 63, 67, 79, 94, 106–107
Mars Orbiter Camera (MOC), 63
Mars Orbiter Laser Altimeter (MOLA), 94
Mars Pathfinder Lander spacecraft, 79, 82–83, 94
Mass flow, 55, 61, 63, 68
Massif, 201
Mauna Loa, USA, 20–21, 88, 196
Maxwell Montes, Venus, 116
Mean planetary radius (MPR), 116–117
Mechanical erosion, 211, 218
Melilite, 210
Mercury, 2, 165–173, 245
Metasomatism, 209, 215
Meteorology, 4
Methane (CH_4), 222–223, 233
Methanol, 222–223
Microbes, 47
Mid-Atlantic Ridge (MAR), 13–14, 123, 125–126, 131, 135
Mid-ocean ridge (MOR), 6, 13–14, 123–125, 127, 131, 136, 201–202
Miranda, 221, 231–234

Möberg, 40, 47, 61, 63
Monan, Io, 188
Moon, Earth's satellite, 2, 4, 9, 143–164, 165, 171–173, 179–181, 196, 207, 219–220, 230–233, 245, 248–249
Moraine, 47, 57
Muscoviense: see Mare
Mount Pinatubo, Philippines, 32
Mount Raincr, USA, 56
Mount Redoubt, USA, 56
Mount Spurr, USA, 55
Mount St. Helens, USA, 55–58, 231, 233
Mudflow, 207, 210
Mulungu, Io, 188
Munro-type komatiite, 216
Mylitta Fluctus, Venus, 134
Myrdalsjökull, Iceland, 47, 58
Mÿrsdalssandur, Iceland, 58

Nairobi, Kenya, 211
Namibia, 210
Natrocarbonitite, 208–211
Near-Infrared Mapping Spectrometer (NIMS), Galileo spacecraft, 186–189, 220
Nephelinite, 208–210
Neptune, 221, 230
Neutral bouyancy, 113, 125–129, 137
Nevado del Ruiz, Columbia, 41–42
Newtonian: see Rheology, Newtonian
Nitrogen (N), 5, 222, 233
Nyererite, 208

Obliquity, orbital, 39
Ocean, 39–40, 130–131, 191, 223–227
Oceanus Procellarum, Moon, 146, 150–151, 159
Oldoinyo Lengai, Tanzania, 207–213, 218
Olivine, 10, 79, 127, 153, 208, 215–216
Olympus Mons, Mars, 42, 64, 67–68, 76–77, 87–93, 95–96, 103
Ontong Java, 29, 217
Opaque index, 168, 170
Opaque minerals, 167–168
Ophiolite, 126
Öraefajökull, Iceland, 57
Orientale basin, Moon, 156–157, 160
Orthopyroxene, 10, 183, 191, 199
Outburst, Io, 186
Outflow channel, 40, 64

Pahoehoe, 18–20, 31, 54, 59, 134, 210–211, 213
Palagonite, 24, 44, 46–47, 61, 63, 79
Paleolake, 63
Palimpsest, 230
"Pancake dome", Venus, 118, 131
Partial melting, 10, 145–146, 162
Patera, paterae, 96–102, 105, 183, 187

Pathfinder: *see* Mars Pathfinder spacecraft
Pavonis Mons, Mars, 76, 91, 95
Pele, Io, 185, 188, 191–192, 194, 196–197, 203
Peleen dome, 23–24
Peneus Patera, Mars, 98
Peridotite, 10, 14, 208, 216
Permafrost, 39, 66–68, 104
PGE, Fe-Ni-Cu-sulfide ore deposits, 214, 219
Phonolite, 209
Phreatic eruption, 24, 40–41, 46, 58–59, 68, 86, 102
Phreatomagmatic explosion, 24, 46, 97, 210
Physical parameters
    lavas, 244
        planets and moons, 2
        planetary environments, 245
Picrite, 220
Pilbara, Australia, 215
Pillan Patera, Io, 184–185, 188–189, 191, 197, 202–203
Pinacate craters, Mexico, 58
Pioneer-Venus spacecraft, 114–116
Plasma torus, Io, 179
Plate tectonics, 12–17, 32, 84, 122–123, 131, 144, 201–202
Pleistocene, 39–40, 43, 46–48, 57
Plinian, 11, 26–27, 31–32, 86, 89–90, 137, 245, 248
Plume, 15–17, 28, 152, 158, 181–187, 191–197, 215–216, 222, 225
Pluto, 234
Polarization, 114
Polygonal terrane, Mars, 63
Pore space: *see* Vesicle
Porosity, 189–190
Potassium (K), 179, 181
Precambrian, 214–218
Pressure, 3, 113–116, 123, 125, 137, 179–181, 190, 192–194, 221–223
Prometheus, Io, 180, 184, 188, 191–194
Proterozoic, 215
Pseudocrater: *see* Phreatic eruption
Pyroclastic, 18, 20, 25, 31, 95, 97, 102, 151–153, 163–165, 168
    airfall, 25–26, 53, 136
    eruption, 25–26, 113
    flow, 26–28, 31, 42, 55–56, 86, 88–89, 97, 137, 194–195
    particles, 3–7, 25, 55, 64, 86, 116, 135–137, 165, 172, 182–184, 210, 247, 249
    surge, 27, 31, 41–42, 53, 55
Pyroxene, 10, 76, 79, 216, 220

Quartz: *see* Silica, silicon

Ra Patera, Io, 196–197
Radar, 114–122, 129–136
Radiating dike swarm: *see* Dike swarm, radiating
Rainfall, 64

Rare earth element (REE), 216
Reflectivity, 114–115, 117
Remotely operated vehicle (ROV), 123
Resonance, orbital, 181–182, 234
Rhadamanthys Linea, Europa, 224, 226
Rheology, 5, 20, 22, 79, 126, 172, 181, 188–189, 218, 223, 233–234, 243
    Bingham, 246
    Newtonian, 55, 210, 245
Rhyolite, 14, 18, 22, 25–27, 47, 84, 244–245
Ridged plains, 98, 102–103, 105, 112
Rift zone, rift valley, 13, 14, 28, 114, 116, 131, 208, 217, 228
"Ring of Fire," Earth, 15
    Io, 201
Rille, sinuos, 103, 118, 147–151, 153, 156, 159, 164, 172, 207, 211, 218–221
Rima, Moon
    Bode, 151–152
    Parry V, 148, 162–163
    Prinz, 147, 150
    Sirsalis, 162
Rock glacier, 40
Rotation rate, 114
Roughness, 3, 114
Rudaki crater, Mercury, 167, 170

Sapas Mons, Venus, 133
Sapping, 214
Salt, 210, 222
Saturn, 2, 221, 230, 233
Scabland, 57–58
Scattering properties, 114
Scoria, 224
Seafloor, 6, 113–116, 121–127, 130–137, 215
SeaBeam, SeaMarc, 123, 125, 130–132, 136
Seamount, 16, 123, 125, 130–132, 136
Sediment gravity flow, 46
Shamash, Io, 188
Sheet flow, sheetlike flow, 49, 87, 95, 103
Shield field, 118, 130–131
Shield plains, 130
Shield volcano, 1, 16, 20–21, 87–96, 99, 118, 150, 156, 163, 169, 182
Shombole, Tanzania, 208
Sigurd, Io, 188
Silica, silicon, 42, 79, 82–83, 118
Silicic volcanism, 43, 47, 79, 135, 182, 234
Sill, 15–16, 150, 216
Sinuous rille: *see* Rille, sinuous
Sippur Sulcus, Ganymede, 229
Skafá River, Iceland, 57
Skjalbreidur, Iceland, 20
Skefilsfjöll, Iceland, 48
Skeidará, Iceland, 58
Skeidarársandur, Iceland, 44, 57

Slope, 3, 5, 88, 93
Smooth plains deposits, Mercury, 143, 145, 165–166, 168, 170
Snake River Plains, Idaho, 16–17, 29, 232
SNC (Shergotty–Nahkla-Chassigny) meteorites, 65, 78, 80–84, 106, 220
Snow, 41, 55–57, 65
Sodium (Na), 179, 181, 208
Sonar, 115, 122–125, 130–131
Soufriere Hills, Montserrat, 234
South Pole–Aitken Basin, Moon, 146, 156–157
Spacecraft: *see individual names*
Spatter, 26, 59, 86, 171
Spinifex, 214, 216–217
Solid State Imaging System (SSI), Galileo spacecraft, 186–191, 220
Stora–Björnsfell, Iceland, 47–48
Stealth plume, Io, 194
Steam explosion: *see* Phreatic eruption
Stellate fracture pattern, 118
Stratovolcano: *see* Composite volcano
Strombolian eruption, 7, 11, 26, 165, 211
Subduction, 13–15, 201
Subglacial eruption, 7, 47, 50–52, 96, 104
Subsidence cauldron: *see* Cauldron, subsidence
Sulfur (S), 5, 46, 117, 179, 182, 185–188, 191–198, 202, 207, 211, 218
Sulfur dioxide (SO_2), 11, 31, 179–181, 191–198, 20102, 222
Sulfuric acid droplets, 32, 117
Sulpicius Gallus, Moon, 149, 151–152
Superheated, 190–192
Superswell region, South Pacific, 131–132
Surtsey, Iceland, 46, 53, 55
Svarog, Io, 186
Synthetic aperture radar (SAR), 114–115, 130
Syrtis Major, Mars, 220

Table mountain, 40
Tanzania, 207–208
Taurus–Littrow, Moon, 151–152
Tempe Patera, Mars, 99
Tempe Terra, Mars, 102
Temperature, 3, 5, 10, 18, 30, 39, 66, 84, 87, 115–116, 143, 181, 185–200, 203, 207, 210, 217–218, 220–223, 232, 245, 247–248
Tephra, 25–27, 31, 42, 46, 53–54
Tessera, tesserae, 122
Tethys, 221, 230
Tharsis, Mars, 41, 64, 76, 87–88, 93, 102–103, 106, 197
Tharsus Tholus, Mars, 88, 96
Thermal
 anomaly, 179
 conductivity, 10, 30
 diffusivity, 211, 234, 245, 247–248

Thermal (*cont.*)
 emission, 79, 186, 189
 erosion, 134, 149–150, 211–220
 expansion, 15, 245, 247
 gradient, 5, 30, 65–67, 78, 91, 117, 144, 196
 inertia, 39, 185
 opacity, 115
Thermokarst, 40
Tholeiite, 10, 86, 116, 208, 216
Tholus, tholi, 88–91, 100
Thrace Macula, Europa, 223
Tidal heating, 179–182, 190–191, 197–199, 220–221
Tindaskagi, Iceland, 48
Titan, 2, 230, 234
Titanium dioxide (TiO_2), 160–161
Tiu–Simud Valles, Mars, 64
Tolstoj basin, Mercury, 167–168, 171
Topography, 114, 116, 230
Tranquillitatus: *see* Mare
Triple band, Europa, 223–224
Triton, 1, 2, 5, 221–222, 230–234
Tsiolkovsky crater, Moon, 146
Tuff, 27, 47, 51–55, 58–59, 210
Tuff cone: *see* Cone, tuff
Tumuli, 18, 59
Tupan, Io, 188
Turbiditite, 53–54
Tuya, 40–41, 46–55, 60–63, 67–68
Tyrrhena Patera, Mars, 76, 97–103, 105

Uganda, 208, 210
Ultramafic, 30, 78, 180–182, 190–191, 199–202, 207–208, 214–220
Ulysses Patera, Mars, 76
Umbria, Italy, 210
Undara, Australia, 22
Uranius Patera, Mars, 93
Uranus, 221, 230, 233
Uruk Sulcus, Ganymede, 228
Utopia Planitia, Mars, 60, 63, 76, 102

Vacuum, 189, 222
Vapor, 222
Vaporum: *see* Mare, Moon
Vatnajökull, Iceland, 42, 50, 57, 61–62
Vega 1 and 2 lander spacecraft, 115–117
Venera 15/16 spacecraft, 114, 116, 129, 131
Venera lander spacecraft, 114–116, 208
Venus, 1–5, 113–137, 198, 207–208, 210, 212–214, 218, 220
Vesicle, vesicles, vesiculation, 3, 11–12, 18, 25–26, 127, 136
Viking Lander spacecraft, 60, 79–80
Viking Orbiter spacecraft, 76–77, 97, 104, 220
Viscosity, viscosities, 12, 18, 22–25, 27, 79, 83, 86–87, 97, 103, 207–217, 220, 223–225, 230–234, 244–249

Volatile, volatiles, 4–5, 10–11, 14, 25, 83, 85–88, 97, 102, 104–105, 107, 152–153, 162, 172, 184, 189–190, 197, 211, 222, 233–234, 245
Voyager spacecraft, 179, 183–187, 189, 191, 195, 221–234
Vulcanian eruption, 26, 151, 163, 165

Wairaki, New Zealand, 66
Water, 3–5, 11–12, 24, 27, 39–68, 79, 82–87, 89, 90, 96–98, 104, 107, 113, 117, 128, 133, 136, 180, 192, 201, 207, 214, 220–234

Wavelength, 114, 134
Wrinkle ridge, wrinkle ridges, 97, 102, 104–105

X_d, height-to-basal diameter, 130
X-ray flourescence (XRF), 79–82, 115, 220

Yellowstone, USA, 16, 27
Yield strength, 22, 150, 244–246
Yilgram, Australia, 215

Zal, Io, 188
Zamama, Io, 180, 188